Agriculture, Geology, and Society in Antebellum South Carolina

Agriculture, Geology, and Society in Antebellum South Carolina

The Private Diary of Edmund Ruffin, 1843

Edited by William M. Mathew

The University of Georgia Press

Athens & London

Designed by Richard Hendel
Set in Sabon
The paper in this book meets the guidelines for permanence and durability of the Committee on Production Guidelines for Book Longevity of the Council on Library Resources.

Printed in the United States of America

96 95 94 93 92 C

5 4 3 2 1

Library of Congress Cataloging in Publication Data
Ruffin, Edmund, 1794–1865.
Agriculture, geology, and society in antebellum South Carolina: the private diary of Edmund Ruffin, 1843 / edited by William M. Mathew.
p. cm.
Includes bibliographical references and index.
ISBN 0-8203-1324-6 (alk. paper)
1. South Carolina—Description and travel. 2. Agriculture—South Carolina—History—19th century. 3. Geology—South Carolina—History—19th century. 4. Ruffin, Edmund, 1794–1865—Diaries. I. Mathew, William M.
F273.R85 1992
975.7'03—dc20 90-24718 CIP

British Library Cataloging in Publication Data available

Photograph of Edmund Ruffin on the title page courtesy of the Valentine Museum, Richmond, Virginia

TO ANGELA WITH LOVE,

FOR STARTING IT ALL OFF

Contents

Preface

The Private Diary of Edmund Ruffin

The diary kept by Edmund Ruffin during his months as agricultural and geological surveyor of South Carolina in 1843 ranks as the most detailed and authoritative report on the state between Robert Mills's *Statistics of South Carolina* of 1826 and Frederick Law Olmsted's account of his travels in the 1850s. Ruffin wished to help save slave society through agricultural reform, and South Carolina offered him his best chance to set about that task with official blessing. Economic depressions of considerable severity had been recurring over the previous quarter-century and were not to pass until the end of the 1840s. The sense of a nadir in Carolinian fortunes had been accentuated by recent memories of slave rebellion and sectional confrontation. Ruffin, therefore, approached the work with some solemnity.

It is necessary at the outset to distinguish the diary from the report Ruffin submitted to the governor, James Hammond, when the survey was completed. They were not one and the same. The report was published at the end of 1843 and, though highly informative, has the stiffness of an official document and a relatively narrow literary and observational range. The diary, by contrast, is imbued throughout with immediacy and vitality, recording Ruffin's social encounters around the state and his observations on the geological and agricultural matters that he had been employed to study.

Like most manuscripts of its kind, it has required some editorial interference for purposes of consistency and lucidity. The text has been both reduced and subdivided. Excisions have been necessary because of length and occasional tedium. Ruffin used parts of the diary for an assortment of preliminary recordings, speculations, and rough sketches, loosely assembled and usually geological in focus. They can be of little interest to historical readers. Ellipses have been limited to this material. Social and economic sections are presented in full. Drawings not considered worthy of reproduction have been ignored completely, with no signs given of their location in the document. Such editing, though reducing the text by some thousands of words, has not been carried to extremes. Ruffin was preoccupied with marling and liming as means of ameliorating acidic, "exhausted" lands in the South, and an officially designated purpose of his survey was to find calcareous strata and encourage their exploitation. Many Virginians had taken his advice in the 1830s, and some tidewater counties of the border state had experienced great increases in crop yields, land values, and incomes. Ruffin's geological work in South Carolina is of interest in this broad context. It shows, too, the hard and physical nature of his labors in the cause of reform, while also demonstrating the abundance of accessible resources. Accordingly, much scientific comment has been preserved.

Subdivisions are indicated throughout by "chapter" headings. The diary in its manuscript form has no such divisions, being broken only by dating. There are powerful arguments for retaining the original format, but it was my experience in the first year or two of editorial work that the rapidly enlarging accumulation of typed sheets had to be split up into dozens of small, labeled parts to help make sense of the account. These breaks proved indispensable for identifying basic geographical and intellectual sequences. It seems pointless now to scrap them in the name of purism. Being mere insertions, they have no effect whatever on the narrative and so serve as guides to a frequently meandering journey. No rearrangements of material have been attempted in the interest of either tidiness or clarity, and the subtitles have been kept brief and neutral to minimize intrusion.

Singulars and plurals, often visually obscure, have been determined according to appearance or logic. Syntactical blemishes resulting from incomplete alterations to sentences (e.g., inserting a new phrase but failing to adjust remaining words and punctuation) have been cleaned up, Ruffin's own aim in such instances being literary precision. Dates take various forms in the manuscript but have been standardized for publication. Obvious blanks in Ruffin's text (no doubt left for later attention) are represented thus: [ER].

Introduction and Biographical Supplement

The introduction describes some of the key physical, economic, and political circumstances of antebellum South Carolina—in particular, the anxieties surrounding depressed agriculture and the institution of slavery. It also considers Ruffin's scientific ideas and reputation, his itinerary in the state, and the outcome of his work. There is little in the way of textual exegesis, the emphasis being on context and issues.

The biographical supplement is offered as a solution to the difficulties posed by footnoting. A journal of this size and complexity requires generous annotation, yet an abundance of notes can detract from the textual flow and disrupt transitions. Many names need much more introduction than the diarist himself provides. It is possible that Ruffin—an energetic writer since the early 1820s—intended his meticulously composed and amended journal to serve as the basis for an extended publication after his return to Virginia at the end of 1843. (As it was, he plunged straight into the restoration of his new Marlbourne plantation, recording his daily business in a fresh volume, which ran to about a quarter-million words). Had the diary ever appeared, it would, in the light of Ruffin's gratitude for help and hospitality in the state and his admiration for many of its scientists and politicians, probably have been fairly discreet in discussing Carolinians. This may explain the large amount of detached personal comment. Only in the case of Francis Pickens did he come close to any expressions of hostility. This, then, is no document for the gossip forager—at once both a minor weakness and a major strength. It is, rather sternly, a diary about issues, written by the most experienced and animated farm reformer in the American South.

The criterion for inclusion in the biographical notes is simply mention in the diary; and the determinants of length of entry are importance to Ruffin, to the survey, and to the cause of agricultural change—these three also being the main matters for comment. Availability of information is another important factor, and usually a distorting one, the worst cases (quite numerous) leading to total omission of a name. The placing of the supplement at the end of the volume, and in alphabetical rather than chronological order, is a way of avoiding clutter around the diary itself, while preserving ease of access for the reader. On all other matters only very selective notation will be presented. Ruffin is usually left as sole commentator on these.

Many of the leading planters and politicians mentioned in the diary remained men of consequence through the Civil War. As a result of his own secessionism and his sharp sense of history, Edmund Ruffin was much involved in the dramatic events in and around Charleston in 1860 and 1861, frequently in the company of men he had visited during his 1843 survey. So there are continuities of issues and personnel. It would, however, be mistaken to impose early 1860s on early 1840s and see Ruffin's efforts as inevitably proto-secessionist. Reform might have been stimulated by sectional anxieties, but major, positive results could have moderated secessionist impulses. The long list of people to whom Ruffin felt specially indebted, as set out at the end of his survey report, included John Bachman, Edmund Ravenel, Whitemarsh Seabrook, James Hammond, Robert Allston, Joseph Johnson, David McCord, David Jamison, William Anderson, John Calhoun, George McDuffie, Mitchell King, and Edward Heriot. They were a mixed group. The diary shows that Ruffin was moving from secessionist to Unionist households without evident discomfort. His best friend in the state, the Reverend John Bachman, was Unionist. Charleston, where he used hotel rooms and an office, had Union tendencies in its professional and mercantile communities. Much of the scientific company that he sought out had little time for secessionist thinking. But of course he also encountered many planters of pronounced states' rights inclination who had supported nullification and continued to have radical convictions. His company, for the most part, was determined by locale and interest.

Ruffin's main preoccupations lay with the difficult agricultural business at hand. Other southern matters could be discussed, agreeably or contentiously, over meals and in the evenings. If he did have priorities concerning the people he wished to meet, it is likely that they were predominantly scientific. Better, in the circumstances of a quick survey, to tour fields and embankments with an apolitical, improving planter than to spend an afternoon seated at some disunionist table bemoaning federal tyranny.

Acknowledgments

My first task, before commencing work on this book, was to determine whether William Scarborough, the editor of the first two (and, since then, third) volumes of Edmund Ruffin's 1856–65 diary, had any plans to extend his splendid labors back to include the Carolina journal of 1843. He had, it transpired, no such intentions. I then approached the Virginia Historical Society, the owners of the journal, for permission to edit and publish. This was generously provided (with some accommodating change in the society's constitution) in 1984, along with an agreement to supply the comprehensive photocopy that I needed on this side of the ocean.

When the bulky packages arrived from Richmond, my wife, Angela, attended to the preliminary transcription. She was probably the first person to go through the journal in its entirety since 1843, troubling over its innumerable obscurities. When I later checked the end product against the original in Virginia, with a particular view to filling in the spinal gaps, I was astonished by the accuracy of her work. It is to her that the volume is dedicated, with love and gratitude.

My visit to Richmond and Charleston in 1985—to examine the journal and undertake research for the introduction and biographical supplement—was made possible by financial support from the Nuffield Foundation. In Richmond I was much helped by the Virginia Historical Society's library staff and by Nelson Lankford from his *Virginia Magazine* office upstairs. I also had the pleasure of meeting James Skelton Gilliam of Hopewell, the most informed and historically dedicated of Ruffin's descendants. I received considerable help too from Edmund Saunders Ruffin, Elizabeth Ruffin, and Inge Walker Sonuparlak, who showed me around the Ruffin properties along the James River.

In Charleston my principal academic debts are to Cam Alexander and

David Moltke-Hansen at the South Carolina Historical Society. Their fine library is downstairs from the room in the Fireproof Building that Ruffin used as his office in 1843. Both dealt patiently with my out-of-town queries and made reading suggestions of the greatest value. I also spent time in Charleston talking about family history with David Ruffin and traveling with him through the old Georgetown rice lands. Other journeys, to Beaufort, Savannah, and Columbia, were less enjoyable, being undertaken alone and by bus.

Back home, I began work on the introductory section, footnotes, and text reduction in the early months of 1986, with the help of study leave from the University of East Anglia. I had recourse not only to recently collected material but also to a mass of items gathered during a decade's research on Ruffin, slavery, and agricultural reform. Two of my children, Jenny and Daniel, gave much help in checking the ever-changing typescript against Ruffin's original.

Before dispatching the manuscript, I consulted Richard Wilson, a historian of British rural society, to see how the material ranked for general scholarly interest and was encouraged by his response. I also had the benefit of two long, detailed, and excellently critical reports, written by William Scarborough and David Moltke-Hansen. I am, finally, grateful to Trudie Calvert for her sensitive and precise editorial assistance.

Agriculture, Geology, and Society in Antebellum South Carolina

Introduction

The South Carolina Agricultural Survey of 1843

Among the best of the early nineteenth-century studies of South Carolina available to Edmund Ruffin as cartographic and literary guides to an unfamiliar state were Governor John Drayton's *View of South Carolina* (1802), Robert Mills's *Atlas of the State of South Carolina* (1825), and the same author's *Statistics of South Carolina* (1826). Drayton was probably one of the first to place topographical images in Ruffin's mind before he left Virginia, and Mills provided him with a report to emulate as well as a set of maps to help him find his way through the tangle of Carolinian land and river highways.

"The shore of South-Carolina," by Drayton's description,

> rises gradually from the Atlantic Ocean. As the approaches become nearer, trees, sand hills, and at length the extension of its shores present themselves. These on the sea coast are continually intersected by inlets, creeks, and marshes; throwing the shores of Carolina into a number of ISLANDS. Some of them present a sandy front to the sea; undulated with conical sand hills, sixteen or twenty feet high: while the sides next the main land, are level and low, and are connected with extensive marshes, intersected by creeks and inlets. . . .
>
> From these islands, the MAIN LAND presents a level country; . . . In some places, the land deepens; and at the distance of fifteen or twenty feet below the surface, it rests upon a bed of small and broken sea shells, and other marine productions. . . . These lands, generally produce extensive pine forests; known with us, by the name of PINE BARRENS; because of their unproductive nature. . . . And a multiplicity of SWAMPS and BAYS, are found throughout the country; which branch out and unite, by an infinity of different meanderings; sooner or later, emptying their waters into some river,

or inlet, from the sea. Natural meadows, called SAVANNAHS, are often seen in this part of the state; some of which cover an area of fifty acres. They are destitute of trees or bushes. . . .

The RIVERS which course along these lands, are bordered with the most fertile soils; and, upon them some of the best RICE plantations are situated. . . . The inundations, and flowings of tides, bear to it, and precipitate thereon, the finest and most subtle particles of manure. . . . From the point to where the tides flow, the lands become hazardous, by reason of freshes, which occasionally pour violently down the river; but the SWAMPS continue deep, and excellent; rising in height above the level of the rivers. . . .

With the SAND HILLS the middle country may be said to commence; stretching in a belt of from twenty to forty miles from Savannah River, to the upper part of Pedee River; and thence into North Carolina. In general, the land hereabouts is barren, or but triflingly productive. The middle grounds between the rivers are the highest, and consequently the most barren. . . . Their soil is of so sterile a nature, that in many places it produces no grass to cover it; and the tracks of any animal passing over it, are discernible, as if they had been upon snow. The low grounds among these hills are either extensive SWAMPS and BAYS, or narrow vallies, into which, the mould from the adjacent high lands have been deposited by the rains which run down their sides. Hence they become suitable for agriculture and pasturage, and are principally those places, near which settlements are effected. . . .

Beyond this belt, and from the first FALLS of the rivers, loose stones appear on the ground, and rocks on the ridges of land, and at the sides of the rivers. Hill and dale alternately rise and fall, as one advances towards the mountains. . . . The swellings of land now rise into more sudden and towering heights. The currents become rapid; are generally fordable; and are often opposed by scattering rocks. The vales are lengthened and embosomed by surrounding hills; and at length the MOUNTAINS spring; whose heights are sometimes hidden by impending clouds; or at others remain superior to the passing ones below. . . .

From this general view of Carolina, it appears, that it may be properly divided into LOWER, MIDDLE, and UPPER Country. . . . Hence, the LOWER country will comprehend all that part of the

> state, from the sea to the sand hills. The MIDDLE country, that part beginning with the sand hills, and ending at the falls of the rivers. And the UPPER country, that part stretching from the falls of the rivers, to the north western mountains.[1]

Around this varied terrain was spread, in Drayton's time, a population of 350,000, of which over 40 percent was black. For whites, the most troubling features of the state's demography were the sluggish growth of the total and the increasing weight of the black component, the latter's advance being particularly notable on the cotton lands of the middle country. By 1820 there were slightly over half a million people in all, 53 percent black and 47 percent white. In 1840, the aggregate was still under 600,000, with blacks topping 90 percent in some coastal parishes. Between 1830 and 1840 the proportion of whites had increased by only 6.7 percent in the lower country and fallen by 1.4 percent in the middle and upper, leaving them as a quarter of the total population in the first and not much more than half in the second.[2]

The state's fortunes had come to rise and fall with those of cotton, the staple being grown not only on the plantations of the coastal and middle areas but on most of the smaller up-country yeoman farms as well. The savage jolt of the Panic of 1837 and the setbacks of 1839 and 1841 provided the immediate economic backdrop to Ruffin's survey, but they were set in a context of more protracted worry. In 1821, the cotton states of South Carolina and Louisiana both exported around $7 million worth of commodities. In 1842, when the invitation was extended to Ruffin, South Carolina still produced $7 million worth, but Louisiana, with the advantage of fresher, richer soils, had quadrupled her figure to $28 million.[3] It is no surprise that, in Lacy K. Ford's observation, "antebellum South Carolinians were extremists in their passion for mobility"; nearly half of the whites born in the state between 1800 and the Civil War took up residence elsewhere.[4] Not only did this movement tend to intensify "Africanization," but it meant, by the arithmetic of representation in Congress, reduced power within the Union.

Given such apparent corrosion, within and without, could slavery hold and the white population survive? These were the anxieties underlying the survey. Ruffin's brief was to look for marl and lime, advise farmers on their use, and pronounce generally on the means by which Carolinian agriculture could be rendered more profitable and competitive. The purpose was, at first glance, practical and mundane. The deeper issues, however, concerned the very foundations of Carolinian polity and society.

Politics, Improvement, and Slavery

EMPLOYING EDMUND RUFFIN. On 24 November 1842, just after his election as governor of South Carolina, James Hammond wrote to Edmund Ruffin—by then the best-known agricultural reformer in the South—to say he hoped "we can make some arrangement to get you here to examine a few localities. If $500 will be an inducement to visit us in the Spring & look over a few plantations, I think we can offer it."[5] Such a survey had been much discussed among planters and had lately been the subject of a committee report presented to the House of Representatives in Columbia.

Hammond's casual tone perhaps owed something to his doubts about the value of any such inquiry. The principal force behind the survey was Robert Roper, a wealthy Charleston rice planter and representative, who later declared that he had nourished the idea as a "wretched foundling, feeble, friendless and exposed to the contempt and obloquy of a thoughtless multitude."[6] If Ruffin had doubts of his own, these were probably confirmed by Hammond's warning that he could expect to encounter "much indifference, much ignorance, much conceit, much obstinacy & some opposition" in the course of his labors in the state. "It will not be the least difficult part to keep your temper."[7]

In this uncertain manner, the survey brought together two of the South's most formidable personalities. James Hammond, then in his mid-thirties, was from a modest family and had found his way into the plantocracy through ambitious drive and an unromantic union with the heiress Catherine Fitzsimons and her ten thousand acres by the Savannah River. He had made his political reputation as an up-country nullifier and had already served as a representative in Washington. Behind a seigneurial facade he was a hypersensitive man of considerable intellectual ability. Ruffin thought Hammond had "the most powerful mind in the southern states."[8]

Edmund Ruffin came from a seventeenth-century James River family of aristocratic Scottish origins and had inherited and successfully worked worn-out lands on the south bank near Coggin's Point.[9] Then approaching fifty, he had won fame as the author of the highly original *Essay on Calcareous Manures* (1832, now in its third edition) and as proprietor and editor of the celebrated agricultural monthly the *Farmers' Register* (1832–42).[10] These were the first publications to draw attention to the prevailing soil acidity of the coastal states and its contribution to low crop yields and depressed land values. They also informed planters that a cure was available in the form of calcareous marl, extensively present throughout the low fluvial lands of the tidewater.[11] Ruffin's biographer Avery Craven has described him as "the greatest agriculturist in a rural civilization."[12] Most improving planters in the 1840s would have agreed. There was no one comparable in view, no one so didactic and challenging on vital matters of theory and practice. The issue of acidity and its correction by marling and liming, as earlier British experience had shown and later soil scientists were to explain, was of fundamental importance for most arable farming in temperate regions of the globe. Proper attention to the problem, it was believed, could reverse "exhaustion" in the Southeast, restore agricultural prosperity, hold the population, and protect slavery.

Hammond, interestingly enough, had been putting Ruffin's reputation to its biggest test yet by starting an enormous program of calcareous improvement at his Silver Bluff plantation.[13] If marl did not help him raise yields and diversify, he insisted, his only option would be emigration to the West. With so much at stake, the work could become obsessive. "The labours of the week," he wrote in November 1841, "have driven politics entirely out of my mind and everything else but marl—marl."[14] The two men also had common political ideas: an intense distrust of the North, an opposition to protectionist tariffs, a strong states'-rights view of the Constitution, a detestation of paper-money banking, and a clear perception of the threat to seaboard slavery posed by economic decline.

Ruffin was Hammond's choice for the survey, despite the possibility that a man of the Virginian's eminence might prefer more comfortable ways of propagating his ideas. As it was, Ruffin was unemployed, without a farm, in 1842. Recent experiences in Virginia had caused him much economic and social distress. He had lost large sums of money in the 1837 crash and its aftermath and had subsequently been involved in wounding exchanges with the Petersburg banking community. He

Portrait of Governor James Henry Hammond, circa 1843 (courtesy of the South Caroliniana Library)

had also seen his *Register* sink in a morass of planters' indifference and bad debt.[15] Eager to return to the privacy of his old life in farming, he had tried, unsuccessfully, to buy his way back into the once-cheap land of his native Prince George County. Ironically, there had been a local rise in farm prices and curtailment of sales consequent upon calcareous improvement. There was, then, no compelling work to detain him in Virginia.[16]

"I am frequently asked if you have accepted," wrote Hammond on 14 January. Two days later, and only a few weeks after publishing the final issue of his *Register,* Ruffin confirmed his wish to take up the "highly honourable appointment." The governor and legislature, however, had wanted a two-year arrangement. This seemed excessive to Ruffin. His wife and his family of nine children, aged from ten to twenty-eight, would remain in Virginia, which was one reason why he refused to commit himself beyond a single year. South Carolina, moreover, was mostly unfamiliar to him, and its low country was notoriously unhealthy. He was apprehensive of the "uncertain & irregular" places he would have to visit. Hammond disliked the one-year proposal but did not press the point. The salary, meanwhile, had been advanced to a wholesome $2,000 per annum.[17]

Despite his obvious worries, Ruffin was probably still reformer enough to be excited at the prospect ahead of him. South Carolina's recent temporary nullification of federal tariffs and its widespread disunionist spirit had won his heartiest approval. His family had favored states' rights, and he had been secretary to the delegates of the agricultural societies of Virginia in 1821 when they petitioned Congress for a reduction in import duties. Agriculturists, he had argued then, should not be required to subsidize industry. The "object of duties on importation . . . ought to be exclusively the increase of the public revenue."[18] He had already enjoyed some direct contacts with South Carolina. The *Farmers' Register* had received communications from several experimental farmers, including James Hammond, and reports had arrived suggesting that the state was unusually well endowed with marl and lime.[19] In 1840 he made his first visit, as a guest of a new Wilmington rail and steamship company, sailing into Charleston Harbor on the morning of 17 April and spending a brief but memorable twelve hours in the city. The weather was splendid and the spring scene quite beautiful. He was reluctant to leave but pronounced himself happy with all

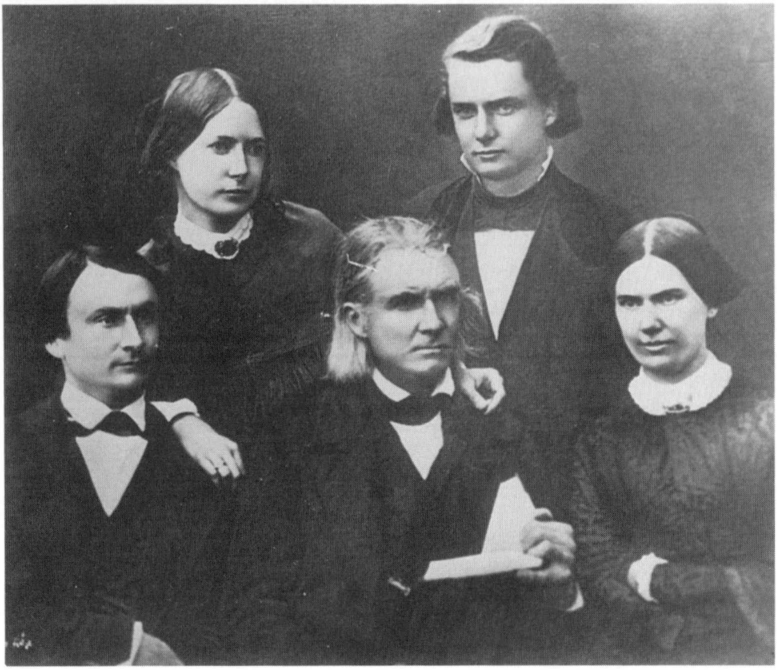

Ruffin with four of his nine children, circa 1850: (left to right) Julian, Ella, Charles, and Mildred (courtesy of the Virginia Historical Society)

that he had encountered. The attentions of numerous "warm-hearted southrons" had been especially flattering. Charleston, by manner and appearance, seemed like "a gentleman born and bred."[20]

To return as an acclaimed scientist and paid employee gave him much gratification. "The act of the constituted authorities of the State of South Carolina which invited me to her service," he wrote to Whitemarsh Seabrook, president of the State Agricultural Society and a future governor, "is the most honorable distinction which has yet been earned by my long continued efforts to advance agricultural interests; and I am impelled by every consideration to desire that the services, sought in my appointment, shall be as faithfully performed as zeal and industry and my imperfect ability and means may permit."[21]

LEGISLATURE AND SURVEY. The idea of sponsoring surveys had been under discussion in the legislature since at least the 1810s, and geological and agricultural examinations in particular were thought necessary by people wanting systematized information of practical import. The first funded efforts, however, were largely cartographic in emphasis.[22] A private geological survey undertaken by Professor Lardner Vanuxem during academic vacations from South Carolina College in the mid-1820s was small-scale and partial, confined to a few up-country districts.[23] Robert Mills's famous *Statistics* of 1826, though of great value, hardly represented rigorous analysis and was spread over "Natural, Civil, and Military . . . General, and Particular" phenomena. By 1842, it was of some concern that twenty of the other twenty-five states of the Union had completed, or had in progress, agricultural or geological surveys or both, usually of substantial range and magnitude. New York had been particularly generous in its provision, voting $104,000 for a four-year undertaking (and giving considerable responsibility to the same Vanuxem).[24]

In November 1839 the matter was debated by a state Agricultural Convention in Columbia. A special committee of fifteen, deputed to prepare business for the delegates, recommended that consideration be given not only to a survey but also to the establishment of an agricultural professorship in South Carolina College, an agricultural school, provision of elementary agricultural education, and a state board of agriculture. The report was opposed by James Hammond, a committee member, and the proposals accordingly narrowed to the professorship and the survey. Even these Hammond wished—unsuccessfully—to see shelved. Further discussion of the survey was deferred on the suggestion that the agricultural and geological parts should be voted on separately. Both, in the event, were supported: the agricultural by a narrow thirty-eight to thirty-two.[25] Despite his later sponsorship of Ruffin's inquiries, Hammond was always highly skeptical of the prospects for improved farming in the state.

The survey idea was further advanced in 1841, when Governor John P. Richardson recommended it to the legislature. Resolutions were also presented by the House of Representatives Committee of Agriculture, under the chairmanship of Robert Roper. The principal one, urging the necessity of a survey because of emigration and the state's seriously weakened cotton trade, was carried. Despite much interest, however, the subject was allowed to lapse on the grounds that no agricultural societies had taken the trouble to submit petitions. The Committee on Agriculture was asked to investigate further and prepare a report for the House. The conclusion, for the moment at least, seems to have been that the survey was little more than a debating topic in Columbia and of no significant concern to planters. Dismayed, Roper, in his other capacity as a vice-president of the State Agricultural Society, persuaded members in the summer of 1842 to support the circularizing of all twenty-six local bodies. They would be asked to organize petitions for the new legislative sessions in November.[26]

Only two obliged: the Wateree and Milton Laurens societies.[27] This extraordinarily poor response after more than three years of public discussion seemed to justify Hammond's reserve. The new governor, however, had by now somewhat moderated his doubts and went so far as to set out the case for a survey in his first gubernatorial Message to the Senate and House of Representatives on 29 November 1842. Interestingly, he failed to urge that a specifically agricultural survey be undertaken. He appealed to the "patriotic consideration" of members, asking them to support "geological researches" of the sort undertaken in other states. His remarks on agriculture were confined to a listing of general aspirations.[28] Shortly afterward, Roper's committee issued its report, unanimously recommending that a two-year agricultural examination be started at once. Apart from soil analysis and searches for marl and lime—very large exceptions—the work was to steer clear of geology "by a consideration of the expense and the remote attendant advantages." The final account, though, should not deliberately exclude significant geological material "incidentally collected."[29] It was an unsatisfactorily ambiguous definition of the survey's range.

The report and its survey recommendation were discussed in the House as the "special order of the day" on 14 December and carried by sixty-four to fifty-four. The vote was hardly an enthusiastic endorsement, and indeed had Charleston been excluded Roper's cause would have been lost. Of the city's sixteen representatives fourteen voted yes and two no.[30] The Senate approved on the seventeenth.[31]

The potential confusion over the survey's scientific content was sensibly resolved by adding "geological" to the surveyor's title.[32] Hammond, a generally powerless governor under his state's constitution, was now obliged to discharge a real function: the selection of a "competent" man to undertake the task. John Belton O'Neall of Newberry had suggested William H. Ellett, the northern scientist who occupied the chair of chemistry, mineralogy, and geology at South Carolina College, proposing that, Vanuxem-style, he do the work during school vacations.[33] It soon became clear, however, that Hammond had already picked his man—an easy choice, given Edmund Ruffin's commanding reputation and the evident success of Hammond's own marling experiments at Silver Bluff. Hammond's study of Ruffin's *Calcareous Manures* in the summer of 1840 (at the suggestion of John C. Calhoun's confidant Congressman Franklin H. Elmore) had been the main factor tempering his pessimism over reform in the state. The idea too of getting someone of Ruffin's distinction to visit and advise Carolinians "disarmed my opposition & that of many others."[34]

Ruffin set out from Petersburg on the evening of 26 January 1843 on the four-hundred-mile rail-and-sea journey to Charleston. By the twenty-eighth he was in the state and by the thirty-first had begun the survey.

OPPOSITION TO THE SURVEY. Given Ruffin's availability and his planting success in Virginia, as well as the popularity of geological and agricultural examinations in other parts of the Union, it might seem odd that South Carolinians took so long to agree to a major survey of their own. The doubting Hammond was not the only impediment. Many voted against the proposal in the 1839 Agricultural Convention; agricultural societies were uninterested; and a large minority of representatives remained unconvinced in 1841 and 1842.

The Roper committee itself contributed to the difficulties by making up a list of duties that invited failure. The "Surveyor pervades every section of the State, examines, as far as practicable, the nature and quality of the soil in every District, detects calcarious [sic] manures, visits farms, corrects injudicious modes of culture, advises as to the best kinds of seed, and communicates the means of affording an increased product; he discloses the improvements of other sections of country, explains the properties and actions of various manures, the necessity of a rotation of crops, and instructs in the general principles of Agriculture."[35] What was more, the whole of South Carolina had to be covered in one effort. The area was similar to that of Scotland—a country well known to American reformers, where examinations had usually been conducted on a county-by-county basis. The plan, clearly, was to do the job superficially and as inexpensively as possible. It would have been sensible to have devised some means of getting two or three years of work from Ruffin, perhaps by agreeing to a judicious phasing of the survey and providing a geological outrider and specimen analyst. The scale of the exercise, as set out, was nonsensical for a solitary man in middle age, working in difficult and unfamiliar terrain and constantly exposed to the elements and to disease.

It was fortunate for the survey sponsors that Ruffin was at such a loose end at the close of 1842 and that his self-confessed vanity made him seek "some just claim to be deemed a public benefactor." Had his *Farmers' Register* still been in existence, or had he purchased the new Virginia plantation he had been looking for, it is likely that he would have stayed put. He could have handled South Carolina's request with pamphlets, short visits, and speeches. He had no special need for supplementary income, most of his considerable wealth having survived the recent crash.[36] And he was not motivated by the desire to seek the stimulating company of secessionists, in prescient anticipation of exciting events in the 1850s and 1860s.

The House of Representatives committee of 1842 set Ruffin some very tall orders, but its deliberations were conducted in a spirit of sober optimism. Governor Hammond, however, had not completely shed his doubts, which he later confessed to Ruffin.[37] His hesitation was founded on his perception of Carolinian planter entrepreneurship. No matter how much success was achieved with marling at his Silver Bluff domain, and no matter how many times Ruffin's principles of amelioration were uttered in mansions and farming societies around the state, an ill-educated, vain, and unresponsive planter community would, Hammond believed, prove as reluctant as ever to engage in bothersome experimentation. During the discussions on state support for agriculture at the 1839 convention, he argued that little should be done until planters were prepared to take a more penetrating, reformist view of their own affairs. A survey, he wrote in his diary, would be a waste of time. "We are not yet prepared to make use of the results."[38] Hammond's view of his fellow planters, indeed, was consistently scathing. He took the trouble to warn Ruffin of the poor reception he could expect. Even though farmers might extend personal politeness and hospitality to the visitor, they would continue to "evince much of that want of enterprise which I fear is characteristic of our state in the main." He continued to abuse them in his correspondence with Ruffin in later years. In 1849 he complained at the South Carolina Institute of their "petty interests, false reasoning, unsound calculation, and perhaps, above all, certain traditional habits of thought and action." The "most injurious" feature of their farming was, perhaps, "the habitual want of personal attention to details."[39]

Such views of the interest and ability of the state's planters have not gone uncontested by historians over the years[40] (and one often senses strong elements of personal and political animus in Hammond's denunciations), yet the fact remains that agricultural practice in the state appeared rigidly unimaginative and unlikely to be altered by the delivery of a report in Columbia. According to David Moltke-Hansen, even the informed planters, conversant with the possibilities offered by reform, "did not always feel the need to pursue such matters vigorously." Theodore Rosengarten notes how the editor of the campaigning *Southern Agriculturist*, John D. Legaré, ended his tenure in 1834 "astonished" by "the almost universal ignorance of elementary science and the hold of habit on the minds of intelligent men."[41] All this has to be stressed, for the planter was the only possible agent of advance and, as such, the person to whom any survey was finally to be addressed. That James Hammond, the sponsoring governor, could remain so despairing gave strong advance notice of failure.

Further reservations were set out by the Reverend John Bachman, naturalist and Lutheran cleric, though these did not prevent his becoming Ruffin's closest friend in the state and a source of considerable support to the survey. Early in 1843 he brought out a pamphlet entitled

An Inquiry into the Nature and Benefits of an Agricultural Survey of the State of South-Carolina. It was based on remarks he had made at a Charleston literary club in December 1842, which he had been pressed to publish by Judges Daniel Huger and Mitchell King. The argument was similar to Hammond's, though less acerbically expressed. Bachman, like Roper and his colleagues, viewed a survey as ideally a pedagogic exercise rather than as a mere tour of inspection. "It points out the errors in the mode of cultivation, and suggests such new improvements as have undergone the test of experiment." Unfortunately, reformers could not go to planters assuming any common basis of language and understanding. The cultivators had no extensive experience with drainage, fertilization, and plant physiology. How were such willfully ignorant people to be instructed? Marling and liming had suddenly become fashionable issues, yet the resources were not newly discovered. Soil analyses had recently been conducted on Edisto Island by Professor Charles U. Shepard, but how many farmers there had enough knowledge of chemistry to be usefully guided by his findings? In what way, wondered Bachman—like Hammond, placing the cart firmly before the horse—"may an Agricultural Survey be rendered beneficial, under present circumstances?" And how could the job be left to the care of "a single individual, unaided by Agricultural Societies and men of science?" One could not reasonably expect it to be properly completed "during the term of a long life." He ended, however, on the inconsistent, appeasing note that the survey might well succeed with the support of South Carolina's body of "enlightened planters."[42]

Criticism could also be voiced by those who wished to see some demotion of agriculture and, with it, the political system dominated by large planters. Diversification of the economy and the body politic would not be best served by continuous concentration of attention on the weak farming sector. A prestigious survey of land and farms might merely confirm the ascendancy of traditional economic and social assumptions. During the late 1830s and early 1840s much writing and speechifying was devoted to the case for extending industry in the state.[43] It was, of course, possible to argue both for more manufacturing and better agriculture, diversified farms being sources of food for factory settlements and industrial activity providing markets and infrastructure for plantation producers. But in political exchange it was often difficult to avoid exclusivity and special pleading.[44]

There was also the reservation that the agricultural problem was essentially one of periodic cotton overproduction and that proposals serving to augment output might be misguided. Many, too, would have no time for a reformer nosing around the state and gratuitously handing out instructions,[45] least of all a Virginian who had just failed in his own Petersburg affairs. What did he know of Carolina? Being admired by James Hammond, too, was not always the happiest of commendations. Even those who favored some occasional experimentation and who had seen Ruffin's writings might have been put off by the evident difficulties involved in finding, digging, hauling, and spreading marl; by the apparent requirement of treating only fields already rich in organic matter; by the suggested need of a specialist labor force;[46] and by the border-state character of the existing arguments and documentation.

LAND AND FERTILITY. The main, immediate purpose of the survey was the introduction of that calcareous amelioration. It was an important objective. In the words of the Roper committee, "modern research attests that all the Atlantic States are deficient in lime as a natural ingredient of the soil, without which, as a constituent, soil cannot be rendered fertile."[47] Lack of lime affected the growth of most plants and in particular impeded the cultivation of leguminous green crops, which were useful both as fodder and as nitrogen-fixing breaks in rotation farming. Acidity also meant the presence of toxic chemicals, the retarded decomposition of organic matter, and the trapping of vital plant nutrients. C. E. Millar, L. M. Turk, and H. D. Foth, in a modern volume on agricultural chemistry, refer to soil amelioration as "the 'backbone' of permanent agriculture in humid regions."[48] There was also the related need for diversification. The rice staple, reported the Roper committee, was geographically restricted, and the state's cotton was under severe threat from cheap production in the West—"from this source of profit," it now seemed, "her palmy days are past."[49]

Analyses apart, much of South Carolina looked awful. Whitemarsh Seabrook, a sea-island planter and future governor, wrote in 1828 of the dilapidation, the choked drainage channels, the deserted mansions, and "the sallow and feverish" faces abounding in the "now dreary region" of the low country. In the language of Ossian, "the fox looked out of the window; the rank grass waved round his head."[50] "A stranger travelling through our country," declared former governor George McDuffie in

1840, "could not be persuaded that it was inhabited by a race of wealthy, hospitable and enlightened planters, so few of the monuments and improvements that indicate a wealthy and prosperous community would meet his eye."[51] This was partly the consequence of extensive wilderness, partly of the fragmentations of a disarticulated slave economy.[52] It also owed a good deal to disease and its spread by human agency. Seabrook spoke of a "progressive deterioration of the climate" resulting from land clearance and the abandonment of inland rice fields.[53] Dr. S. H. Dickson, in a lecture to the State Agricultural Society on malaria, insisted that "improvident clearing, imperfect cultivation, and neglected drainage of land, have been the curse of our middle and lower districts. . . . The rule should be to leave all low spots which cannot be perfectly drained and brought under permanent tillage."[54] The specter of malaria stalks the Ruffin journal. It is cited as the cause of much abandonment, and in the end persistent summer indisposition forced Ruffin to quit the state.

Soil exhaustion was much referred to at the time and has been a feature of analyses ever since. It need not be examined here and is, of course, much complicated by subjectivity and economic relativism. The very concept was something that Ruffin's dynamic soil science set out to challenge. There had, however, been much destructive cultivation since the earliest European settlements came to the Ashley River, and much former farmland had been abandoned. Absenteeism was rife. It was locally observed in 1831 that "agriculture was . . . scarcely even a secondary consideration. . . . Indeed, some who consider themselves the better class of society, appear to act as though the tilling of the soil was an abject and merely mechanical calling, nearly useless and only allowed to exist by sufferance."[55]

Manuring had only recently been practiced on any noteworthy scale, and because of the sparse numbers of livestock in the tidewater, planters had to rely heavily on vegetable matter for their composts. The nutritional value was often low and the management crude and wasteful.[56] Beneficial effects were also constrained by the absence of preliminary calcareous amelioration. But the difficulties of supply could make the exercise almost exotic. A certain "Summerville" wrote in 1831 of his mixtures of salt; wood ash; dung from hogs, sheep, cows, and pigeons; human and other urine; burned shells; clay; rotten wood; mud from ponds, creeks, and bottoms; grasses and rotted couch grass roots; brine of salted meat; salt fish; soap suds; leaves of trees; kitchen garden offals; rice and other straws, hay, corn shucks, and cornstalks; and a sprinkling of lime to activate decomposition—collected for the most part by "small negroes, and those who are past hard-labor."[57]

Frederick Porcher's 1844 *Report on Manures,* submitted to the Agricultural Society of Black Oak—an advanced parish on the Ashley just behind Charleston—has the tone of a pioneering document, concerned with novelties of uncertain value. He was unable to record much progress. "It is but a very few years since . . . the benefits of manures were mooted at every social meeting; and even now, though no one is so outwardly heretical as to question their utility, there are yet many who have derived so little practical benefit from their application, that their faith in their efficacy is rather a confidence in the testimony of others."[58] Ruffin had his own views about local manuring practices, which he expressed in his diary. In his report, he suggested that there were two main categories of material, poor in quality and difficult to collect: woodland leaves on most mainland plantations, and marsh grass and mud on the farms of the treeless sea islands (with a small additional dressing of animal manure). "The labor given to these different manuring operations is great, and perhaps in some cases the cost is carried beyond the profit."[59] The fertilizing work that he observed—most notably on the islands—did not affect his dominant perception of a state still busily engaged in the destruction of its soil. He later publicized his complaints about "land-killer" planters in an address in Charleston. "Ruin" and "destitution," he declared, were staring Carolinians in the face.[60] Some might solve their problems by emigrating; the resident majority, however, had to make amends at home.

THE VOCABULARY OF REFORM. In the view of the men who set up the survey in the early 1840s, the path to recovery in South Carolina lay through a sequence of land amelioration, improved efficiency, expanded incomes, and demographic stabilization. Despite the importance of marling, however, and the general validity of Ruffin's ideas,[61] it might well be argued that the reasoning was simplistic and uncoordinated with other social and economic realities. Bachman, Hammond, and Porcher had articulated their doubts: what prospects for amelioration could be held out in a country where the most elementary habits of land care had barely been established? And how could progress be achieved when the very terminology of exhortation was obscure and sometimes contradictory? These were questions for the proponents of reform themselves to answer.

It is significant, for example, that the notion of a survey as a means of promoting crop-boosting agents ran counter to the expressed wish of most of South Carolina's leaders in the 1830s and 1840s to see cotton production constrained, thus alleviating the problem of depressed prices. Individuals such as Hammond might seek recompense in larger, marl-assisted crops, but the collective goal, as he himself insisted, was fewer bales, fetching higher unit returns and, perhaps, larger aggregate receipts as well. Ideally, this would be combined with increased cultivation of nonstaple crops and, perhaps, a spread of effort into livestock farming and industrial activity. Calcareous amelioration could powerfully contribute to *either* this diversity *or* an enlarged staple output.

The awkwardly contradictory options, however, were never clarified or worked into any of the official arguments presented for the survey. Only a few seem to have been alert to the dangers of advance, one of whom was George McDuffie. "It will be . . . perceived," he told the State Agricultural Society in 1840, "that I have made no disclosure or recommendation of any improvement by which large cotton crops may be made. . . . I have intentionally abstained. . . . Indeed, if I could now reveal a process by which our common soils could be made to produce two bales of cotton to the acre, I should have great doubt whether the revolution would be a blessing or a curse to that great interest."[62] The doubt was valid, though it could have been answered that because South Carolina's days as the dominant short-staple cotton state seemed to be past and planters' decisions were atomized, individual restraint was a foolish irrelevance. McDuffie's point, however, had undoubted force in general terms.

The issue was not just one of correct judgment but also of confusion in the collective mind of the state's agricultural leadership, a difficulty, certainly, if unresponsive planters were to be persuaded to change their ways. Further blurring, according to Theodore Rosengarten's study of the *Southern Agriculturist,* attached to "something so fundamental as the meaning of overproduction," which was measured by different criteria. Many sought to explain it, and some denied its existence.[63] Diversification was also ill-defined. Distinctions were rarely drawn between the idealist diversification of the theoretical reformer and the pragmatic diversification already practiced by farmers in the low country as well as up-country: working that "co-existence of market involvement with the ethic of self-sufficiency" described by Lacy Ford and analyzed by Gavin Wright and Robert Gallman and Ralph Anderson.[64] This category of diversification might be termed *subsistence* and contained a strong food-grain element. The *reformers' version,* despite its self-sufficiency components, was usually forage and stock, with more animals and a positive return from the new items to the old by way of fertility improvements. Three other variants were *staple* diversification, with some combination of cash crops; *area* diversification, with geographical rather than sequential mixes; and *sectoral* diversification, with increased emphasis on nonagricultural activity, within as well as beyond the farm.[65] Diversification, in short, was a confusing slogan unless carefully qualified. For many, loose instructions to diversify as a means of easing overproduction could be devoid of meaning and, therefore, of effect.

The reformers' diversification did not stress forage as much as it might have, and this was another limitation. It was barely recognized that neutralized land, even in the South, permitted the extensive cultivation of leguminous fodder crops, which would result in an increase of nitrogen nutrients in the soil, the enlarged production of animals and manure, and the substitution of expensive imported hay. Because legumes—clover, "artificial" grasses, and the like—could not grow well on acidic soils, they had been excluded from local calculations, partly through unfamiliarity and partly through the assumption that they were victims of climate rather than soil.[66] Ruffin, with millions of subsequent farmers, knew otherwise, and a well-conceived inquiry would have been required to address the matter.

The 1840s were the decade when the need for reform was felt most acutely, in consequence of past and current problems in cotton production. There was to be no sense of secure recovery until the 1850s. An attempt could have been made to evaluate the opportunities for extended self-sufficiency for both plantations and the state. What were the most promising regions, crops, and livestock for diversification practices? How could cattle rearing cease to be, in James Cuthbert's words, an activity planters were "disposed to neglect" or to "condemn as troublesome, expensive, and unnecessary"?[67] What advances were needed in labor skill and organization? What strain might there be on planters' purses? Could the highly defective transport system cope with an increased range of items moving in and out of plantations?[68] What portion of the old production should be retained? Amelioration, the modish reform, was discussed too much in isolation by the survey's advocates, and its points of contact with the multiplicity of agricultural possibilities in the state were seldom mentioned. Its significance, as Ruffin knew, lay within a compound of reforms.

One point that might have been stressed and reiterated was that planters had no need to make sharp, unattractive choices between staple farming and mixed farming. As a great many cultivators above and below the falls were already demonstrating, they could have both, and with marling and liming, they could have more of both. On ameliorated and manured soils cotton yields could be enlarged within reduced acreages and the land freed for other crops and for livestock. If the bigger cotton outputs on the smaller cotton acreages continued to give excessive quantities and low prices, that would matter less because increased areas would be devoted to other useful products. The planter would have widened his range of market options, applying both upward and downward adjustments to cotton and other acreages and probably also altering the overall balance between commercial and subsistence production. *There would be no need for any drastic curtailment or abandonment of the staple.* This cannot be emphasized too strongly. Old habits and old creditors could still be appeased. And the liberating factor would be amelioration and manuring: not inflating the cotton crop but setting it in a flexible and compensatory context.

Ruffin said as much but held back from any insistent argument.[69] Others also dropped the odd hint, but briefly and casually. Whitemarsh Seabrook and John Belton O'Neall, of coast and piedmont respectively, observed in 1845 that farmers could "profitably curtail the quantity of land devoted to the cotton crop. An abandonment of the present extremely defective mode of culture . . . would insure a larger quantity of cotton than would be lost by divers[if]ying the products of industry. In other words, his cotton crop would be larger, his corn, wheat, rice, oats, barley, horses, mules, hogs, cattle, sheep, butter and vegetables, would be the produce of his farm." Former governor James Hamilton, Jr., the manager of the old nullification campaign, declared himself convinced that by manuring and marling on cotton fields, "half the land now in cultivation will give the same return." This advanced system helps lift "half the tax on our lands, by allowing us to fallow half, or enables us to double our product in provisions and stock, if desirable."[70] The question remained: how were planters, in large numbers, to be encouraged and instructed to act? The preliminary answer—to refine the pertinent concepts and choices—was not formulated.

INSTITUTIONAL ANXIETIES. Beneath the concern about farming practice there lay much general unease concerning the prospects for slavery. The institution seemed menaced from internal unrest and abolitionist assault and from the weakening of many of its old economic and political supports. Concerns over cotton competitors, emigration to the newer states, loss of congressional representation in Washington, provocative tariffs, postnullification isolation within the South, and growing economic dependence on the North were partly triggered, partly intensified, by alarm over the prospects of slavery.[71] If slavery were to collapse, or even be eroded, from internal and external pressures, it was believed that the economy, body politic, civilization, and system of race control would be destroyed.

A good many people sensed that perhaps the only way to save slavery within the Union lay through the fundamental, local processes of farm reform. Rhetoric was mostly vacuous, and politics at the national level was becoming arithmetically askew. Agricultural improvements presented a difficult means of rescue, but as southern slavery was essentially a plantation institution, the productive power of the land was utterly vital. Had the proposed advances been likely, by skill, technology, and capital requirement, to render slavery redundant, the equation of course would have collapsed. As it was, the improvements of amelioration, manuring, and diversification were all broadly workable with a low-technology slave labor force. In Rosengarten's words, "reform in this time and place did not challenge slavery." That, of course, is not the same as saying that slavery managed to cope with the new ways.[72] Drew Gilpin Faust suggests that agriculture was the *universal* issue and codifying solvent in time of crisis; most troubling matters could be dealt with by ritualized references to the world of farming. Slavery was certainly not everything therein, but as the economic base crumbled and a reactive agricultural evangelism intensified, there came to be articulated a growing sense of institutional vulnerability. "Agricultural reform," writes Faust, "could be equated with the survival of both slavery and the South."[73]

Whitemarsh Seabrook told his fellow planters as early as the 1820s that "the preservation of the domestic institutions of the South, depends on a radical change in the habits and practices of the tillers of the soil." Hammond, a profoundly conservative reformer, observed in his first Governor's Message in 1842 that Carolinian "domestic institutions are of a character so immutably agricultural, as to vibrate with all its reverses and vicissitudes. They would perish, or depart, in proportion as the profits of this great branch of industry were neglected." Ruffin's ideas on farming were radical at a strictly agricultural level, but in propagating them, he was motivated by a combination of reformism and conservatism very similar to Hammond's.[74]

South Carolina was not anxious about slavery and its economic base because of some accidental paranoia. The state's agricultural difficulties were very real because it enjoyed neither the temperate opportunities of the border states nor—rice and sea-island cotton apart—the remunerative staple farming of the fresh lands to the south and west. Cotton, in the words of the Roper committee, could no longer continue as "our wealth and pride . . . : a fairer bloom opens on other lands, and every day imparts a more ominous warning that the sceptre has departed."[75] There was, too, the great numerical weight of the black population. In 1840 slaves, with tiny numbers of free blacks, were heavily predominant in the low-country districts of Beaufort, Colleton, Charleston, Georgetown, and Williamsburg. In the remaining twenty-five districts of the interior, only about half had a white majority.[76] Local imbalances along the coast were greatly exaggerated in the summer, when whites moved inland to escape malaria, leaving behind what William W. Freehling terms "a massive, concentrated Negro population." The five seaboard districts also tended to be the most disunionist politically, though J. P. Ochenkowski has recently warned against oversimple correlations.[77]

"The South Carolina gentry," writes John Barnwell, "foresaw the end of slavery leading inevitably to the end of law, the end of property, the end of material and moral progress." Steven A. Channing considers that a "base line of race fear stretched unbroken across the state, embracing patrician planter and yeoman farmer in an unyielding agreement upon the necessity for the rigid control of blacks." Some historians, such as Michael O'Brien and David Moltke-Hansen, have doubts about the more apocalyptic assertions but hardly deny the power of race anxiety. There was nagging concern over insurrection, a number of plots being detected in the state in the early years of the nineteenth century, and

Nat Turner's Virginia uprising appeared for a time as possibly part of a much wider southern movement. The 1822 Denmark Vesey conspiracy in Charleston had been a close call and a decidedly traumatic experience for low-country whites. It was followed by numerous cases of arson in Charleston and the discovery of a plot in Georgetown. By the 1830s the state had become generally quiet, but unrest and suspicion had left their marks. Slavery was the root of the trouble, but by the logic of apprehension its preservation seemed all the more necessary.[78]

Any weakening of slavery was inconceivable to white Carolinians, especially those in the middle and low countries. This had to do with material function as well as physical and legal constraint. A combination of climatic, political, economic, and social deterrents had made it impossible to replace slaves with any system of free white labor. Slavery also played cultural and political roles. The institution, some argued, created an elite and allowed it to operate by the highest standards of social behavior. In March 1858 James Hammond made his notorious "mud-sill" speech in the United States Senate, asserting, by figurative allusion to the lower classes of society, that large numbers of unintelligent people had always to be available to perform dirty and menial tasks so as to liberate a cultivated minority for the trials of social and political leadership. The South secured this structure through slavery.[79] Back in 1841 Hammond had expressed similar views to the South Carolina Agricultural Society.[80] Staple culture, he contended, had created "a large class elevated above the necessity of any kind of labor, many of whom have devoted themselves to letters, travel, and to public affairs. Even those whose choice it has been to reside on their plantations . . . are accustomed to take enlarged and manly views of every thing; to govern masses; to sway, comparatively, a broad mass of territory. . . . Such characters are essential elements of a high state of civilization."[81] Robert Roper widened the issue to include the sheer age and maturity of the seaboard states. "Society with its refinements weaves a light but tenaceaous [sic] web, luxury presents its allurements, and an enlarged sphere of action is afforded by the demands of civilization."[82] Such cruel class pomposities were widely entertained. Hammond and Roper show that they were not confined to the self-satisfied and complacent. They were justifications of a sort common in various species of tyranny. Edmund Ruffin also propagated them. In so doing he, like Hammond, contradicted his own repeated vituperations against the culture and practices of real planters.[83] Southern thought was blemished by a strong reluctance to square circles.

Down from the rarefied atmosphere of fine art and manly culture, there was a very tangible and visible world of rough and self-serving political contest. Slavery's function there was to help maintain a status quo by which rich, predominantly low-country planters held a tight grip on public affairs.[84] James Hammond had been much concerned by the tactics of James Hamilton and others in recruiting popular support for the nullification struggles and by the difficulties of putting the genie back in the bottle when political life returned to seminormality.[85] The population at large might, in their ignorance, respond to peddlers of progressive ideas. Slavery, Hammond argued in 1849, would be "weakened by every accession of administrative & executive power to the masses *even here*"[86]—and with slavery, of course, James Hammond and the planter class. There was also a fear of party, which could endanger the known rules of planter factionalism.[87] Once again, there are close affinities between Hammond and Ruffin. Ruffin had no time for popular democracy. The sponsor of Jacksonian democracy was to him an "ignorant & vulgar despot."[88]

The Compromise of 1808, by which the up-country population negotiated its way into a share of political power, still left control of the state senate in the hands of the low country, and representation tipped in the latter's favor in the lower house. Suffrage had been extended to virtually all white adult males two years later, but such liberality was sharply constrained by the stringent property qualifications demanded of governors, senators, and representatives. The legislature was run by the large slaveowners. "The people," pronounced Hammond, "should not hold any power."[89]

Internal dissension complicated the defense of slavery and worsened the unease. Divisions between Unionists, cooperationists, and secessionists—to name but three—were very wide. Every period of argument and conflict, such as that preceding nullification, produced intensified animosities, increased rigidities of thought and posture, and inflammatory memories for use in subsequent crises. In October 1832, shockingly, gang warfare came to the decorous streets of Charleston. On one particularly rowdy evening hundreds of Unionists and nullifiers confronted each other, clubs in hands; Joel Poinsett and William Drayton were struck by bricks. Although slavery was, for most white people, unquestionably worthy of protection, the means of support were sources of a disturbing divisiveness.[90]

Ruffin never discusses slavery as a political issue in the diary. The institution is there as a social datum and is not allowed to intrude on the strictly geological and agricultural preoccupations of his text. But as he would have well appreciated, it had been slavery and the various anxieties surrounding it that had underlain his invitation to survey the old nullifying state. The institution and its uncertain foundations set the context.

Geology, Land, and Scientists

STRATA AND SOILS. Carolinians were familiar with the broad features of the state's geomorphology, and something was also known about geological formations and the position of calcareous strata—rich in lime and usually formed from ancient shell deposits. Back in 1765, John Bartram visited the huge Shell Bluff exposure on the Savannah, which James Hammond later used as his source of marl supply. "Took five or six hands and two battoes and provisions down the river twelve miles to the great oystershell bank, above 150 feet high. . . . This perpendicular bank is formed of oystershells all broke to small particles like coarse sand. . . . One may cut it very easily with a knife. It reacheth down to the water. . . . This perpendicular bluff continueth above 200 yards."[91] Governor John Drayton observed in 1802 that there was much "marle" and seashell deposition in the low country. James Hammond had been able to tell Ruffin, before the survey began, and with fair accuracy, that marl deposits stretched from the Atlantic to a line extending from Mars Bluff on the Pee Dee to Bean Creek on the Savannah. This was similar to Franklin Elmore's 1838 suggestion that the inland limit passed about forty miles to the southeast of Columbia.[92] A planter on the Pee Dee informed Ruffin's *Farmers' Register* of extensive, thick, shelly banks in Marion District. Marl, according to a correspondent of the *Southern Agriculturist* in 1838, was to be found "in Barnwell, Colleton, Orangeburg, Charleston, and Sumter districts" and probably "in all that section of the State, called the middle and low country." It was so abundant on the Ashley River that it was "brought to Charleston, and used for filling up the streets." It also appeared on the banks of the Cooper, the Santee, and the Wateree.[93] Despite this knowledge, though, few had turned the resources to good effect. The authoritative Dr. Joseph Johnson of Charleston estimated in 1840 that there were only about half a dozen planters in the entire state who were making notable use of marl and lime.[94]

Geologically, the patterns of calcareous strata were very simple. What was needed by 1843 was not general description but details of exact locations and quality. Marl, being shells, whole or fragmented, set in various, normally soft, matrices, was exclusively a coastal plain and riverine deposit (Tertiary, with Secondary and Quaternary). Limestone was usually older, consolidated rock (Secondary and Tertiary), located both above and below the fall line (the largest coastal plain deposits being the Santee limestones running east from Orangeburg District to the sea). Both had fossils that helped in identification. Marl usually was more accessible because, being confined to the fluvial lands of the low country, it was often visible on banks and bluffs and approachable in river craft.

The character of the soils that lay on the varied structures was determined by climate in the main and accordingly was fairly standardized. Lands above marl, in short, were not automatically calcareous: the strata might be a good way under the surface and calcium might be lost through leaching. Rainfall totals of between 40 and 80 inches per year would usually ensure the speedy removal of calcium and other bases from the soil, imposing acidity and accompanying iron, aluminum, and manganese toxicity.[95] All Carolina soils are classified as ultisols. H. D. Foth and J. W. Shafer describe these as "intensely weathered and leached": wet, sour, unpopular, and badly in need of marling and liming. W. R. Paden and W. H. Garman, examining South Carolina cotton production in 1947, advised farmers of "the importance of liming acid soils in order to have the maximum amount of cotton ready for harvest at the earliest possible date." E. N. Fergus, Carsie Hammonds, and Hayden Rogers later suggested that if cotton was grown in a yield-boosting rotation with legumes the soils should have a pH as high as 6.5. "This means that most of the soils on which cotton is grown should receive lime." Once the acidity problem was solved, environmental benefits could come into full play. "Ultisols," write S. W. Buol, F. D. Hole, and F. J. McCracken, "represent a vast potential for agricultural production. They develop in climates that have long frost-free seasons and an abundance of rainfall." Only acidity and infertility held them back.[96] Most of such basic ideas were first formulated by Edmund Ruffin. He is still an accredited pioneer, sometimes called the "father" of American soil science.[97]

SCIENTIFIC GUIDES. The quantity of presurvey geological information owed much to the localized analyses and passing observations of numerous scientists. These were of particular service to Ruffin for purposes of fossil identification and dating. Lardner Vanuxem, a Philadelphian and the first professor of geology and mineralogy at South Carolina College, undertook a small, sporadic survey in 1825 and 1826. The outcome was a brief report on the up-country districts of Abbeville, Pendleton, Greenville, Spartanburg, and York. In 1829 he co-authored an article with Samuel George Morton on the Secondary, Tertiary, and alluvial formations of the coastal plain and was a source of much handy information on Quaternary marl deposits along the coast. His later, better-known work was with the New York survey.[98] Morton, another Philadelphian, and a man of wide interests, produced a study of Cretaceous formations in the early 1830s.[99] T. A. Conrad, a local naturalist, had looked at Shell Bluff in the early 1840s and judged it to be Eocene. He also studied strata of the Miocene, another of the Tertiary subdivisions.[100] Lewis R. Gibbes was a Charlestonian who enjoyed a variety of literary and scientific pastimes and whose work included useful examination of Tertiary fossils and of Quaternary deposits under Charleston.[101] Edmund Ravenel began his working life as a doctor and academic, subsequently retiring at an early age to the Grove plantation off Cooper River to engage in shell collecting and brickmaking. He was a useful man for a surveyor to visit, given his sharp eye for fossil variations.[102] Ruffin spent two nights in his company in mid-February.

The greatest of all the scientists in and around the state at the time was the celebrated British geologist Charles Lyell. He was widely regarded as the founder of modern historical geology, holding that present and ancient conditions on the surface of the earth were essentially similar. Such views were in explicit opposition to neptunist and catastrophist notions, which stressed dramatic discontinuities, sometimes of a seemingly biblical character.[103] In the early 1840s, Lyell was just coming to international prominence, though his was very much a passing presence. He had recently completed his first American journey, part of it in the Southeast, and a two-volume account of his travels was published in New York in 1845.[104] Most of his articles also postdated the survey, though the *Principles of Geology* was available (in two editions),[105] and Ruffin did make occasional reference to him on the matter of Tertiary fossils. On the first day of 1842, Lyell and his wife took a cotton boat down the Savannah from Augusta to James Hammond's marl excavations at Shell Bluff and decided, like Conrad a little later, that the beds were upper Eocene. During one of his stops at Charleston, Lyell sailed up to the Grove to meet Edmund Ravenel. He also visited John Bachman. Any observations of interest were no doubt passed on to Ruffin when he made his own calls on Ravenel and Bachman a few months later.[106] Others, such as William Ellett, Robert Henry, Mitchell King, Henry W. Ravenel, Frederick Porcher, and Louisa McCord, were also involved in geological exchanges, though some of their concern derived less from disinterested curiosity about an advancing science than from the urge to see how geology and religion could be reconciled at points of conflict.[107] It was not a matter that ever attracted Ruffin's serious attention.

Ruffin's account may sometimes remind the reader of Charles Darwin's first *Beagle* journal, published in 1839. Both recorded, with obvious fascination, the oddities of landscape, geology, zoology, economic practice, and social custom. Ruffin cannot, however, be accused of stylistic plagiarism, the similarities being merely those of naturalist convention. Darwin's volume did not come into his hands until 1864; he called it "a very interesting book, & new to me."[108]

Ruffin's Itinerary

STAGE ONE.* "I may presume to claim," Ruffin later wrote of his time in the state, "that no person employed by any government, from discretionary duties, ever labored more constantly & assiduously than I did."[109] The presumption may have been excessive, but the work was unquestionably hard and seriously undertaken. There is little similarity between Charles Lyell's lordly passages with his wife and the isolation and physicality that characterized so much of Ruffin's effort. The writ-

*The stages here correspond to the editorial subdivision of the diary.

ing, too, is very different: Lyell brief and anecdotal, Ruffin intensely involved with his subject matter.

The outlines of the journey can be roughly charted in the diary's subheadings. There follows here the barest of summaries.

Ruffin set out from Charleston on 31 January 1843, shortly after his forty-ninth birthday. Most of February was spent in the vicinity of Charleston, where he examined promising ground along the Cooper and Ashley rivers, traveled around the Santee Canal, and had a look at James Island. Distances of up to forty miles were recorded for individual days. He also attended to office arrangements and specimen analysis in Charleston itself. His headquarters were the Charleston Hotel and Robert Mills's Fireproof Building, a few hundred yards apart on Meeting Street.[110]

At the end of February he set off for Beaufort in the southern corner of the state. His route was a deliberately slow one past the renowned long-staple cotton fields of Edisto Island. Arriving in Beaufort, on Port Royal Island, he arranged trips to Lady, Distance, and St. Helena islands; his work in these geologically less bountiful parts took about two weeks. From there he moved round the headwaters of the Broad River and down to the Savannah at the former Swiss settlement of Purysburg, subsequently striking back east by a lonely and troublesome inland route along the southern boundaries of Barnwell and Orangeburg districts.

On 18 March he arrived at the Santee River near Eutaw Ferry, the journey from the Savannah having taken just four days. The surveying, clearly, had been only cursory. He moved around the richly calcareous lands by the Santee and the socially congenial Charleston hinterland until the end of the month before taking off eastward for Georgetown and the great rice plantations at the mouth of the Pee Dee. He had two weeks there in helpful company before striking upriver to examine the finest marl exposures in the state. He got as far as Mars Bluff near the Marlborough District boundary and then, after less than a week up the Pee Dee, turned speedily back to Charleston across Black Mingo Creek and the Black and Cooper rivers. He was in the city by 24 April and set sail the following day for Wilmington and a three-week break in Virginia. His work thus far had been confined to the tidewater, with the main emphasis on the sea islands and the fluvial country behind Charleston. The one coastal area he ignored was Horry in the northeast, his assumption being that there were no knowledgeable people to

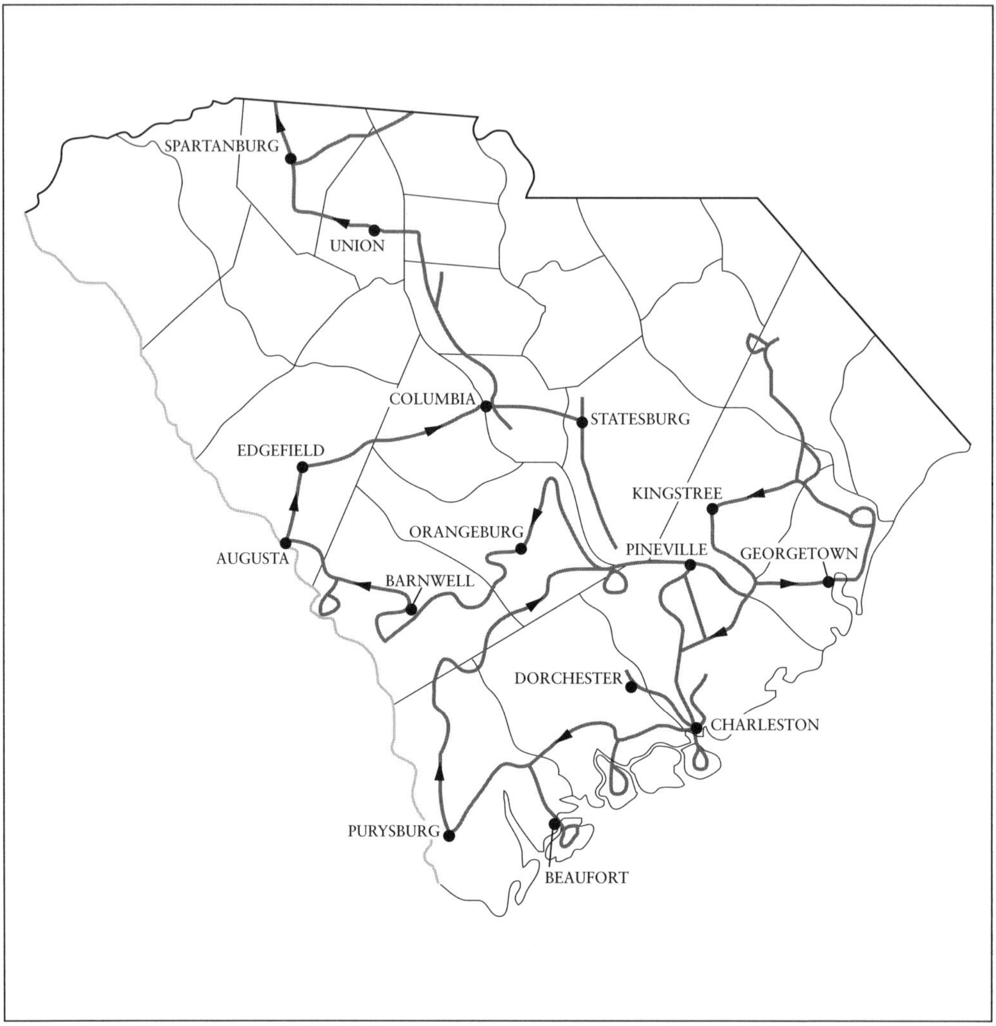

RUFFIN'S ROUTE THROUGH SOUTH CAROLINA
Source: Base maps drawn from Robert Mills's *Atlas of the State of South Carolina*, 1825.

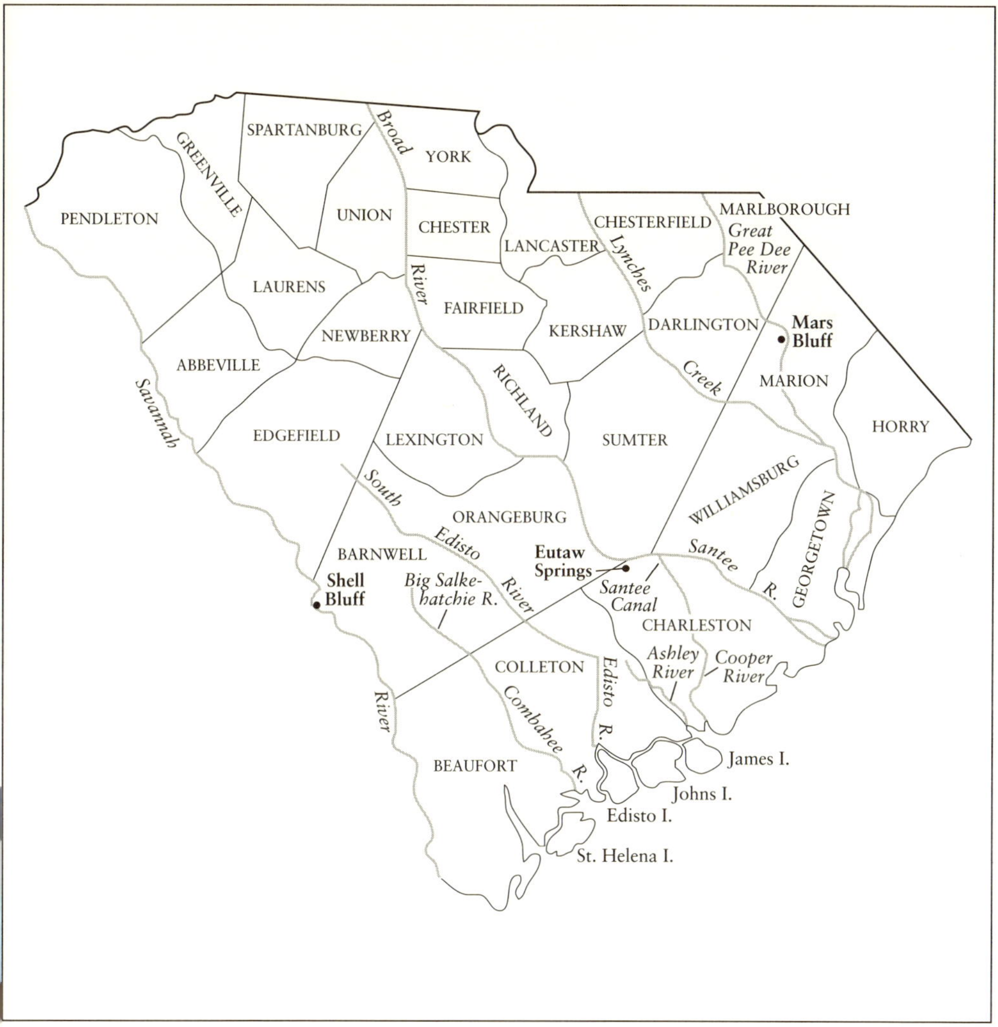

SPARTANBURG
GREENVILLE
PENDLETON
UNION
YORK
CHESTER
LANCASTER
CHESTERFIELD
MARLBOROUGH
Great Pee Dee River
Broad River
LAURENS
FAIRFIELD
KERSHAW
DARLINGTON
Lynches Creek
Mars Bluff
NEWBERRY
ABBEVILLE
RICHLAND
MARION
HORRY
Savannah River
EDGEFIELD
LEXINGTON
SUMTER
WILLIAMSBURG
South Edisto River
ORANGEBURG
GEORGETOWN
BARNWELL
Eutaw Springs
Santee R.
Shell Bluff
Big Salke-hatchie R.
Santee Canal
CHARLESTON
COLLETON
Ashley River
Cooper River
Edisto R.
Combahee R.
BEAUFORT
James I.
Johns I.
Edisto I.
St. Helena I.

be found in that "dark corner." It was a district with few plantations and an emphasis on subsistence agriculture.[111]

STAGES TWO AND THREE. After his return to South Carolina on 10 May, Ruffin retraced some former routes along the Cooper and Santee, pushing on into Orangeburg and arriving at Columbia on the twenty-first to meet planters assembling for a state convention. He spent a few days surveying in the neighborhood of Orangeburg village and then headed west into Barnwell for a busy two weeks of exploration on the Savannah and its eastern creeks. The main feature here was James Hammond's Silver Bluff plantation. Governor and surveyor visited the great marl exposure at Shell Bluff together, spending their evenings in specimen analysis and games of billiards.[112]

On 16 June Ruffin left to return to the capital, traveling by Augusta and Edgefield. During the last week of the month he looked at marls and limestones in the western parts of Sumter and Clarendon, staying close to the Wateree and Upper Santee river system. At the beginning of July he took a few days' rest in Columbia, followed by a trip up-country to Fairfield, a notable piedmont cotton district, for another convention. By the tenth he was back on the coast at Charleston, packing for a second journey home to Petersburg. For some weeks he had been bothered by ill health. Because this had interfered with his work and was causing him some anxiety, he had decided to seek advice from doctors in Virginia. His concern, no doubt, was that he had lingered too long in the heat and damp of the low-country summer. This time he was gone for six weeks.

Ruffin returned to Columbia by rail on the evening of 4 August and by the following day was back in his buggy, riding north again to Fairfield. He had not made a full recovery and was soon afflicted again by biliousness and headaches. He passed through Monticello in a straight northwesterly line to Unionville and a place near the Cowpens battlefield known then as Limestone Springs, with diversions to Spartanburg and the Nesbitt ironworks on Broad River. By now he was in the northernmost part of the state, and after five days of illness and confinement at the springs he passed into North Carolina and on to Flat Rock in the Blue Ridge "to seek a perfectly healthy climate" and a week or two of rest.

He had had enough though, despite all the personal attention still available. Arriving in Flat Rock on 13 September, he closed his diary.

"These notes of late have been so little agricultural in their character . . . that I shall cease taking the trouble to continue them." He found his way home again, returning to South Carolina in the fall to see out his contracted year and finish his report. Odd jottings and letters suggest that he visited some areas in the western piedmont and spent time in Charleston and Columbia. On 4 November he wrote from Charleston to Colonel David McCord[113] to say that he was too busy to pay a last call on him and his wife, Louisa, at their plantation in Orangeburg District and that after the completion of the writing he proposed to "set out to use the little remainder of my time of service to the best purpose in further marl investigations." He wished to examine as yet unvisited areas. Illness remained a problem. "I have been sick half my time since June," he told McCord, "and very sick at the last attack, after frost, in Fairfield. I am far from well now—& do not expect my damaged health to be re-established for a good while."[114] Much of the fieldwork over these last weeks of 1843 was given over to his son Julian.[115]

What was initially conceived as a two-year survey had turned out to be a series of excursions concentrated effectively into a mere five months. Despite the intensity and authority of the work, it was hardly a comprehensive evaluation. Ruffin's resignation at the end of November was a disappointment to James Hammond, who seems to have hoped that Ruffin would still be persuaded to serve the full two years. Hammond acknowledged Ruffin's "important" services in a brief note, pointing out how much more useful his efforts would have been if he had seen them through to completion.[116]

Fortunately, Ruffin had suggested a successor: "an obscure & unknown teacher of a school in Petersburg" by the name of Michael Tuomey. He was an Irishman by birth, a trained geologist who had met Charles Lyell and was a person high in Ruffin's esteem.[117] The new appointment was made at the beginning of 1844. Tuomey was to "at once report to the back country" to engage in mineralogical investigations. The transfer, though, was hardly speedy. "Where is Mr. Tuomey?" Hammond asked Ruffin in March. The work got under way by early summer, and the governor soon declared himself well pleased. The emphasis now was to be on physical science, in accordance with Tuomey's qualifications and Hammond's original preference. Tuomey completed Ruffin's second year and then won extensions totaling two years and four months before moving on to Alabama.[118]

Ruffin's report was finished and submitted to Hammond on 1 December 1843.[119] It was published by the state printer before the end of the year, and long extracts appeared soon after in the *Southern Agriculturist*. A supplement based on Julian Ruffin's efforts along Lynch's Creek, the Waccamaw River, and other sites in the Southeast was sent down from Virginia in mid-January 1844.

The Survey and Its Achievement

CONTEMPORARY EVALUATION. In the letter accompanying his report to Hammond, Ruffin apologized for much "superficial and incomplete" work and for some missing and delayed items. Comment, however, also seemed necessary on the difficulties of surveying within a benighted farming community. There was, he alleged, an insufficiency of elementary "facilities" for his geological investigations, resulting from planters' ignorance of calcareous exposures and the lack of excavations of their own. It was, he declared, "impossible for me, by my own direct labors, to supply all that would be wanting for a general Survey and Report." His partial solution to the latter problem lay with enlisting help from agricultural societies, mainly by the distribution of questionnaires.[120] Ruffin was highlighting two of the most fundamental flaws of the survey: its reliance, in the field, on guidance from scientifically illiterate farmers; and its being the responsibility of a solitary individual who had no familiarity with the state and who fell victim to its dangerous climate.

No doubt recalling the many promises of good intent that had been made to him during his tour and the interest he seemed to have aroused at clubs, societies, and conventions, Ruffin suggested that many diggings had, "since my visits, been commenced for marling operations, and will now be rapidly increased in extent and number."[121] Such optimism continued into 1844. He observed then, from the distance in Virginia, that although Carolinians had come late to calcareous manures, "most of the intelligent planters . . . are now well awakened to the value of this improvement, and many have already commenced marling, and some of them are making unexampled progress."[122] But he was relaying hopes, not observations. He could be excused these for a year or two because, after his Virginia disappointments and retreats, he had a powerful personal interest in success and some horror of further public failure. When the undeniable letdown became apparent and he remembered all the insistent protestations of his alleged disciples, expectation gave way to contempt. This was to signify the end of his career as a reformer and the beginning of his switch to private farming and radical disunionism.

If Ruffin, in Virginia, could dream for a bit, what of the more informed South Carolinians, who picked up the farming gossip and were well aware of major new departures if and when they were occurring? Were their optimistic comments merely polite, or were they part of the same self-delusion and posturing that, in part, had got the survey under way in the first place? The *Southern Agriculturist,* seeing salvation through the serialization of gospel, rejoiced that the state had the richest marl known and that Ruffin had "awakened a spirit of experiment and inquiry," which had "given a new impulse to agriculture." Ruffin's travels, thought a Charleston planter, led to "so many experiments which could not have existed but through his persuasion."[123] Henry W. Ravenel believed, a little more circumspectly, that "a vast mine of agricultural wealth" had been revealed, requiring only "the energy and industry of the planter to be made available."[124] James Hammond, writing to Ruffin, conferred on him an exaggerated and probably self-regarding compliment, but one nonetheless likely to please a man in search of "reputation." How lucky he was, said Hammond, to be "one of the few benefactors of mankind whose services have been appreciated by the world, while living."[125] Robert Roper in 1844, with his own sponsorship very much in mind, invited his fellow planters: "Receive with favour the Agricultural and Geological survey of the State. . . . It is now young, but promising, and will not fail at maturity to reward your nurture." Roper died shortly after. "The most fitting tribute we can pay to his memory," declared Joel Poinsett, "is to carry forward his views and to complete the great work of fully surveying . . . every District of the State, in order that its agricultural resources may be discussed and developed."[126] Other sources of hope and flattery included Mitchell King, Frederick Porcher, and John Ramsay.

R. F. W. Allston, the celebrated rice planter and future governor, may have touched on Ruffin's main but unmeasurable achievement when he wrote that a survey "serves to discover the best systems of culture and of management perfected by gifted individuals . . . and communicate them to distant parts of the State." This had been done through the report, documenting sea-island cotton growing, Georgetown rice cultivation, up-country legume farming, Hammond's improvements, Samuel Porcher's embanking, and a variety of other agricultural achievements that may have helped spread an idea or two around, even if there were no takers for marling and liming. Ruffin himself was aware that his main efforts had been directed toward the discovery and use of calcareous manures and that it was in that area that his principal claim to fame lay. This had been, he wrote, "the one department in which I counted upon rendering the most early and important services to the improvement of the soil and agriculture of South Carolina."[127]

He knew for sure that he had confirmed the existence of rich marl beds throughout the tidewater, even though his successor, Michael Tuomey, had only faintly praised him "for the correction of errors as regards localities" and "the directing of public attention to the economical value of . . . those formations." Ruffin had advanced a clear categorization into the Pee Dee beds (upper Cretaceous, eastern, and rather poor); the Great Carolinian/Santee (middle Tertiary, extensive central and western, and rich); the Great Carolinian/Ashley and Cooper (middle Tertiary, Charleston area, less rich than Santee); and coastal (Quaternary, patchy by sea, loose with many whole shells). Marl, he had demonstrated, underlay most of the low country and was much exposed along river courses. South Carolina had possibly ten times as much accessible marl as Virginia, and it was of markedly superior quality. No other known region possessed "half as great advantages and resources for agricultural improvement, or more needs the employment of these means."[128]

These, plus his own accumulated theories and past recommendations, were the offerings on which Ruffin staked his Carolinian hopes. By 1845 he knew that in practical results it had all been futile. Experiences in the state in 1843 had already hinted at the possibility of such an outcome. Many planters had not known what marl looked like and had called him out on a number of wild goose chases. One confused it with petrified wood and many others with lime rubbish around old indigo vats. Ruffin was in South Carolina for eight weeks before he saw anyone engaged in a marling operation, which caused him some dismay, but the greater disappointment came from the apparent hollowness of his reputation in the state. Many people were familiar with his name but not with his ideas. "It is," he wrote in his diary on 5 March, "both astonishing & mortifying to me to find such general ignorance in regard to the use of calcareous manures, upon which I have so long been writing & publishing & I have sent so many copies of my publications to this state." The *Charleston Mercury* reported during his visit to Edisto Island (in an item which Ruffin pasted into his diary) that planters there believed they "had attained the full measure of agricultural knowledge. The appearance, then, among them, of any individual who proposed to teach them anything new as essential to their agricultural prosperity, was looked upon, to say the least, with incredulity." Near the Santee Canal, several former *Farmers' Register* subscribers worked right above extensive masses of untouched marl. Audiences at meetings did seem curious, often promising to start amelioration at once, but after Ruffin moved on the interest usually died away. They had, he assumed, been humoring his "supposed marling mono-mania." On a return visit to a farm behind Charleston, he was pleased to note that marl was at last being excavated, only to discover that the earth was wanted solely for embanking.

Ruffin surmised at the beginning of his survey that James Hammond alone had possibly marled six times as much land as had all the rest of South Carolina's farmers together. With the exception of a few deviants, for a few years, that, broadly, was the way matters were to remain. From beginning to end, Hammond was the only large-scale, determined marler or limer in the entire state. Likewise, Shell Bluff was to remain the only marl exposure well known before the Civil War—and that was on the Georgia side of the river. Such extreme narrowness of response is surprising, given the abundant resources, the scale of fertilizing in the postbellum decades (much of the material coming from the phosphatic marl beds under Charleston),[129] and the wide and productive resort to calcareous applications in the present century. It was, however, what Bachman and Hammond had forecast.

By the start of 1845 Hammond had marled about twenty-three thousand acres of field and drained swamp. His initial labors, of course, predated the survey by some years. Tentative experimentation had grown into a massive program. "I thank you & Shell Bluff for my conversion to Marl," he told Ruffin. "If we live 3 or 4 years longer," he wrote in October of that year, "I want you to come & see the *revolution* at Silver Bluff." By the end of the 1840s he estimated that his net annual gain from seven years of marling was around $5,000 or 50 percent, despite much trouble with the weather. But no one nearby seemed to be responding or even talking about agricultural experiments. "Altho' everybody now acknowledges the value of it, *nobody* seems to calculate on using it." People exaggerated the costs and the labor requirements, and, what was more, they were disposed to mock. Hammond found himself "only laughed at by my neighbours, who would enjoy every failure as a good joke." Carolinian planters were "so very plain." They had little interest in new ideas. "In our days we may perhaps get the people to read agricultural papers & to listen to an occasional lecture," but it still seemed a long way off.[130] His improving work, of course, was by no means entirely calcareous in emphasis. Drainage and diversification were major allied programs, and, wealth accumulation apart, the principal goals were the security, economy, and independence conferred by plantation self-sufficiency. Faust sums up: "In his planting efforts, he discovered a new arena for personal accomplishment and public eminence, and his agricultural leadership encouraged and reinforced his emerging political authority as a defender of the South and its institutions."[131]

This one instance of success, then, concerns the highly intelligent and fiercely ambitious *arriviste* from the obscurity of Newberry, who married into land and money and who in his younger days had prayed, "Oh God. . . . Make me great among men."[132] He was, arguably, the most atypical planter in the state. His achievements never stimulated easy imitation.

"I fear," Ruffin wrote to Hammond in 1845, "that on you, almost alone, still will rest the whole burden of establishing by practice in S.C. the truth of my doctrine." In his private notes a few years later, Ruffin observed that "the results have not yet amounted in value to the state to one-thousandth part as much as I had expected." There is little evidence to support Avery Craven's appraisal that "the interest he awakened in better farming, though not revolutionary in results, was the beginning of a better day in the eastern cotton belt." The great marl beds lay almost totally untouched. Despite the context of economic and institutional anxiety, the survey had ended up as an affair between gentlemen, having as much to do with status, gesture, and seigneurial custom as with farming practicalities. To John Brisbane, president of the St. Andrew's, Ashley, and Stono Agricultural Society, even James Hammond was suspect. He had, knowingly, "memorized his name" by means of the survey.[133]

CHARLESTON ORATION. Thus far, South Carolina had been spared an oration from Ruffin. The state had plenty of men among its own leaders ready to spring to their feet at the smallest anniversary, with portentous lamentations and exhortations. Ruffin probably always had suitable sentiments ready for delivery, but he had a nervous loathing of public speaking as well as a contempt for exhibitionist speechifying so he had kept his silence on the grander issues. In November 1852, however, following an invitation to give the agricultural address at the South Carolina Institute during its annual fair in Charleston, he made up his mind to tell his former hosts in Hibernian Hall (just across the street from his old office in the Fireproof Building) what he really thought of them. It is recorded that "great interest was manifested" during its delivery.[134]

He at once reminded them of his "personal sacrifice of interest, of ease, and of domestic comforts" in 1843, in his "zealous effort to promote the agricultural improvement of your country." The outcome, however, had been ruined expectations. Since retiring from his office, "it has continued to me a subject of painful reflection, that I failed to earn the great and most grateful recompense I had hoped for—the inducing the people of this State to begin, in earnest and properly, to use their wonderfully abundant and valuable resources for the improvement of their lands—and in rendering this benefit to them, to acquire for myself some just claim to be deemed a public benefactor." Ruffin, too, had self-regarding weaknesses, though he was honest about them, declaring that only such vainglorious ambition could have drawn him from his home.

He proceeded to remark on the typical Carolinian's energetic pursuit of maximum, immediate profits; the exhausting culture; the tendency to resort to the clearing of forest land at costs generally higher than those of marling his worn fields; the evidence from Virginia of the great benefits of calcareous amelioration; the high profits that still, perversely, could be secured under primitive farming; the role of slavery, which could be highly destructive under defective planters or "a great and valuable aid" under intelligent improvers; the unpopularity of green, manuring crops and fallowing; the expensive importation of northern lime; and the inadequate drainage of swamplands.

Having made these familiar points with some eloquence, he then raised the level and generality of his attack. Many in the farming community were, he considered, "ready enough to accept and to apply to themselves and their fellow planters, the name of 'land killers.' But while thus admitting, or even assuming this term of jocose approach, they have not deemed as censurable or injurious, their conduct on which this reproach was predicated. . . . Their error, in regard to their own interests, great as may be, is incomparably less than the mistake as to other and general interests not being thus affected. . . . Individual planters may desert the fields they have exhausted in South-Carolina, and find new and fertile lands to exhaust in Alabama." Carolinians left behind, however, had the miserable job of coping with "the enormous evils produced"—evils to the community and to posterity. The South as a whole had already degenerated to frightening lengths. The political weakening in Congress and the scale of antisouthern legislative defeats were clearly on the record.

Yet not everything was lost. Radical changes of heart and practice would yield an "increase of political power, both at home and abroad, which at this and the near approaching time, would be especially important to the well-being and the defence of the Southern States, and the preservation of their yet remaining rights, and always vital interests." But how to get planters to alter their ways? That was the old and fundamental question, and Ruffin's speculative answer was at once desperate and provocative: "If it were possible that, for all lower South-Carolina, the system of improvement could be directed by one mind and will, as much as the operations of any one great individual estate, the most magnificent results could be obtained with great and certain profit, and in a few years." This, from the master of Marlbourne, who had turned an acidic farm in Virginia from profitless sterility to high, calcareous profit in half a decade, must have drawn a few long glances from the "killer" audience in Hibernian Hall.

"Planters of South-Carolina," Ruffin ended, warming to the role of consul in the senate, with Hannibal at the gates

> —I have offered to you in plain and unvarnished language, and, possibly, it may be in ungracious and distasteful terms, the last advice and admonition that I can expect to utter to you or any similar audience. My burden of years, and infirmities much greater than even suited to my age, admonish me that my labours must soon close. I would deem it a reward of more value to me than will be the short

> remainder of my life, if you and your fellow-labourers . . . would heed my words, and fully profit by them. It is but little that a private individual can do, to warrant to a great commonwealth or community, the beneficial results predicted upon stated premises and conditions. . . . Choose, and choose quickly! And remember, as my last warning, that your decision will be between your purchasing, at equal rates of price, either wealth and general prosperity, of value exceeding all present power of computation, or ruin, destitution, and the lowest degradation to which the country of a free and noble minded people can possibly be subjected.

MARGINALITY. There are bound to be elements of success and achievement in any man's intelligent conversing, writing, and journeying through a state of the size and importance of South Carolina. In the case of the Ruffin tour, however, these are minor, personal, or elusive, and no time will be spent identifying and evaluating them. Some will be obvious by now. Others will become clear as Ruffin's account is read.

The survey left agriculture almost exactly where it was when Ruffin arrived and gave succor to the complacent and the cynical. Only the anonymous forces of the cotton market could, with a future lift in prices, give superficial relief. Such apathy is noteworthy, considering the profound weaknesses of this most exposed of southern states and James Hammond's massive and convincing demonstration of local farming possibilities. The survey also put an end to Ruffin's sui generis career as a full-time, quasi-professional reformer. The buying, draining, marling, and expanding of the Marlbourne estate in the 1840s and 1850s marked a return to the life he had led back in the 1820s on the south bank of the James River.[135] He was very clear in his writings about his desire to quit the farming public, although the inclination dated from Petersburg in the last years of the *Farmers' Register.* There can be little doubt that it was confirmed by the failure of his Carolina survey. This did not match the *Register*'s collapse, but it was bad enough—if only because the second galling setback confirmed the message of the first.

Ruffin had initially seen South Carolina as the redeeming state, the place that *did* appreciate him, the community with the sympathetic grandees, and the part of the Union that had shown a grasp of vital political realities. Failure there must have been a terrible letdown, and it may explain why he came back to Charleston in 1852 to list the omens of disaster. It seems to confirm that he had been experiencing formatively disturbing mental processes carrying him out of peaceful, scientific reform and into sententious, belligerent disunionism. Ruffin, South Carolina, the South, and the cause of intelligent farming were the losers.

In explaining failure and marginality, there is no need to engage in tedious repetition of illusions, inconsistencies, and fallacies. The general southern issues have already been examined in my *Edmund Ruffin and the Crisis of Slavery in the Old South*. It is clear that one man was not enough (and he said as much himself); that, with short periods and intermittent illness, there was inadequate time; that he had physical difficulties in moving around; that the scientific help, on the whole, was unprofessional; that there was distracting planter hospitality; that entrepreneurial interest was close to minimal; that transport defects in the disarticulated slave economy yielded severe impediments through high carriage costs for marl;[136] and that the legislature's instructions ignored vital issues of cotton overproduction and plantation diversification. Perhaps, too, mental disjunctions set in over the gap between the principal, immediate issue and the dominant, fundamental issue: marling and slavery, respectively.

There were three other factors of note, the last two of which Ruffin was very well aware. The first was the language and presentation of reform appeals. John Bachman had warned of scientific illiteracy in the state, a problem that had, somehow, to be accommodated rather than dismissed if the exercise were to be more than merely patrician. As it was, however, even the large planters were unresponsive. If the prime targets had been missed, why bother with other, less promising ones? Abstractions from books and journals, moreover, and examples from the border states, were unlikely to be persuasive in South Carolina. "I do not take Mr. Ruffin's work," said one farmer shortly before the survey began, "it is too far north for me." "You cannot influence us," wrote another in 1842, "so long as you speak only of its [marl's] results on corn, wheat, and clover."[137] Ruffin, moreover, while allowing his *Calcareous Manures* to run to five increasingly fat and desultory editions, never took the trouble to convert it into a clear and systematic pocket book for busy planters in the field.

The second problem, which he was at pains to explain in a newspaper letter of February 1843, was that calcareous applications like marl needed organic matter to work on in the ground. Since they hastened decomposition they could, in a worn-out field, merely accelerate the depletion of nutrients. Marl and lime had to be directed to new fields, or fallows, or acres to which vegetable material had been added.[138] This requirement, for some, was hopelessly complex. According to his Hibernian Hall account in 1852, Ruffin's failure derived not so much from being totally ignored as from errors and hasty abandonment by the few who did follow his general lead.[139] He was certain what had gone wrong. "I infer that the main and usual cause . . . has been the same prevailing bad practice, before denounced, of incessant, or, at least, much too frequent tillage, which does not permit the fields to receive and retain organic matter from their own growths especially. This cause had operated on nearly all the trials of marl made previous to my service in South Carolina." In the report he reiterated his own earlier observations on the subject: "that without vegetable matter to combine with, calcareous manures will be of little value." Conversely, there could be no failure "on land in proper condition."[140]

The third problem is related, for it may be combined with the second to produce what can be termed the "bother" factor. Many planters simply did not wish to take the trouble to ameliorate and diversify. The crude, inexpensive, preliminary tasks of digging, hauling, and spreading marl could easily repel farmers, especially when combined with the need to attend to organic supply in the soil.[141] These did not have to be ignorant men of low motivation. On a visit to Point Comfort plantation on 24 April 1843, Ruffin had a chance to see how his grandiloquent sponsor, Robert Roper, was coping with the bother. The encounter was neatly ironic.

> Walked with Mr. Roper to see his lately discovered marl, & of the labor of digging which he complained heavily, & seemed to think difficulty enormous. The marl . . . lay only 3 or 4 feet under the surface. . . . When I before visited this place, Mr. R. had no idea that his land had marl. But I was sure it must be there & accessible enough, & advised his digging for it. This he ordered a few days after. . . . And upon examination, the bottom of the ditch was of this rich marl, which, hard as it is, is so convenient to use, that I would desire nothing better to cover the nearest 500 acres. However, so much are the difficulties of all new operations magnified by

their newness, that Mr. Roper thinks the job excessively laborious & costly.

Notes

1. John Drayton, *View of South-Carolina,* pp. 6–11. (Ed.'s capitalization)

2. Coclanis, *Shadow of a Dream,* pp. 112–13; Ford, *Origins of Southern Radicalism,* pp. 39–45.

3. Coclanis, *Shadow of a Dream,* p. 123.

4. Ford, *Origins of Southern Radicalism,* p. 38; see also Smith, *Economic Readjustment of an Old Cotton State,* pp. 19–44.

5. Hammond to Ruffin, 24 November 1842, Ruffin Papers.

6. Roper, "Address," pp. 239–40. For his importance, see *Farmers' Register* 10 (September 1842): 453–54; Seabrook, *Memoir on Cotton,* p. 27n; Poinsett, "Agricultural Address," p. 241.

7. Hammond to Ruffin, 5 February 1843, Ruffin Papers.

8. Ruffin, *Diary* 1:75 (20 May 1857). For a fine biography, see Faust, *James Henry Hammond.*

9. For biography see Craven, *Edmund Ruffin;* Mitchell, *Edmund Ruffin;* see also Mathew, *Edmund Ruffin.*

10. See Ruffin, *Essay on Calcareous Manures;* Allmendinger, "Early Career of Ruffin"; Mathew, "Edmund Ruffin and the Demise of the *Farmers' Register.*"

11. Mathew, *Edmund Ruffin,* pp. 84–89.

12. Craven, *Edmund Ruffin,* p. vii; see also Gray, *History of Agriculture,* 2:780.

13. The best early account is in *Southern Cultivator* 1 (March 1843): 9–10.

14. Hammond Diary, 1841–46, 13 November 1841, Hammond Papers.

15. Mathew, "Ruffin and the Demise of the *Farmers' Register.*"

16. *Southern Agriculturist* n.s., 4 (February 1844): 48.

17. "Incidents of My Life," 2:155, 159; Hammond to Ruffin, 14 January, 5 February 1843 (and enclosed legislative document of 25 January 1843), Ruffin Papers. Family details have been taken from the introductory material supplied by Scarborough for Ruffin, *Diary,* vol. 2.

18. *American Farmer* 3 (January 1822): 324–25.

19. *Farmers' Register* 10 (July 1842): 366; (December 1842): 522. Correspondents had included Whitemarsh Seabrook, William Murray, J. Jenkins Mickell, Joseph Johnson, Lardner Vanuxem, Thomas Cooper, and Franklin Elmore.

20. Ibid. 8 (April 1840): 243–44.

21. Ruffin, *Report of Survey,* Appendix 1.

22. Bachman, *Inquiry into the Agricultural Survey*, p. 9; Gene Waddell, Introduction to Mills, *Mills' Atlas of the State of South Carolina*, pp. i–xii.

23. Millbrooke, "South Carolina Surveys," pp. 27–28.

24. *Journal of the House of Representatives, 1843*, p. 91; Millbrooke, "South Carolina Surveys," p. 28.

25. *Proceedings of the Agricultural Convention*, pp. 13–16 (26, 27 November 1839).

26. *Farmers' Register* 10 (September 1842): 453–56.

27. *Journal of the House of Representatives, 1843*, p. 88.

28. Hammond, *Governor's Message*, p. 10.

29. *Journal of the House of Representatives, 1843*, pp. 88–92.

30. *Courier*, 17 December 1842, p. 2.

31. *Journal of the House of Representatives, 1843*, p. 93.

32. Hammond to Ruffin, 14 January 1843, Ruffin Papers.

33. O'Neall, "Agricultural Address," p. 201.

34. Faust, *James Henry Hammond*, pp. 114–15, 115n, 237.

35. *Journal of the House of Representatives, 1842*, p. 88.

36. Ruffin, *Address on Exhausting and Fertilizing Systems*, p. 3; "Incidents of My Life," 2:142–46, 158–59, Ruffin Papers.

37. See diary below, 18 February 1843; Hammond to Ruffin, 21 December 1846, Ruffin Papers, in which Hammond remarks, "[I] don't value consistency a copper."

38. Faust, *James Henry Hammond*, pp. 213–14.

39. Hammond to Ruffin, 5 February 1843, 7 July 1844, 22 July, 4 November 1846, 11 March 1851, 24 October 1853, Ruffin Papers; Hammond, *Address before South-Carolina Institute*, pp. 12, 14.

40. See Fogel and Engerman, *Time on the Cross*, passim; Moltke-Hansen, "Expansion of Intellectual Life," pp. 22–45; O'Brien, ed., *All Clever Men*, pp. 1–25.

41. Moltke-Hansen, "Protecting Interests," p. 172; Rosengarten, "*Southern Agriculturist*," p. 292.

42. Bachman, *Inquiry into the Agricultural Survey*, pp. 5–9, 11–40.

43. Hammond, "Anniversary Oration," pp. 189–90; Roper, "Address," pp. 232–35; Poinsett, "Agricultural Address," pp. 252–54; Lander, *Textile Industry*, pp. 50–55.

44. Robert Roper complained in 1844 of the "present antagonistic positions" affecting relations between agriculture and industry and of a "perversion of sentiment and sound policy." The planters, as he saw it, were principally to blame (*Proceedings of the Agricultural Convention*, pp. 225–26). James Hammond later agreed: "The ancient and illustrious calling of agriculture . . . is

apt to engender a haughty contempt of all Mechanic Arts" (*Address before South-Carolina Institute,* p. 12).

45. See, for example, the reaction of Edisto farmers to Ruffin, as recorded in a March 1843 newspaper clipping inserted in diary.

46. Problems discussed in Mathew, "Slave Skills," pp. 171–87.

47. *Journal of the House of Representatives, 1843,* pp. 89–91.

48. Millar, Turk, and Foth, *Fundamentals of Soil Science,* p. 146.

49. *Journal of the House of Representatives, 1843,* pp. 89–91.

50. Seabrook, *Address at United Agricultural Society,* pp. 9–11.

51. McDuffie, "Anniversary Oration," p. 106.

52. Mathew, "Agricultural Adaptation," pp. 136–37.

53. Seabrook, *Address at United Agricultural Society,* p. 10.

54. Dickson, "Essay on Malaria," p. 171; see also Freehling, *Prelude to Civil War,* pp. 10, 12–13, 33.

55. *Southern Agriculturist* 4 (March 1831): 121.

56. Smith, *Economic Readjustment of an Old Cotton State,* pp. 89–96; Ruffin, *Report of Survey,* pp. 82–85. See also Rosengarten, *Tombee,* pp. 51–52.

57. *Southern Agriculturist* 4 (May 1831): 247; see also (April 1831): 215–17.

58. Porcher, *Report on Manures,* p. 1.

59. Ruffin, *Report of Survey,* p. 83; see also pp. 82, 84–85.

60. Ruffin, *Address on Exhausting and Fertilizing Systems,* pp. 9, 22, 26.

61. See Akehurst, *Tobacco,* pp. 28–29; Clay, Orr, and Stuart, *North Carolina Atlas,* pp. 193–94; Donahue, Shickluna, and Robertson, *Soils,* pp. 279, 289; *Farmers' Register* 3 (April 1836): 748; Fippin, *Address,* pp. 9–10; Foth and Schafer, *Soil Geography,* pp. 27, 38, 177–81, 183, 186, 192; Genovese, *Political Economy of Slavery,* chap. 5; Hart, *Southeastern United States,* pp. 30, 36–40, 43, 45–46; Midgely, "Lime," p. 329; Smith, *Economic Readjustment of an Old Cotton State,* p. 70; Stuck, *Soil Survey of Beaufort and Jasper,* p. 47.

62. McDuffie, "Anniversary Oration," p. 100.

63. Rosengarten, "*Southern Agriculturist,*" p. 282.

64. Ford, *Origins of Southern Radicalism,* pp. 52–81; Anderson and Gallman, "Slaves as Fixed Capital"; Wright, *Political Economy,* pp. 55–74.

65. For further discussion see Mathew, *Edmund Ruffin,* pp. 123–26.

66. Ibid., pp. 13, 179–80.

67. Cuthbert, "Address," p. 594.

68. Coclanis, *Shadow of a Dream,* pp. 146–47; Ford, *Origins of Southern Radicalism,* pp. 61–62; Mathew, *Edmund Ruffin,* pp. 153–68.

69. Ruffin, *Address on Exhausting and Fertilizing Systems,* pp. 5, 9–19, 22.

70. O'Neall and Seabrook, "Report on Cotton," p. 324; Hamilton, "Address," pp. 305, 310.

71. Barnwell, *Love of Order,* pp. 46–48, 85, 131, 179, 182, 207 n. 38; Chan-

ning, *Crisis of Fear*, pp. 125, 167; Faust, *James Henry Hammond*, pp. 167, 250; Freehling, *Prelude to Civil War*, pp. xii, 24, 47, 86, 179, 206, 222, 226–27, 234–35, 249, 254–59, 346–54, 356–57, 360; Lander, *Textile Industry*, pp. 29–32; Low, *Protection in the United States*, pp. 9–12, 15–18; Ruffin, *Address on Exhausting and Fertilizing Systems*, pp. 19–20, 22–23; Smith, *Economic Readjustment of an Old Cotton State*, pp. 20–22; Wiltse, *Calhoun, Nullifier*, chaps. 1–7.

72. Rosengarten, "*Southern Agriculturist*," p. 294; Mathew, *Edmund Ruffin*, pp. 204–13.

73. Faust, "Rhetoric and Ritual," pp. 541, 555, 558–59, 561.

74. Seabrook, "To the Planters," p. 426; Hammond, *Governor's Message*, p. 10. See Mathew, "Agricultural Adaptation," pp. 129, 131, 139–42; Mathew, *Edmund Ruffin*, Parts I and V.

75. Roper, "Address," pp. 225–60.

76. Barnwell, *Love of Order*, p. 18.

77. Freehling, *Prelude to Civil War*, pp. xii, 11; Ochenkowski, "Origins of Nullification."

78. Barnwell, *Love of Order*, p. 196; Channing, *Crisis of Fear*, p. 26; Moltke-Hansen, "Protecting Interests," pp. 170, 179–81; O'Brien, ed., *All Clever Men*, pp. 20–22; Barnwell, *Love of Order*, pp. 14–17; Freehling, *Prelude to Civil War*, pp. 53–65, 109, 250.

79. Eaton, *Growth of Southern Civilization*, pp. 307–8; Faust, *James Henry Hammond*, p. 348.

80. In contemporary usage the name of the organization was given alternatively as South Carolina and State.

81. Hammond, "Anniversary Oration," p. 183.

82. Roper, "Address," p. 227.

83. Mathew, *Edmund Ruffin*, chap. 3.

84. Collins, *White Society*, p. 99.

85. Faust, *James Henry Hammond*, p. 148; Freehling, *Prelude to Civil War*, p. 241.

86. Quoted in Faust, *James Henry Hammond*, p. 272.

87. Freehling, *Prelude to Civil War*, pp. 90–91.

88. Craven, *Edmund Ruffin*, pp. 42–46; Mathew, *Edmund Ruffin*, chap. 3; Ruffin, *Diary*, 2:119, 650 (1 September 1861, 14 May 1863).

89. Quoted in Barnwell, *Love of Order*, p. 28; see also pp. 20–31, and Freehling, *Prelude to Civil War*, pp. 23, 204.

90. Freehling, *Prelude to Civil War*, pp. xii, 2, 86, 240, 253, 260–61, 314–21, 330. The best overall account is in Channing, *Crisis of Fear*.

91. Cruickshank, ed., *John and William Bartram's America*, p. 116. See also Bartram, *Diary*, p. 64.

92. Drayton, *View of South-Carolina,* pp. 6–11; Hammond to Ruffin, 11 January 1843, Ruffin Papers; *Farmers' Register* 5 (February 1838): 693, 9 (August 1841): 469–70.

93. *Farmers' Register* 6 (May 1838): 111; *Southern Agriculturist* 11 (June 1838): 297–98.

94. *Farmers' Register* 8 (May 1840): 202.

95. Chernov, *Nature of Soil Acidity,* pp. 5–6; Millar, Turk, and Foth, *Fundamentals of Soil Science,* p. 126; Brady, *Nature and Properties of Soils,* p. 382; Midgley, "Lime," p. 329.

96. Hunt, *Natural Regions,* p. 236; Foth and Schafer, *Soil Geography,* pp. 178–81, 186, 192, 195; Paden and Garman, "Yield and Composition of Cotton and Kobe Lespedeza," pp. 309–10; Fergus, Hammonds, and Rogers, *Southern Field Crops,* p. 598; Buol, Hole, and McCracken, *Soil Genesis,* p. 297. See also Stuck, *Soil Survey of Beaufort and Jasper,* pp. 47, 167–72.

97. See, for example, Foth and Schafer's observation that he "ushered in a new era of increased agricultural productivity for the abandoned ultisol soils" (*Soil Geography,* p. 180). See also Millar, Turk, and Foth, *Fundamentals of Soil Science,* p. 146. Others such as Craven and Collings have called him the "father" of American soil science (Craven, *Edmund Ruffin,* p. 60; Collings, *Commercial Fertilizers,* p. 288).

98. Millbrooke, "South Carolina State Geological Surveys," pp. 27–28; Vanuxem and Morton, "Geological Observations"; *DAB,* 10:218.

99. Longton, "Some Aspects of Intellectual Activity in South Carolina," pp. 180, 193, 406; *DAB,* 7:265–66; Tuomey, *Report on the Survey of South Carolina,* p. iv.

100. Arden, "Lyell's Observations on Southeastern Geology," p. 134; Tuomey, *Report on the Survey of South Carolina,* p. iv.

101. Longton, "Some Aspects of Intellectual Activity in South Carolina," pp. 155, 403; Tuomey, *Report on the Survey of South Carolina,* p. iv.

102. Discussed in diary and Biographical Supplement, below. See also citation by Tuomey for his special contribution, through fossils, to Eocene studies (*Report on the Survey of South Carolina,* p. iv).

103. Greene, *Geology in the Nineteenth Century,* pp. 8–9, 25–27; Arden, "Lyell's Observations on Southeastern Geology," p. 131; *DNB,* 12:319–24.

104. Lyell, *Travels in North America in the Years 1841–2; with Geological Observations on the United States, Canada, and Nova Scotia.*

105. First published as three volumes, 1830–33, reaching its twelfth edition in 1876.

106. Arden, "Lyell's Observations on Southeastern Geology," p. 134; diary, below.

107. Longton, "Some Aspects of Intellectual Activity in South Carolina."

108. Darwin, *Journal of Researches into the Natural History and Geology of the Countries Visited During the Voyage of H.M.S. "Beagle" Round the World,* (7th ed., 1890); Ruffin, *Diary,* 3:374 (23 March 1864).

109. "Incidents of My Life," 2:159, Ruffin Papers.

110. The hotel has been destroyed and replaced by an ugly motel; the Fireproof Building survives as one of the most beautiful and idiosyncratic buildings in the state, presently housing the South Carolina Historical Society and its collections.

111. I am grateful to David Moltke-Hansen, director of the South Carolina Historical Society, for suggesting this explanation to me.

112. Faust, *James Henry Hammond,* p. 126.

113. See diary and Biographical Supplement.

114. Edmund Ruffin to Col. D. M. McCord, 25 November 1843, Manuscript Room, Library of the University of South Carolina, Columbia.

115. His second son, Julian Calx Ruffin, owner of the Ruthven plantation in Prince George County, Virginia. Born in 1821, he was killed at Drewry's Bluff in 1864.

116. Hammond to Ruffin, 1 December 1843, Ruffin Papers.

117. Ruffin, *Diary,* 1:55 (11 April 1857). Tuomey was somewhat more experienced than Ruffin's words convey, having, for example, accompanied Charles Lyell on a field trip in the Petersburg area during the British geologist's brief visit to Virginia in 1842. See Arden, "Lyell's Observations on Southeastern Geology," p. 135.

118. Hammond to Ruffin, 4 March 1844, Ruffin Papers; *Southern Agriculturist* n.s., 4 (February 1844): 50; Ruffin, *Diary,* 1:55 (11 April 1857), and accompanying editor's note; Millbrooke, "South Carolina State Geological Surveys," pp. 30–32.

119. Ruffin, *Report of Survey.*

120. Ibid., p. 4.

121. Ibid.

122. *Southern Agriculturist* n.s., 4 (February 1844). 47.

123. Ibid. (February 1844): 50; see also ibid. (September 1844): 353; quote ibid. (July 1844): 253.

124. *Southern Cultivator* 3 (February 1845): 21.

125. Hammond to Ruffin, 14 February 1844, Ruffin Papers.

126. Roper, "Address," p. 240; Poinsett, "Agricultural Address," p. 241.

127. Allston, *Essay on Sea Coast Crops,* p. 1; Ruffin, *Report of Survey,* p. 3.

128. Tuomey, *Report on the Survey of South Carolina,* p. iv; Ruffin, *Address on Exhausting and Fertilizing Systems,* p. 18.

129. Rogers, "Phosphate Deposits," pp. 183–220.

130. Hammond to Ruffin, 7 July 1844, 10 October 1845, 22 July, 4 November 1846, 22 July, 21 December 1846, 10 July, 9 November 1849, Ruffin Papers.

131. Faust, *James Henry Hammond,* p. 134.

132. Ibid., p. 5.

133. Ruffin to Hammond, 7 September 1845, Hammond Papers; "Incidents of My Life," 2:160, Ruffin Papers; Craven, *Edmund Ruffin,* p. 82; *Southern Agriculturist* n.s., 4 (July 1844): 252–53. See also diary and Biographical Supplement. Brisbane here might have a sense of what William Robert Taylor calls the South Carolinians' "struggle to see themselves in history" (*Cavalier and Yankee,* p. 261); see also Faust, *Sacred Circle,* pp. 33, 46–47.

134. Ruffin, *Address on Fertilizing and Exhausting Systems,* South Carolina Institute's introductory note on p. 1. The *Address,* passim, is the source of quotations in the following paragraphs.

135. Mathew, *Edmund Ruffin,* chap. 2, sec. 4.

136. Ibid., chap. 10.

137. *Farmers' Register* 8 (June 1840): 341; 9 (May 1841): 287.

138. Ruffin, *Address on Exhausting and Fertilizing Systems,* pp. 24–25, 24n; also newspaper clipping from *Mercury,* 10 February, pasted in diary and reproduced below.

139. Ruffin, *Address on Exhausting and Fertilizing Systems,* pp. 24–25.

140. Ruffin, *Report of Survey,* p. 54.

141. Mathew, "Slave Skills."

Private Diary of Edmund Ruffin, Agricultural Surveyor in South Carolina, 1843

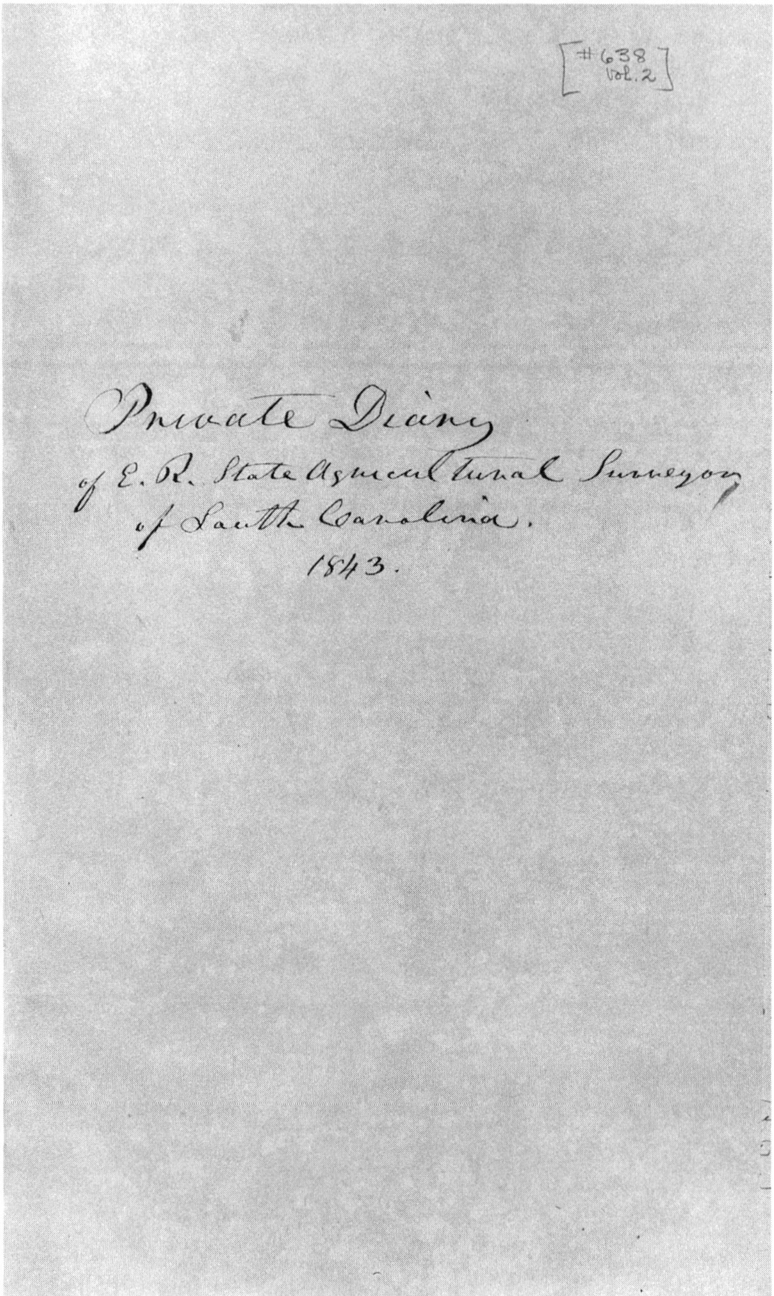

Title page of the diary kept by Ruffin during his agricultural survey of South Carolina (courtesy of the Virginia Historical Society)

I

Winter & Spring

THE LOW COUNTRY

1. Arrival in Charleston (26 January–30 January)

In the last days of December, 1843 [sic], by a letter from Gov. Hammond, I learned that he had chosen me to make the agricultural survey of that state. Highly honourable to me, and therefore highly gratifying as was this appointment, & pleasing as would be its duties, if suitable to my private interest & domestic relations, these & other considerations also caused delay of my determination until further correspondence with the governor had served to remove some of my objections & difficulties in regard to the execution of the trust. On the 16th of January, I wrote to announce my acceptance & that in ten days thereafter I would set out on my journey to Charleston.

26 JAN. Left Petersburg in the car of the railroad, at 7.30 P.M., & reached Weldon, on the Roanoke by 2 A.M. by Blakely, & thence 2 miles up the river by steamboat to Weldon.—63 miles.

27 JAN. Was transferred immediately to the car on the Wilmington railroad. An obstruction on the track caused a delay of three hours, notwithstanding which we reached Wilmington by 3 P.M., where we embarked on the steamboat for Charleston, & arrived there the next morning before sunrise. 180 miles by railroad & 200 by sea from Weldon.

Put up at the Charleston Hotel.[1] In the forenoon, Mr. R. W. Roper called, & conferred with me in regard to the first movement to be taken. This gentleman was chairman of the Committee of Agriculture in the House of Representatives of S.C. & had proposed & so zealously advocated the institution of the agricultural survey that to him

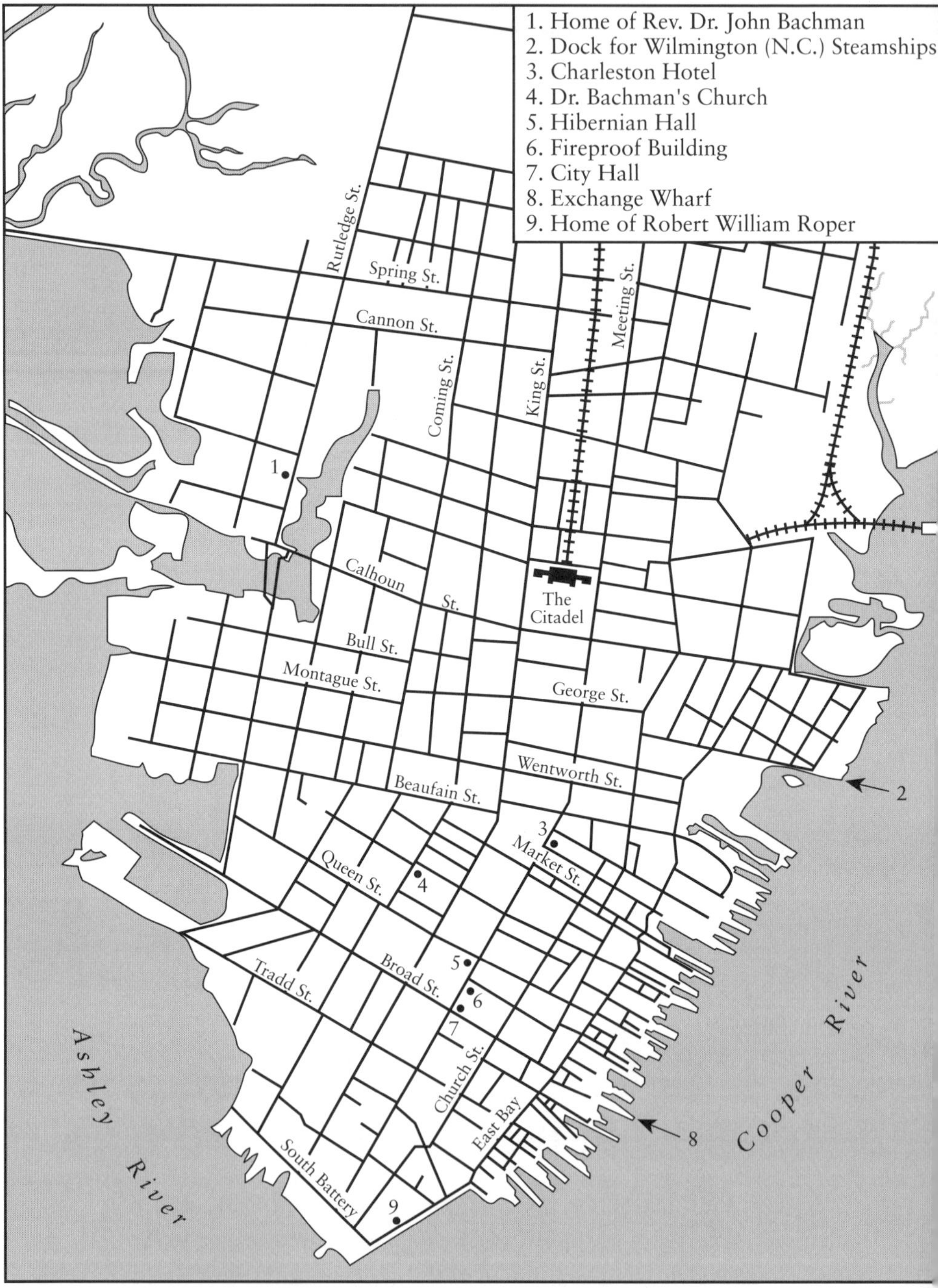

CHARLESTON, S.C.
Source: Drawn from J. H. Colton's 1855 map of the City of Charleston

is justly due the principal credit of the adoption of the measure. He most kindly offered, for himself & other planters every aid to facilitate & advance my labors. The marls & other calcareous deposites of the state I announced to be the first and main objects of exploration & observation—& in the lower districts first, to be enabled to leave them before the first of May, when they begin to be unhealthy to strangers, or even to residents who expose themselves. It was agreed that I should accompany Mr. Roper on the next day but one, (the next being Sunday) to his plantation on Cooper river, whence with the aid of his company, I should seek for the marls reported to be visible in that vicinity. In the evening, the Rev. Dr. J. Bachman & Judge M. King called, but being out, I missed seeing them.

1. The hotel was at 200 Meeting Street, some blocks north of Ruffin's office in the Fireproof Building; it had recently been rebuilt after a fire.

29 JAN. Attended the German Lutheran church, of which Dr. Bachman is pastor, & heard him preach. Since I had the pleasure of becoming acquainted with this highly gifted & interesting man, he has been grievously afflicted by family misfortunes, the heaviest of which is the death of a second daughter, who like her elder sister whom she so speedily followed, had not long arrived at womanhood, & had just before been happily married. As I heard the service of God performed by the minister, & the father who had had so many blessings to be thankful for, & such heavy afflictions to bear up under, I was forcibly reminded of the character & sorrows of La Roche as pictured in Mackenzie's admirable & pathetic story.[1] Dined & spent the evening with Mr. Roper, by previous invitation, with his family, met several intelligent gentlemen.

1. *The Story of La Roche* (1793) by the Scottish writer Henry Mackenzie concerns a Swiss pastor whose daughter died shortly before marriage; he continued preaching, finding comfort in his faith.

30 JAN. The day commencing with steady slow rain, & closing with very heavy rain. Of course our designed departure did not take place. Availed myself of the delay to call on Judge King & Dr. Bachman, & with the latter, upon his most kind & urgent invitation, I dined & spent several hours.

With Dr. Joseph Johnson I have had several conversations, & he has shown much interest in my objects, & by his kindly offered suggestions & aid to my arrangements, has done much to forward my views.

2. Scenes of Desolation on Cooper River (31 January)

31 JAN. From Charleston to Point Comfort, 28 miles. According to the previous arrangement, I set out with Mr. Roper & in his carriage for his plantation on Cooper river. The route for 18 miles is upon the "state road," to Columbia, &, for the few first miles from Charleston, near the track of the railway. The surface of the land is low & level, & cut through near the city by many narrow depressions into which the tide rises, & which may be called creeks. Farther on these creeks, & their low grounds are fewer, but of more extent, & mostly are beyond the influence of the tide.

The state road is on the middle or ridge between the heads of the creeks flowing into the Cooper river on the one side & the Ashley on the other; & therefore, as might be inferred, the poorest land in the peninsula borders the road. It is generally very poor, & also very sandy, & mostly also on a sandy subsoil. Some, however, has a sandy clay subsoil. The trees are nearly all pine, & generally of second growth, the land having been formerly cultivated & afterwards turned out.

After leaving the main road, & turning to the direction of our destination, there soon appeared some improvement in the soil, in a closer texture; & when more nearly approaching the river, clay so much predominated in the soil as to be objectionable, the high land being very stiff & intractable under tillage, as well as poor. Some oaks, usually of small growth, mix with the pines of first growth on all this stiffer land. The pines of original forest are mostly of the "long leaf" species,[1] & many of the great size & beauty for which that kind is distinguished. But whenever of second growth, whether after culture, after mere cutting down the first growth for fuel, the second growth pines are of the "loblolly"[2] or "old-field" kind, of mean sized appearance. The growth of all is so thin that the land beneath is generally also sprinkled by a thin growth of broom grass.[3] But mean & forbidding as certainly is the upland soil, I saw none on the whole route which would not be well improved by calcareous manures, & be made enough productive to be profitable, if these manures can be applied without great expense. But the bad soil & actual poverty of region are much less striking to the eye, & painful to observation, than the appearance of desolation & abandonment of all attempts to improve or even cultivate the land, which is seen throughout. This scene of abandonment and desolation commences immediately on leaving Charleston; & where, from proximity to that rich & populous city, the land might have been expected to be under the most perfect culture, & almost as rich & productive as garden ground, the fields are mostly lying untilled, & naked, & the houses all show decay & dilapidation. Farther on there is still less cultivation, indeed almost none, until reaching the plantations on Cooper river. For much the greater part of the journey, the country appears like the former residence of a people who have all gone away, leaving their lands tenantless. And such is the fact to a greater extent than would have been supposed possible. Many plantations, once cultivated, & the favourite residences of their proprietors, are now entirely deserted & on all others, the extent of tillage has greatly diminished. This is owing to the sickliness of all this part of the country, which now compels all its white inhabitants who have the means, to flee to the city or to some other healthier places, to reside for five months of the year. It is true that if sleeping, or being at night in the country be carefully avoided & exposure to wet at all times, the most sickly plantations may be visited by their proprietors during the day with safety. And many proprietors, who are within a few miles of the very healthy pine ridges, have summer residences there, & can see & attend to their plantations every day. But this resource cannot be generally available. And there can be no doubt that the regular or general absence of the proprietor for several months in every year, & of the time of most important operations, must be greatly injurious to business & profits, & to the agricultural interests of any country so situated.

Our journey extended through the several parishes of [ER]. Goose Creek parish especially, which is now as much a scene of desolation as any, formerly furnished residences & plantations for many of the most wealthy planters. We passed two abodes of former magnificence as well as wealth, as appeared from the still beautiful remains, in avenues of noble & venerable live-oaks.[4] One of these places was the property & residence of one of the Middleton family. It is now owned by a Jew, who does not cultivate at all, or to very small extent. The avenue of large live oaks leading to the mansion appears to be about a third of a mile in length, is 60 feet in width, across which the branches of the trees interlock, so as to present to the eye an unbroken irregular arch—every vacancy between the crooked & irregular limbs of the trees being filled by the upper covering of evergreen leaves, & of the long & pendant masses of the dark gray moss,[5] which hangs from these as from all other trees in the low & humid region. The other place is Spring Grove. There a similar avenue, of less length, has other live oaks on each side so as to form a grove of these trees of 80 or 100 yards in width extending the whole length of the avenue. These trees are generally as much as two feet in diameter, & some of them are three feet. The acorns were planted by a lady. This plantation of about 900 acres, was once under profitable culture. Now it has neither resident proprietor nor cultivator. It was bought not long since at $3000, & is held merely as a resource for timber for another place.

Mr. Roper having lately bought a tract of poor wood land through which we passed, enabled me to learn its market price, which was $800 for 360 acres or about $2.25 the acre. It had a small dwelling house & some cleared land, but was of most value for its timber trees of long leaf pine, which here are of the large size & length of trunk for which that kind of pine is noted.

The lateness of our arrival left no time after dinner for anything except a hasty view of Mr. Roper's rice pounding mill, of which the motive power is the tide-water collected on some 40 acres of his rice land, which thus is made to serve as a mill-pond during the winter months, & the water is then drawn off, to give place to the regular tillage of the crop.

1. *Pinus palustris,* indigenous to the eastern United States and the main source of tar, resin, pitch, and turpentine, is common on barrens and dry, sandy lands in the low and middle country. It is the dominant pine over most of the coastal plain in "old-growth" areas after fire or hardwood felling, reaching seventy feet in height.

2. *Pinus taeda* or old field pine, indigenous to the southeastern United States, is of little value. It is common on dry, sandy land and is of low or reduced agricultural worth in the low country (notably pine barrens). It reaches eighty feet and often may be several hundred years old.

3. *Andropogon scoparius.*

4. *Quercus virginiana,* evergreen, indigenous to southern United States and

Mexico, grows to great height and girth. It provides fine durable timber, especially prized for shipbuilding. The best growths are in the sea islands and near salts. It is dominant in maritime forests, mixed with palmettos and slash pines, and can be low and scrubby on dunes and in sandhill areas.

5. *Tillandsia usneoides*, Spanish or beard moss (though not a moss at all but a bromeliaceaous plant), grows, ephiphytic on trees in low, damp situations along the coastal plain and is especially common on live oaks. It is useful as a cattle food supplement and substitute for hair stuffing in mattresses.

3. By the Cooper Rice Lands to Mepkin and Mulberry Bluffs (1 February–4 February)

1 FEB. A violent & piercing N.W. wind, & an intensely cold day for this latitude. At 1 P.M. the thermometer, in the house, was at 38° & at 8 P.M., 28° on the outside.

Rode through several of the adjacent plantations in passing to & from that of Col. Wm. A. Carson, 4 miles distant. He cultivates 700 acres of embanked tide swamp in rice, & has commenced embanking another body of 300. The work is now in progress with the labor of 43 Irishmen. The rice lands here & at Point Comfort, even at this time of the year, furnished to my unaccustomed sight a scene as beautiful as it was novel. The wide extent of perfectly level land, as black & as rich as soil can be, dry & firm, though several feet below the level of the tide at flood, the long straight drains & canals, & the good & strong condition of the embankments, & flood-gates, all were admirable. There can scarcely any where be found more elaborate & perfect culture, & more exercise of skill, intelligence & industry, than in the rice culture of this region. And it is strange that one branch of such admirable culture should exist, where all others seem to be of entirely opposite character. But rice-culture is every thing on this part of the river, & nothing else seems to be cared for, beyond the furnishing provisions to the plantation. And the immense value of this culture, & the quantity of rich tide swamp land devoted to it may be inferred, from the fact of the great quantity of rice exported from lower S.C. notwithstanding the general barrenness & waste condition of most of the other land, & the bad cultivation & poor products of that which is not waste & useless.

Although the quantity of tide swamp land bordering on this river,

& of inland swamps leading to it, form a large proportion of its entire basin, (as generally in this whole region,) still but a small part is deemed profitable for or is actually under rice-culture. Nearly all the inland swamp lands formerly were under it—but have been thrown out, & are now under water. The lower tide swamps where the water is usually brackish in summer are unfit, because the water is essential to flood the crop, & any salt therein will kill it. And higher up the river, where freshes[1] affect the river much, rice cannot be safely made, because the freshes may prevent the fields having the water drawn off, when essential to the crop. Thus, it is only a comparatively small intermediate space of the river swamps on which rice culture is generally, but not always, safe from one or both of these causes of disaster. This belt of tide-marsh suitable for rice culture is about 16 miles across on this river, along its crooked course; & even of this, the lower plantations are precareous [sic], owing to exposure to "salts",[2] & the upper, to freshes.

As soon as arriving at Mr. Roper's mansion, he showed me specimens of white earth very lately dug in deepening a canal on the land of his next neighbor above, Mr. James Ferguson, & part of which was said to be chalk & was gathered up to be used as such by carpenters & others requiring chalk for marking. The application of muriatic acid[3] from my test vial showed at once that all the earth was highly calcareous. This day we went to the locality which is the canal at the junction of the firm land with the rice tide swamp. The digging had penetrated the whole stratum for some 50 or 60 yards of its course, & for 6 feet depth where it rose highest. The general appearance is of a friable clay, not untuous [sic] or adhesive, white slightly tinged with gray. In this mass are thinly scattered numerous small & distinctly separate masses or nodules of the pure white earth, which has the qualities of very pure & fine chalk, soft enough when dry to be rubbed away by the fingers. There can be no question of the value of the whole body of this peculiar marl for manure. . . .

At Col. Carson's, some hard whitish lumps were shown . . . which were also calcareous. These were found scattered in the ordinary clay in digging a canal. . . . No calcareous deposite has been found so low down Cooper River as this place, though presented in plenty, as I soon after saw, but a short distance above. 12 miles.

1. Or freshets, a rush of freshwater into saltwater, often causing flooding and damage to planting operations.

2. Saltwater entering the river from the sea.

3. Commercial hydrochloric acid; any acid or strong vinegar added to earth in water would if, producing effervescence, identify the calcareous content.

2 FEB. Though the cutting wind had subsided, still severely cold through the day. The Thermr. at 22 at sunrise, outside & against the north wall of the house, & in a situation more sensible to the temperature of the air, Mr. Ferguson's was at 19°. Ice half an inch thick had formed last night over the large sheets of water flooding some of the rice grounds.

Under the guidance of Mr. Roper, & soon after of Mr. Ferguson & Dr. S. W. Barker, proceeded to examine the already known exposures of marl or calcareous rock along the river in this neighborhood. On Dr. Barker's plantation, 3 miles above Mr. Roper's, along the course of the river, a deep ditch in the high land had reached small calcareous lumps, & some marl fossils. Deeper digging would very surely reach the marl. Adjoining is Mulberry Bluff on Mr. Thos. Milliken's land, & commencing where the river at high tide washes against the high land. The face of the low bluff, consist of marl, & which by the digging of a canal from here, is exposed to view for 300 yards in extent. This is said to be the general appearance of the calcareous formation of this whole neighborhood, (& was subsequently found to be true,) & I at once recognised in the softer parts the same very rich marl of peculiar texture & appearance, of which specimens had formerly at different times been sent to me in Va, from the Santee & elsewhere, without very precise statements of localities. The marl is of a pale buff color, or nearly a dingy white, of very uniform appearance & texture—compact & firm, & for the greater part is dug easily enough with picks or grubbing hoes. But parts are of almost stony hardness, & would require being burnt to lime for manure. I am not yet enough informed to know what causes the greater hardness of these masses, & to what extent exposure to rain etc. will serve to pulverize the softer lumps. At this bluff the invisible height of the stratum is some 10 feet above high tide, & how deep below the river has never been ascertained.

The same marl was seen at various places as high as the head of the river, where entered by the Santee canal, which was the highest point reached in this day's exploration. There is a noble exposure at Steep Bluff, (on Blamine's land), half a mile below the Santee canal. Here the marl presents a perpendicular face, 15 feet or more above high tide, & descending as much below as could be examined. At Lewisfield there is another excellent exposure, though the marl there seems harder. Elsewhere on that plantation, & also on Mazyck's, the marl rises to the surface in the fields, & the rocky masses are scattered over the surface.

In the road to Monk's Corner, near Broughton's Swamp, a ditch, deepened by washing of water exposes a very easily accessible body of marl, of very different appearance & texture, being like white crumbly clay, & is manifestly the same deposite as is exposed at Mr. Ferguson's canal several miles distant. A mile higher up, in the same road, adjoining Moss Grove plantation, the ordinary marl of the neighborhood lies naked at the surface. The whole day, was spent in visiting & examining these & other localities so that it was some time after dark before we returned to Point Comfort, cold, hungry & weary, & in the best condition to enjoy the creature comforts that were there awaiting us. 25 miles.

3 FEB. Crossed the river & visited the extensive exposure of marl at Mepkin Bluff, in sight of & below Point Comfort. The land is much higher on the eastern side, & the prospects beautiful, extending widely over the rice-fields, the river meandering through them & embracing many of the neighboring mansions & settlements on the higher lands. Mepkin was the property & family seat of Henry Laurens,[1] the President of Congress, whose remains & those of his son Col. John Laurens, who was killed at the close of the revolutionary war, now repose at Mepkin. Henry Laurens was the grandfather of Mrs. Roper, whose hospitable attentions I am now enjoying.

The marl at Mepkin is exposed for several hundred yards in a perpendicular but not high cliff, washed by the river. It seems softer than this peculiar kind is usually found, & it could scarcely be more acceptable than it is for use. It is also exposed at Mr. John Harleston's shore, a mile below; & the lowest place where it is visible on this river is at Strawberry Ferry, between one & two miles from Mepkin. Went there to trace it, but it could not be seen, being so little elevated there as to be covered by the tide, which was then high.

Fossils are very scarce in this marl, and the few seen are mostly much decayed & difficult to obtain perfect or to preserve. At Mepkin I found the best specimens, & yet very few of any. The marl is of secondary[2] formation, & is very different in its fossils, as well as in appearance, from any marl yet seen by me in place, except that on N.E. branch of Cape Fear River, 15 miles from Wilmington, N.C. which is manifestly of the same kind. Though it is now nearly three years since I visited that locality, & though at that time I had paid no attention to the shells, & suspected nothing of the different ages of different formations, I could have no difficulty in identifying that body of marl with this, from general resemblance, & the like peculiar qualities. And from the remarkable uniformity, I feel certain, in advance of analyzing, that all these specimens are very rich in carbonate of lime. Numerous & different as are the varieties of marl in Va, there is none that could be mistaken for this. The nearest approach to it in appearance is the yellow tertiary eocene marl at Coggins Point on James River. The sharks' teeth found yesterday & today, have marks of difference from those of both eocene & miocene marls,[3] as the teeth found in both these are different from each other. Still the difference is not so marked that I could certainly distinguish those of either one from the others. But all the shells seen here are entirely different from those known of the tertiary marls.

This body of marl below the Santee canal, seven miles in a straight line from one extremity to the other, as manure has as yet been untouched. Not one of the proprietors has as yet made the smallest experiment of its use. Yet several of those whom I saw are very intelligent & well educated men, & zealous & profitable planters; & several of them who either own marl or have abundantly cheap access to it, have been among the subscribers to the Farmers' Register, & readers of my various publications on marl & marling. For their benefit, to what little purpose have I then labored! But all who were present yesterday to be importuned promised forthwith to begin & to marl a few acres, or at least one acre, this winter. Still, I believe the promises were made more to gratify me, & in good-humored indulgence to my supposed marling mono-mania, than because of being impelled by my arguments or proofs of the value & profit of marling.

But though no marling has been tried on Cooper river, so far as I have heard, it has been near the Santee; & to that point I shall direct my route, though not expecting to proceed there when I left Charleston.

There is one use to which the marl of Mepkin & of Mr Harleston's immediately below has been put. The bluffs have been dug down & the earth taken by lighters to heighten the embankments of the rice-grounds; & as the stratum of marl formed the greater part of the perpendicular section, so it has formed the greater bulk of the earth so carried. Even the thus digging & boating it, & thus experiencing the facility of both operations, have not induced the least application of the marl as manure.

I have not been altogether inattentive to the rice culture & other cultures of the lands in this vicinity; but so novel are they to me, that I wait to be better informed before expressing any opinion thereupon.

The Santee canal[4] of which I saw the southern extremity yesterday, is an improvement which promised to be of great value in advance theory, but which, from unlooked-for circumstances, has been much less useful in practical operations. It is 22 miles long, from the Santee to the head navigable water of Cooper river; & to boats from the upper Santee, serves to avoid the circuitous navigation of that river to Charleston, & the dangers of the passage along the sea coast. The rail road from Columbia has served to divert much of the cotton to that newer route; & unprofitable as is the canal to its shareholders, the tolls are so high, that many proprietors of boats take the old river & coast-wise navigation rather than pay the canal tolls for its short distance into the narrow & safe waters of Cooper river. 6 miles.

1. Laurens (1724–92), South Carolina merchant, planter, and revolutionary statesman of Huguenot extraction, succeeded Charles Pinckney as president of the First Provincial Congress, 1775, and John Hancock as president of the Continental Congress, 1776–77.

2. The oldest category of marl, normally hard, of Cretaceous subdivision, running into Tertiary at 63,000,000 B.C.

3. These are younger, usually softer than secondary marls. Eocene is the second longest (58,000,000–36,000,000 B.C.) of the five main divisions of the Tertiary (63,000,000–1,000,000 B.C.); Miocene, the second last (25,000,000–13,000,000).

4. The canal ran from Biggin's Creek (on the Cooper thirty-four miles from Charleston) north-northwest to the Santee and middle country.

4 FEB. After breakfast set out for Mr. Fred. Porcher's plantation, designing thence to follow the calcareous formation to Eutaw Springs & return by the Railway to Charleston. My host furnished me with letters of introduction, & with horses to the intended stopping place. The route chosen was along the course of the Santee canal, and there-

fore it was necessary to go backward two miles to cross the Cooper at Strawberry ferry. On the level high land just across the ferry is the site of Childsbury, formerly a village of some considerable population, but of which not an inhabitant now remains, nor a house, except the old colonial brick Church, which is kept in repair & used as the place of worship of the neighborhood. The church yard is a beautiful & solemn place, with its tombstones overshadowed by very large & wide spreading live-oaks, from whose out-spread limbs hang the dark moss, as if designed & well suiting for ornamental funereal drapery. The view from this place, as from most other of the eminences on this side & on the river, is commanding & beautiful.

The road up the country passes back of the river clearings & even much higher land, & a more undulating surface than on the other side of the river. The wood-land poor, but soil more of medium texture, & well-improvable if beginning by marling. Some red clay subsoil seen. The quantity of short-leaved pine[1] prevailed instead of the usual long-leaved pines. About 6 miles above the ferry, in passing through the plantation of Mr. Gaillard, saw a cotton field of last year, & as yet the only appearance of this crop seen, except at Fairlawn, which is just opposite this place across the river.

In this farm, where the road descends a hill, the marl is bare to the surface, & rises several feet on both sides of the road, with very little overlying earth. It could be carried upon the adjacent cultivated grounds at the lowest cost—perhaps for 50 cents an acre, exclusive of the labor of spreading. But still not a shovel-ful has been applied. At the next place, Sportsman's Retreat, turned in to search for the deposite, having heard before that it was exposed there in great abundance. But the overseer (the only white resident) was not at the house, & no one was seen to direct to the place. A negro gunner whom I met there & inquired of, & who seemed much more intelligent than usual with his class, was with difficulty made to understand what I was seeking for; but when understanding, said that the rock was exposed in the road turning off to the Parsonage ahead where I sought & found it naked across & each side of the road,—where the surface has but a gentle slope. Doubtless this calcareous formation extends under the whole of this part of my route, & plentifully exposed on most of the plantation as was seen on the opposite side of the river. With such great plenty of exposures, & uniformity of texture & appearance, there is no use in seeking for more views of the marl, unless where some use of it has been made, or is designed to be commenced.

1. *Pinus echinata* is most prominent in the piedmont and Blue Ridge.

4. Along the Santee Canal to Black Oak (4 February–6 February)

At [ER] miles from the ferry stands Biggin Church an ancient & venerable brick building, which was burnt & the wooden portion destroyed by the British soldiers in the revolutionary war. The walls remained uninjured, & it has since been rebuilt, the ancient form being preserved.

Between the two churches mentioned, & not far from the former one, another large & modern wooden edifice was seen, which seemed as if designed for worship. Upon inquiring of my attendant, learned that this was the church built for the negroes of the neighboring plantations, where regular services of religion were maintained at the expense of their masters.

At Biggin Church left the main road & arrived at the canal at the same point where it had been reached a few days before. From the elevated bridge which crosses just above the lower lock, the whole lower section of the canal is in view, having a straight course for nearly three miles. My route was along its margin, for 5 miles to Wantoot. For that distance, & some higher, the canal passes through the very large Biggin Swamp, which furnishes the head-waters of the Eastern Branch of Cooper River. Low-lying high on firm land in some of this distance is on one side of the canal, but the wide-spread swamp always on one side, & for some miles is on both sides of the canal. The lower section of three miles is dug very shallow, & the water raised to a sufficiently high level by the banks on each side. For about two miles continuously, & indeed generally wherever firm ground reached to the canal, the marl formation was exposed in the sides of the banks, thrown from the bottom, & by the rocky lumps scattered outside. It is evident that the stratum is of very uniform level & is near the surface. After a large interval of swamp, these hard lumps were again seen at the third lock. If needed for use, the marl could be excavated very cheaply for transportation along several miles of the portion of the canal seen.

Abundant evidences of the improvement caused by the marl thrown out of the canal is seen, even at this season, in the growth of white clover & wire-grass[1] (or joint grass as called here,) which plants so especially delight in calcareous soils. But even the lesson thus afforded, has not induced the throwing any of the marl farther than accident has spread it into the cornfields that reach to the bank of the canal.

Upon reaching the residence of Mr. F. Porcher was much concerned, & my plan of procedure frustrated, by finding him confined to his bed by recent & severe sickness. However, he insisted on seeing me, & conversed as long as I would consent to subject him to such exposure. And though kindly & urgently pressed to stay & to receive every attention & accommodation that hospitality could offer, I proceeded immediately after the late family dinner to the house of Mr. Wm. Cain, which was not a mile distant. It was only by such resistance that it requires some obstinacy to exert, that I avoided being sent in Mr. Porcher's carriage, or at least on horse-back this short distance. And at last, that I might walk, I had to compromise by accepting the services of a man-servant to carry my scant change of clothing, which was wrapped in a handkerchief & the parcel weighing less than 5 pounds.

I learned particularly from Mr. Porcher (what the report of had brought me to seek him,) that he had applied the calcareous rock in 1840 to 22 acres of land, at the rate of 100 bushels to the acre. The deposite under his surface, (which was not visible, but found by digging a canal) is more hard than generally lower down or higher up the country. About half was soft enough to apply as dug, & the balance, of hard lumps, was burned to lime.[2] The two kinds were applied to different ground, & if there was any difference of effect, it appeared to be in favor of the softer or raw marl. The total result was altogether satisfactory. The application was on cotton land, which was cultivated every other year in that crop, & rested the intervening years, & not grazed except in winter. The season of 1840, the year of the application, was very bad for the crop, & the 22 marled acres made nearly all the product of his plantation, the balance of his cotton land not marled, producing scarcely any crop. In the next crop, 1842, Mr. P. thinks that the greater product of the marled land over the other & otherwise equal land, was 40 percent.

Upon meeting with Mr. Cain, the announcement of my name, in advance of his seeing the letter of introduction, secured his immediate &

hearty welcome. Mr. Cain is a member of the senate of S.C., & a zealous friend to the institution of the survey, & therefore feels additional interest & gratification in forwarding my labors.

In this neighborhood all rice culture . . . has been abandoned, except for home consumption, & it is bounded by part of the eastern line limiting the culture of cotton. The regular & large culture of this crop was first seen on these two plantations. The kind cultivated is not the green-seed,[3] but the long-staple Santee, a black seed variety which is the product of sea-island seed, but somewhat degenerated from the fine & long staple of the sea-island product.[4] The long staple Santee made hereabout sells usually for three times as much as the ordinary or green-seed upland cotton. On the other hand, land planted in the latter kind is about 50 per cent. more productive.

The mode of cultivating corn & cotton here, as of corn on the upland of Cooper river, is in high narrow ridges, & principally by the hoe. What is most novel & strange in the practice of all this low country is the small use made of the plough. Mr. Cain works 60 laborers of all sorts, equal to 50 or 51 able men, & only 5 mules for his whole ploughing & hauling labors. He does not use oxen at all, & therefore these few mules have to perform not only all the little ploughing given to all his crops, but all the hauling of fuel, of manure, of crops from the fields, & finally delivering all for sale on the canal 3 miles distant.

The general mode of cultivation hereabout is the same. Where most care of the land is used, the crop whether of corn or of cotton, is repeated every second year, & the land rests under its volunteer growth of weeds in the intervening year. If manured, then the year of rest is not given. The field is left in high ridges. When about to be tilled, in winter, the hands with broad hoes scrape the surface of every ridge, with all its vegetable covering, into the alley where it is covered by the earth thus drawn down. When manure is applied, whether from the stables & cattle pens, or only of unrotted rice-straw (on the river) or of leaves raked up in the woods, it is laid in the alley in mass, before this operation of "ridging down", & covered with earth by the hoes, as are the weeds & rubbish. Thus the corn (& also cotton,) is planted over this list of putrescent matter, which has been so carefully & laboriously kept from being intermixed with the soil, & which effect I should much object to, even if not more laborious. The little ploughing given,

& repeated hand-hoeings together serve at every repetition to put more earth to the new ridges, by drawing from those previously existing.

The cotton crops of this neighborhood are extensive, & pretty good. The soil is sandy, much of it very good, & the surface gently undulating. A fine soil & surface for improvement & for tillage. 19 miles.

1. This may have been *Poa compressa,* a close relation of Kentucky bluegrass.

2. This was a purer, lighter source of calcium carbonate than marl, though usually requiring processing (as here). "Marl & lime" is a proper pairing, though lime's main uses lay outside agriculture, most notably in mortar for building, thus the combination of widespread production and importation with limited farming use.

3. *Gossypium peruvianum,* indigenous to South America, reaches a height of 9–15", with strong 1–1½" staple; so-called upland.

4. *Gossypium barbadense,* indigenous to tropical America, was the most valuable cotton and was grown on islands and coasts of South Carolina, Georgia, and Florida, gaining from saline conditions; it produces a fine, long 1½–2¼" staple.

5 FEB. Sunday. Accompanied my host's family to Black Oak Church near the canal.

Mr. Cain had never seen any marl on his plantation, nor had searched for it, but supposed certain springs flowed from the body, from their waters being impregnated with lime. Our conversation induced him to have a slight examination made by one of these springs & near the edge of the swamp, & in less than 18 inches the spade brought up the ordinary softer marl. I trust that he will proceed to make use of it. Much of his & other wood-land in this neighborhood, has intermixed with the pine, good sized red-oaks & some hickory.[1] Such land is of superior quality; & it is the opinion of both Mr. Cain & Mr. F. Porcher, that this growth indicates marl being at no great depth below the surface. I learned from Mr. Cain that the visible marl extends 6 miles from this towards the north west, to a plantation of Mr. Dubose, 45 miles from Charleston. Then some pine land intervenes, & then the marl again shows, & becomes very abundant on the surface about the Eutaw Springs. The examination of this extension I must defer to another time.

1. *Quercus rubra,* widespread in North America; and *Carya* spp., a nut-yielding deciduous tree, with tough, elastic wood and inner bark providing

yellow dye, was traditionally used to identify good farmland. In piedmont and inner coastal plain forests oak and hickory as dominants, in a wide variety of sites.

6 FEB. Left Mr. Cain's house, & rode across 23 miles to the Charleston & Hamburg railway. After two miles reached the "pine barrens".[1] This name is applied to land of peculiar & in some respects very valuable character. It furnishes healthy residences in the summer & autumn, when every planter even higher up the country than this place, is compelled to quit his plantation, & if he remains in the lower country, to reside, (or certainly to sleep every night) either in Charleston, or the sea side, or in the "pine barrens". The latter, when within a few miles of the plantations, afford very convenient retreats, as the planters can see & attend to their work every day, & as long as they please, & safely, provided they avoid getting wet, & always sleep in a healthy place. Mr. Cain's summer residence is between two & three miles from his plantation, & with those of six other planters form a scattered little village of cottage-like buildings. Several other of such villages occurred in other parts of the same body of pine barren. This land lies midway between the swamps, & is probably the highest as it certainly is the dryest [sic] land; but there is no perceptible ascent to it from the plantations bordering on the swamps. The surface is very level, the soil sandy & dry. The growth is almost exclusively of long-leaved pine, there being besides only a very few black-jack oaks,[2] & no undergrowth of pines or any other trees. The pine trees are generally small, very few being as much as two feet & not many 18 inches through, & therefore they are destitute of the magnificent appearance which these trees present when of their full size. Still, the pine growth of the barrens has a peculiar beauty in the very straight & clear & long & slender trunks of the trees, their general uniformity of appearance, & the absence of all smaller growth. Although the pines stand quite thickly, objects are visible at distances very rarely thus exposed in forests. No gray moss is seen—the prevalence of which always indicates unhealthy ground. After passing through some 8 miles of this land, the surface lost its uniformity of level, small swampy places began to show, & next we reached the Monk's Corner road which I had travelled on to Cooper river.

1. This is the colloquial term for sandy, infertile land on which pines form the dominant growth.

2. *Quercus marilandica,* indigenous to the southeastern United States, is one of the principal large trees on the piedmont ridgetop forests and sandhills; it also grows on drier habitats along the coastal plain.

5. Return to Charleston (6 February–7 February)

The railway was reached by 12, where I discharged the horses & servant, & waited two hours for the arrival of the train, in which I reached Charleston. A letter from home awaited my arrival at the hotel. It brought me news both to alarm & to relieve. My little granddaughter,[1] whom I had left sick, had been worse & in danger—but the crisis was passed & the disease much alleviated.—40 miles.

1. Most probably Nanny Ruffin (b. 1841), eldest daughter of Edmund Ruffin, Jr.

7 FEB. The greater part of the day spent in trying to hire two oarsmen by the month or day, & in vain, though aided then & also by previous inquiries of Dr. Joseph Johnson. A suitable boat for exploring the rivers was found & engaged conditionally, but not a man could be found. While expecting to be thus kept idle, & my designed expedition up the Ashley for tomorrow to be frustrated, met with Dr. Bachman, who, before hearing of my difficulty, told me that he had last week been in St. Andrew's Parish, had announced my designed visit, & that it was urgently wished for; & further, he proposed to take me there in his carriage. It was the very place that I had designed first to go to by water, & the first stopping place was the very house to which an invitation had met me on my first arrival in Charleston—Mr. J. Brisbane's. I consented most willingly, & Dr. J. who was to have gone with me in my boat, if hands could have been procured, agreed to join us. I had made an auger to bore for marl 10 feet long, & a sounding rod. The former will be too inconvenient for land carriage, but the rod as well as my box of apparatus for analysing marl will be taken along.

An apartment in the "Fire-proof building", had been placed at my disposal, & this afternoon my baggage & boxes of books were moved there. In this building are the state offices, as the Treasury, Comptroller's Office etc. It is one of the many beautiful public edifices of this city. My apartment has all the necessary furniture for the purpose. It will be very convenient to deposite & exhibit my specimens, to analyse marls, & to keep my books & heavy baggage while I remain in the lower country, as frequent returns to Charleston is [sic] made necessary by the character of the country; & the routes for travel by land or by water. After making this arrangement, spent the evening with Dr. Bachman. Walked today not less than 7 miles.

6. Ashley River Plantations (8 February–9 February)

8 FEB. Left Charleston immediately after breakfast with Dr. Johnson & Dr. Bachman, in the little carriage of the latter, & soon reached the residence of J. Brisbane esq. on the Ashley river, 9 miles from the city. The marl, similar to that on Cooper river is here visible, rising only to between ordinary tides. This is the most eastern exposure above low tide known on this river; but at a mile or two only from Charleston it forms the bottom of the river, where a bridge across was formerly constructed, & which was destroyed & carried off to sea in a few months because of the piles being driven no deeper than a very little into a stratum, (which is no doubt this marl,) & therefore having no sufficient hold. It would have furnished an excellent foundation for piers of masonry, but was altogether unfit to receive piles, which were prescribed in the belief that the bottom was of mud.

My companions were compelled to return in a few hours, after which I rode with Mr. Brisbane to examine the stratum at several of the neighboring plantations above. These had been first shown to me at Mr. Brisbane's, as a novel feature, which I found afterwards to accompany the marl throughout our ride, & as I infer, forms a sure indication of its presence below. This is in masses of stony hardness, full of hollow impressions of shells, but of which none remain, the mass seemed to be but slightly calcareous. . . . In an inland rice swamp of Mr. Brisbane's nearly two miles back from the river, we saw where plenty of these stony masses had been thrown up in digging ditches of two feet deep.

I found in my host a fund of interesting & general information in regard to lower S. Ca. where he was born, & in several different parts of which he has long lived. His conversation was as agreeable & instructive to me, as was his hospitable & kind aid serviceable to my pursuit. 17 miles.

9 FEB. After breakfast, with Mr. B's company & guidance, embarked in a boat rowed by two of his negroes, to examine every exposure of marl on the Ashley. The river is deep, narrow, & very crooked in its course. Large tide swamps form its borders generally on both sides, & always on one side. The margins where the high lands are washed by the river, generally forming low bluffs, are inconsiderable compared to the extent of swamp forming the shores. At every bluff, the marl was visible, & generally as far as the high land formed the shore. The stratum was examined, & specimens taken at the lands of J. Brisbane, O'Neale, Cohen's & Webb's, & Oak Forest on the left (or eastern) bank of the river, & at Bee's ferry, Drayton Hall, Magnolia, Pringle's, & Cattell's, on the other side. In some places it is abundantly soft for easy digging, in others hard enough to be difficult to dig with a pick-axe. At Brisbane's & Bee's ferry, it was the softest. . . .

It was much to be regretted that I had no implement suitable for examining the bottom of the river, better than a common boat-pole, pointed with iron. With this, & with much ado, succeeded in getting up a little earth from the bottom 12 feet below low water mark, opposite the marl exposure at Magnolia, which to the eye, & also by the application of acid, was manifestly the same marl. It rose here in the bluff 8½ feet above low water, showing altogether 20½ feet of the section exposed by the river. . . . The boatmen who were propelling their vessel by long poles, when questioned, reported that the bottom of the channel, even where deepest, was hard. Thus, there is every reason to infer that this deposite extends from near to Charleston continuously up the river, to 25 miles (along its course) where my day's search terminated, & above which Mr. B. believes that it does not appear in sight. The river is navigable about 3 miles higher, for boats; but as the tide was about to ebb, & the day was far advanced, we turned back, when near to the now deserted site of the former town of Dorchester. It was two hours after dark when we reached Mr. Brisbane's residence.

The lands on the left bank of this river, which I saw yesterday in riding, & as reported on the river generally, are of pretty good quality, & very capable of being improved by the marl which is so abundant beneath every plantation, & so cheaply to be applied. Yet here, as on Cooper river, not an acre has been tried, & an utter disregard or ignorance of the value of the manure is universal.

In addition to the value of the land, for cultivation & still more for improvement, the river banks offer many beautiful sites for residences, which were preferred as such by the early settlers, & for a long time the Ashley river plantations were the most highly appreciated & productive lands in the colony. Now these lands are almost left untilled, are rarely inhabited by the proprietors (themselves almost altogether a new race,) & the whole presents a melancholy scene of abandonment, desolation & ruin. Along the river there are seen what were once large & costly houses & some even princely establishments, & all (except Middleton Place which its owner still resides on part of the year & maintains in all its ancient splendor,) are either in ruins, or seem to be going to ruin. Very little land is cultivated. But little rice is made, & only by a few persons. One occupant only on the left bank cultivates cotton for sale, & he only a few acres; some corn & even hay, imported from the north are bought by a large proportion of the proprietors; & there seems to be no thought of doing any better. The principal business now pursued is cutting wood to sell in Charleston.

In most parts of the country, it is not easy to learn what are the fair market prices of lands, because sales are too rare to establish rates. On the Ashley river, sales have been so frequent that there is no difficulty in knowing the price at which more than half the plantations have actually been sold for: though the continued & great decline of value makes every successive sale at a lower rate than the preceding. In answer to my inquiries on this head, I heard from Mr. Brisbane of such low sales on this navigable river, & close to Charleston, that they seemed as if the land, for cultivation was counted for nothing. In his neighborhood, say from 7 to 12 miles from Charleston, for river plantations, 3 to 5$ the acre was as much as could be expected to be obtained from a purchaser & sometimes much less had been obtained. Some cases will be stated more particularly. Mr. Brisbane's plantation on which he resides, of upwards of 500 acres, he bought 7 or 8 years ago for $1300. He has since built a good two storey dwelling house, which with other buildings could not have cost less than $2000; & he thinks that he could not now sell the plantation for what he paid for its purchase. Adjoining this, & above, he has a plantation of 300 acres, for which he gave

$1500. He has advertised it for sale, & would take $1000, but no one has even so much as inquired the price. Next above is the plantation called Lauristin & of 1000 acres. It has a large wooden mansion house of 3 stories, which with the out houses must have cost fully $8000. It was sold for $4000. Of the next plantation (formerly Williman's), the size & price when sold were not remembered; but the value believed to have been rated no higher.

The next plantation presents a still more remarkable case. It consists of 1400 or 1500 acres, of which 900 were of wood-land, & much of it very good soil. The mansion house was of brick, & of three stories, & 4 good rooms on a floor—& constructed in the best manner. After Charleston had suffered by the great fire in 18[ER],[1] there arose such a demand for building materials, that bricks sold for from 18 to 20$ the 1000; & this plantation, then being in market, was bought for the purpose of demolishing its superb old mansion & selling the bricks at the reduced price at which those once used would command. The speculation was a profitable one; for it was understood that the buyer got back his purchase money in one year by the sale of the bricks, & of wood from the land. He died before the work of demolition had been quite completed, owing to which it ceased, & part of one of the walls still stands as memorial of the remarkable evidence of a declining agricultural community.

The next plantation above is Cohen's, above 800 acres, & having a large wooden mansion house. It was bought for $750. These plantations are along the river, in the order named, & occupy the whole space. The nearest to Charleston is but nine miles distant, & the most remote about 13, & the water route but little more. All well supplied with wood, & all the most abundantly with marl, of rich quality, & very cheaply obtained, but which latter does not add one cent to the appreciation.

Again:—two miles lower down, & much nearer to the city, not on the river, but close to the rail-road, a tract of 250 acres of the best land in the neighborhood was bought in 1832 by Mr. B. at $750. It was mostly virgin land & much of it well wooded. Next to this & on the river, is a tract of 230 acres, having a very handsome & good brick mansion 50 feet by 25, & of three stories. The house has a beautiful situation & prospect, Charleston being in full view over the broad & straight stretch of the river. This Mr. B. bought for $1450. The mansion alone had probably cost in being built, $10,000.

Oak Forest, formerly the highly embellished seat of Sir James Wright,[2] a loyalist, is even now a beautiful place, although the land is deserted & gone to waste, & the house falling to decay & destruction. This place was near the upper termination of our expedition, & we stopped there & viewed the place. It was to me, connected with its history, a subject of melancholy interest—& of more painful interest to my kind host & guide, who now saw this place for the first time since he had ceased to be its proprietor, & when he had suffered its last sacrifice in sale. There are 1460 acres of land. $16,000 had once been offered for it to a former proprietor. Mr. Brisbane subsequently bought the plantation for $8000; & when after 10 years residence he was compelled to sell it, he obtained for it only $2900. It now is again for sale, & the sale has been delayed only because it was expected not to pay the debt for which it is bound. Mr. B. said that he should not be surprised if it should sell for less than $1000.

I have rarely seen a situation of more beauty, even in its desolate state, than Oak Forest. The house stands on the table land of perhaps 40 feet above the river, & within 30 or 40 yards of the river which washes against the steep bank of which the lower part is of solid & firm marl, & as precipitous as a wall, & to which the surface slopes regularly & rapidly. Just before the house & near to the descent to the river stand wide spreading live oaks, of which one seemed fully 5 feet in diameter, & the others very little less. On the other side of the house the entrance is through an avenue of these noble trees. Nothing is wanting but the essentials of improved agriculture & improved healthfulness to render this place as profitable as it has been beautiful. After stopping here long enough for us all to eat of the cold provisions brought along, we re-imbarked [sic] at 3 P.M. before the tide had quite begun to ebb, & returned. A light rain fell at intervals, but not enough to be hurtful.—24 miles.

The Ashley is frequently fresh as low as Mr. Brisbane's, & sometimes brackish nearly to the head of tide. Therefore the tide marshes, when embanked & put under rice culture, cannot rely upon the tide for irrigation, as any salt in the water is destructive to the rice plants. The rice fields formerly embanked had resources for flooding in inland streams which were obstructed by cross dykes, so as to form ponds, or "reserves" of water as they are termed. These reserves were emptied & their bottoms exposed to the sun in summer; a more sure producer of malaria could scarcely be devised. However, all the old river rice fields have long been abandoned to the tide, & the reserves are no longer kept up, though probably in no case drained for dry culture. Two bodies of tide swamp have recently been embanked, & the work not yet completed; & these present the only signs of rice-culture now on the river.

The marshes bordering on the river are very extensive. They are covered by two kinds of tall & coarse grass, growing usually separately. The one known as "marsh", furnishes good & early grazing, & cattle were already upon it. It supplies a coarse hay, & elsewhere is much used for manure. The other growth is the rush, a longer & taller grass which is not considered of any value for any purpose.

The head of this river is an extensive inland swamp, the waters of which, dyed with vegetable matter, do not lose their dark color as low down as I saw the river. This is contrary to what I would have supposed, the water being in contact with so much calcareous earth, which according to my views combines with the dissolved or suspended vegetable extract, & thereby should make the water clear. But if the color of the water, still remaining would seem to oppose my doctrine of the "black waters" in one respect, that which it has lost strongly sustains it in another. All along the places where the marl is perpendicular, much of it . . . is covered by a thin coat of black matter, which must be the vegetable extract of the water combined with the calcareous earth.

1. There were notable fires in 1833, 1835 (twice), and 1838—the last the worst.

2. Wright (1716–85) was the last British governor in chief of Georgia. He was ordered to leave in 1782. Wright had formerly been attorney general of South Carolina, ca. 1739–60.

7. Conversation and Work in Charleston (10 February–11 February)

10 FEB. Returned to town with Mr. Brisbane, & deposited my load of specimens in my office.

Dr. Johnson has succeeded in engaging for me a good boat & pair of oarsmen; & I hope to be able to set out tomorrow morning on my first cruise in my own vessel. The accommodation is obtained however at a pretty dear rate; & there is reason to fear that there may be more objections to my crew than their high wages. I am to pay $1 per day for each man (they finding their own provisions) & 50 cents for the boat—9 miles.

11 FEB. A heavy rain last night, & in the morning still drizzly & threatening worse weather. Deeming the designed water trip too disagreeable to be undertaken, abandoned the intention for the present. But after having so determined & discharged the boatmen until Monday morning, in half an hour the weather cleared, & we might have had a pleasant voyage.

Had my office well cleaned out, & arranged & labelled the specimens from Ashley river. Also specimens of the marls of Virginia which I had brought with me. . . .

Gov. J. H. Hammond arrived in the evening, & will remain some time on official business. I called on him & was received with great kindness & courtesy. We lodge in the same house (the Carolina [sic] Hotel,) and he pressed me to join as one of his party, in their sitting room & at meals, in which however each one pays his own expenses. This however I shall decline, both because of my short stay, & because I prefer more privacy, & less of form & ceremony.

This being Saturday evening, laid in my stores (namely, a box of crackers & some cheese,) & made every other arrangement for an early start on Monday morning. Col. McCord, of Orangeburg district sought me out today, & in making personal acquaintance, I have found an intelligent & agreeable companion. He has promised to accompany in my first cruise.

Dr. Bachman had inserted in one of the city papers a short notice of my explorations, & the extent of marl seen, & urging the planters to forthwith begin the use, by applying it to one acre, if no more; & called on me to furnish directions for such applications by new beginners. This piece had appeared just before my return—& I responded to the call today by writing an article & sending it to the paper. I also made use of the leisure time to bring up my journal, as there had been no time to make an entry for several days before.

[A copy of the newspaper piece is inserted later in the diary.]

To the Editors of the Mercury—

In your paper of the 10th, a call is made upon me by your correspondent, J.B. [Dr. John Bachman], to which it is my duty to respond. J.B. urges every planter having access to marl, to apply it to at least one acre of land for the crop of the present year; and he requests of me to give directions as to the manner of the application.

Of the lands which as yet come under my personal observation, there is but a small proportion to which the application of marl would not be both beneficial and profitable. The only exceptions to be stated are such soils as are naturally and sufficiently calcareous, and other lands not sufficiently drained, and suffering from wetness. So, no one need wait because not knowing where to marl. But if there be choice of lands for experiment, the best and earliest effects of marl may be expected on soils containing the most vegetable or other putrescent matter—or which are the most *acid,* the latter quality being indicated by the greater propensity of the land to nourish the growth of pines and sorrel.[1] Land which has been some years lying out of cultivation, or has been permitted to rest at frequent intervals, and kept from being grazed, when under a regular course of culture, will be much more benefitted by marl, than would the like kept under very exhausting tillage. But even if any such unfavorable condition of the land should prevent the full and proper effects from marl on the first crop, it will show them afterwards—and the more strongly according to the greater quantity of nutritious matter that the land may be enabled to receive, whether from its own growth or from other sources. Whatever early effects may be produced by marl will never be lost, under judicious mild culture, (which is always the most profitable culture,) and the early product will continue to increase for twenty years or more.

There is no little choice in regard to the kind of crop to be marled, as of the kind of land. All known cultivated crops are thus greatly benefitted. Perhaps a benefit additional to that of *increased bulk* of product may be found by marling for the long staple cotton, in hastening its maturing, and thereby making much of the product white, which usually is stained, because not ripe before cold weather. Of rice too, I should infer that calcareous manures will increase the product and still more certainly improve the quality of the grain.

As to the manner of application, I would advise (generally,

and without knowing the circumstances of any particular case,) that the quantity applied should be about 100 bushels to the acre. Of course, it ought to vary according to the degree of richness of the soil, and the nature and condition of the land. But 100 bushels, even if the marl is very rich, as most of it is, will be no where injurious by its too great quantity, and will also be a sufficient dressing for ordinary poor lands. Should the marl be much poorer, and therefore the dressing be too light, still it will be serviceable in proportion to its strength, and may be added to afterwards, when such deficiency has been correctly ascertained. The marl should be scattered as evenly as may be convenient, over the whole surface, so as to be thoroughly intermixed with the soil, and equally diffused by the subsequent tillage processes. It should never be accumulated in the alleys, or under the lists, or in any way kept in mass, (no matter how small the masses,) beneath the stalks of growing plants.

For continued and regular marling to be made most profitable, of course far more full directions would be needed. But these alone will suffice to make applications always more or less profitable. I most earnestly second the proposal of J. B., with one amendment thereto; viz: that instead of *one acre* every planter who has it in his power will marl ten acres or more for the coming crops. The *first beginning* of every new operation is attended with much trouble, and so will be the first attempts at marling. But the greater part of this trouble is in *getting ready;* and when that has been done and one acre has been marled, ten more may be covered with perhaps less trouble than was required for the first acre alone. No man will fail to profit by a judicious application; and there is but little doubt that for all the labor and expense thus directed, there will be returned, in increased products, on the general average, a permanent net profit of at least 50 per cent per annum on the out lay.[2] Every acre marled this year will by its effects shown, probably induce the marling of twenty the next, and a hundred the year after.

EDMUND RUFFIN)

1. *Rumex acetosa,* sorrel dock or common sorrel, thrives on sour soils and was used in sauces, salads, and as medicine. Its roots provided olive-green dye.

2. Nine farmers in Prince George County, Virginia, estimated in 1840 that the increased annual value of their marketable crops, after marling, stood individually at between $426 and $3,200. Ruffin calculated in 1842 that eighty-nine marling planters in King William County had enjoyed additional per capita gains of $5,427.

8. Visit to Edmund Ravenel at the Grove (13 February–15 February)

13 FEB. By the appointment of my intended boatmen, 6 o'clock in the morning had been fixed upon to set off, & in a few minutes after I was ready & waiting for them. But nothing was heard of them until 9, when they made their appearance & I discharged them instantly. There would have been but little use in trying the further service of rascals who commenced by such neglect. I had before sent to engage another pair of hands & another boat, which was done by 10. But they had to fix themselves & prepare their provisions, & before they finally returned I had been kept until there had passed, altogether, 8 hours of waiting, in continual expectation of setting off. I should have felt less annoyed by having to do hard work for thrice the length of time. At last, at a little after 2 P.M., all was ready to cast loose from the wharf. The first thing my best boatman did was to fall overboard & to lose or spoil by the wetting all his stores of rice, sugar & ground coffee. He suffered nothing more, except his cold bath, & having to row in his wet clothes. He is however a smart active young fellow, & his haste only caused his accident, which he treated as a mere joke.

My destination was the residence of Dr. Edmund Ravenel, 24 miles distant, & on a creek of the Cooper River, & which I would have distrusted reaching after so late an hour of departure, but for relying on the knowledge of my boatmen of the river. I soon found that they knew nothing about it so far from the city, & I had to be guided by Mills' map,[1] which is very incorrect as to the river's course.

The margins of the river are mostly of broad & extensive marshes & the few intervals of firm land are very low, say from 3 to 6 feet above tide. After rowing some 12 or 15 miles, the wind permitted the use of the sail. At 18 or 19 miles, strange as it may appear, I lost the route of the main river without knowing it until I found myself in a very crooked creek.

The passage taken at first by mistake had formerly been the course of the river which it had left by cutting through a narrow neck of marsh, & thus making a more direct route. When it was ascertained that we were lost, without there being any idea of where the boat was, night was approaching. There was nothing to do but to row on, taking such of the crooked passages of the water through the marshes as seemed by their direction & size as most likely to direct to my destination. And so crooked was the route that it was doubtful whether we were rowing around a large & irregular island of marsh, or were going any where but the right course. At last, when not expecting to touch dryland, without returning the course for miles, we reached a pine flat, & landed. After walking half a mile, I reached the negro houses, & found to my agreeable disappointment, that Dr. Ravenel's was the next plantation, & his house about a mile distant. I hired a guide to show me the way, & also to pilot the boat around, which was 8 miles, it having gone up the wrong creek, & it being necessary to go out into the river to ascend the right one. I was doubly glad to see Dr. Ravenel, who received me with the hearty welcome which I have found from every stranger whom I have sought or made my host.

Dr. Ravenel was formerly professor of chemistry in the Medical College at Charleston. He is not only a man of scientific education, as his station would imply, but is devoted to the natural sciences & especially to conchology. He has a very extensive collection of shells, recent & fossil, but very few of the fossil shells of the tertiary, & none of those of the Va marls. I hope by giving these to him to add much to his gratification & to the value of his fossil collection. I had heard that Dr. Ravenel had probably made more full examinations of the marls of S.C. as a geologist, than any other person, & therefore would give more correct information as to the characters and localities of different deposites. This would have been sufficient to induce me to seek his acquaintance & instruction. But, what increased the inducement was my hearing from Dr. Johnson that the marl of Dr. Ravenel's estate (the Grove) was different from what is usual, abounded with fossils, & moreover was of the tertiary formation like that of Va.

Dr. Ravenel informed me that his marl, & not only his but all of lower S. Ca. is tertiary, & as Mr. Lyell said when viewing it last winter, of the eocene division of the tertiary. To such authority of course I yield at once my previous opinion that it was secondary. Indeed I had myself no ground of my own for that opinion except what still remains in full force, after having seen all Dr. Ravenel's fossils of this marl, in addition to those I had gathered myself on the Ashley. Among them there is not one that is certainly the same as any tertiary shell I have seen of the eocene (or miocene) of the Va beds. And those I found being new to me, without knowing them, I concluded they were of an older period, & therefore of the secondary formation as was supposed by geologists of the north. They may be tertiary—but I still think they are different from any tertiary known more north. Dr. Ravenel says that the only chalk (secondary) or N.J. fossils found in S.C. were at Eppingham's mills on Lynch's creek, 20 miles from Camden. These belemnites[2] & other fossils of the same age were found by Mr. Blanding, & with other fossils of the ordinary marl, sent to northern geologists, & as Dr. R. supposes, the labels being misplaced, induced the still prevailing opinion that all the calcareous formation of the lower country is secondary.

On these & various other subjects I had much interesting conversation with Dr. R. & received much general information from him as to the localities of marl known to [sic] on the middle portion of the Santee & on the tributaries of the Cooper river. He has not any personal knowledge or particular information as to the other rivers or their neighboring lands.—24 miles, besides the additional lost distance.

1. District maps in Robert Mills, *Mills' Atlas of the State of South Carolina,* published in 1825.

2. *Belemnitellinae* are dart-shaped mollusc fossils of the Cretaceous.

14 FEB. The marl found on Dr. Ravenel's land is not visible on the high-land bordering on the creek, but is reached by digging a few feet deep in a swampy piece of ground a mile back. None is now exposed: but some large lumps were left of the last digging, which afforded sufficient samples of the marl, & the fossils were seen in Dr. R's cabinet. The marl is of granular texture, the grains coarse, & the masses easily reduced to the separate grains. It is soft enough in the bed to be dug up by the spade without much difficulty. The fossils are not casts as those of the Ashley, but shells & mostly in good presentation. . . .

Dr. Ravenel has made some use of his marl as manure, & with very satisfactory results. Still he has intermitted such efforts, not because of want of confidence in their utility & profit, but because all his energies

& nearly all his disposable labor are directed to other than agricultural pursuits, which have consequently been comparatively neglected. In 1837, before having discovered marl on his land, he tried shell lime, at 20 bushels to the acre on 20 acres, with good effect from so small a quantity. Afterwards he put 60 bushels of marl per acre on the same land. The effects have been good, & have continued to increase. He has now a small experiment in progress of newly marled land in rye & clover. This I am especially pleased with, believing that it will be seen in S.C. as it was in Va, that marling will make clover grow well,[1] & make profitable returns where it could scarcely live before, & impossible to make a crop. He also in 1839 marled 9 acres of embanked salt marsh, which was designed for dry culture. But the oats died, evidently because of too much salt in the land, & the same result was found after marling. This obviously was no experiment of marl, because the excess of salt prevented its action.

Dr. Ravenel's main business in [sic] the making of bricks by a machine which is worked by a steam engine. This machine had just been invented, & like most other new invented machines does not perform near as much as was promised for it. Still, it makes 20,000 bricks a day, with the labor of 24 hands, most of whom are women, & boys or girls, & the bricks are put up in the kilns as fast as moulded, & are ready immediately for burning. This is a great gain over the ordinary hand-moulding & long drying of bricks; & nothing is now wanting, but a sufficient demand for bricks, to reward Dr. R. richly for his outlay & expenditures for this new business.

The clay is used as dry as it is usually dug in moderately dry weather. It is taken up from the lower to the upper (third) floor of the house by elevators, (as grain is raised in mills,) there ground by suitable means so as to reduce all the lumps to a moderate state of division, then it runs down & is conveyed into deep moulds, & pressed by stampers with an enormous force. The bricks are passed out as soon as so pressed, & are of remarkable compactness. They are taken immediately to the kiln, & as soon as a kiln is completed, fire is put to it. On account however of the greater degree of moisture which these bricks contain, a gentle heat is only given at first, & for some days, until the water is expelled, when the necessary high degree of heat is commenced.

There is on this plantation an embanked rice field of tide marsh of 100 acres. As the saltness of the river water here forbids its being used to flood the rice, streams from the interior were used for this purpose;

& these, by their super-abundant supply, destroyed the crop frequently. To remove this excess, & also for the other purposes which will be named, Dr. Ravenel has had dug a canal 3 miles long, & 12 feet wide generally, & 20 feet at the lower end. This intercepts & draws off, when necessary, the excess of water which before injured the rice crop, & also furnishes excellent still water navigation for the wood which is required for the brick-kilns. The earth excavated in digging the canal, is good brick earth throughout, & was left heaped on the sides to be so used. It is thrown into the large flat boat, & brought therein to the house where the bricks are moulded.

Though other pursuits have thus almost broken up everything like a regular & proper system of culture on this plantation, still there are operations seen which are far in advance of what is the custom of this part of the state. There is far more use of the plough, & the use found as much preferable as might have been expected. Dr. R. agrees with me that the general disuse, or extremely limited use of the plough, & of horse labor is the great defect of the tillage of lower S. Ca. He also deems drainage wanting very generally on the high & dryest lands, owing to the general character of the subsoil, which is clayey & impervious to water, even when the soil is very light, as it is here & where I saw it in Mr. Cain's neighborhood. From the ditches I saw recently dug here, & the water they held, I am satisfied of the correctness of this opinion.

The operation of mowing the rushes on the marsh was going on to furnish litter[2] for the stock. This seemed to me altogether too late, & I still think it would have been better, for the value of the manure, if mown & heaped in autumn. But on examination, I found that most of the stem of the rush is yet green; & the marsh grass[3] [ER] though the last year's growth is dry, has already shot out young green spines of two feet in height. This supply of litter is inexhaustible, & it ought to furnish immense quantities of manure, if properly made use of. The rush (juncus [ER])[4] is good for nothing as food. It grows on the higher parts of the marsh. The marsh grass prefers the lower, & by tempting the cattle at this time of the year to seek its tender growth near the low edges of the marsh, they are frequently missed, & drowned by the next rise of the tide. This danger takes much from what would be otherwise the great grazing value of these very extensive salt marshes. Would it not be best to mow them closely at the close of autumn, for manure, & then again in early spring for food for cattle?

This evening rowed down the creek to the river 3 miles distant, to try to get up some of the hard bottom of marl. The instrument I had contrived for this purpose, is a straight & round rod 32 feet long, 2 inches in diameter at the middle, & tapered to 1½ inches at each end. A straight gauge or convex chisel, made to form three-fourths of a cylinder, is fastened by its socket to one end of this unwieldy implement. I could not succeed in getting through the very firm & compact plastic blue clay of the bottom of the creek, nor the softer brown clay or mud of the river. But on examining these, I was surprised to find that the latter effervesced strongly with acid. Off Cote-bas, which is just across the river, Dr. R. had found hard bottom, which he did not doubt to be the marl as deep as 42 feet. . . .—10 miles.

1. In correcting acidity, marl enabled the essential, root-based bacterial processes of nitrogen-fixing to occur in all such legumes, benefiting soil as well as plants.

2. The litter provided straw, leaves, and other undecomposed plant parts for animals to lie on and to absorb urine and dung, serving later as manure.

3. Not botanically specific, this could be any grass growing in marshy land.

4. Possibly *Juncus roemerianus,* or black needlerush.

15 FEB. The morning was stormy, with strong wind & heavy rain for a short time, & slow rain afterwards, so that I feared my designed return to Charleston would be prevented. However, at 10 A.M. a little before the beginning of the ebb tide, the rain ceased for a while & we set off, though slight rain fell afterwards at different times. Though the tide soon was favorable, the wind was pretty strong & exactly against our course. Supposing the length of the tide to be much more than sufficient to row 24 miles, some time was used in endeavoring, & in vain, to reach & bring up marl from the bottom of the river, as it could be seen nowhere along the banks. Having heard that it was visible on Clouter's Creek, I took that passage to search for it. This passage, improperly called a creek, is the narrow part of the river formed by a very large island of Marsh. In crossing the main river to reach this passage, it was so rough that it was a difficult job; & the head wind was now so strong that it became doubtful whether the flood tide would not begin before the end of the voyage, when it would be manifestly impossible to proceed against both wind & tide. While thinking of the probability of passing the night in the boat, in the marsh, or not much better off if ashore, the wind suddenly veered about, & could not have been in

a better direction. The sail was hoisted; & without once shifting it, or altering the steering the least from the desired course, the boat sailed 8 or 9 miles as rapidly as it could well cut its way through the water, to the Exchange wharf, almost to the junction of the Cooper with the Ashley. Though it is a ticklish sort of conveyance, which should not be attempted by any one who has not confidence in his swimming powers, the fast running of a good sail boat is to me more delightful & exhilirating [sic] than any other mode of travel on water or on land. Still I have had very little experience of it, & know but little of the conduct of a boat, though I am now suddenly raised to the dignity of "Captain" & "Boss", as my men call me. However, in this run, I yielded the management to them, & was content to be merely a passenger.

This boat, as nearly all the shore boats I have seen here, is such as used by fishermen, & who go out to sea in them about out of sight of land. They are in fact canoes, being dug out of the trunk of a single cypress tree.[1] They have keels, are of good form, & are altogether excellent boats both for rowing & sailing.—24 miles.

1. *Taxodium distichum,* bald cypress, is the largest, thickest tree in the state, with a swollen base; with tupelos it is the dominant growth in permanently flooded sites along rivers, in "bays," and in deep swamps.

9. Ashley River and the Lost Village of Dorchester (16 February–20 February)

16 FEB. It is a general custom in this district for the planters of each neighborhood to form a club & meet to dine & spend the day together sociably at stated times once in one or two weeks. Each member furnishes the dinner etc. in his turn. There are sumptuary regulations which prevent the operation of the besetting sin of members of all such associations, that of each entertainer aiming to give something better than usual, so that the expenses of the entertainment increase until they become onerous as well as unsuitable & ridiculous, & serve to defeat the objects of the association. The day is spent in conversation, & in temperate conviviality. The plan of these meetings is much to be applauded. When I was up the Ashley, several of the planters whom I hoped might be operated on to commence marling were absent from home, & my limited time prevented my waiting for them, or going off from my route to seek others. Mr. Brisbane urged my returning after-

wards & joining the neighborhood club at their next dinner, which I promised to do if in my power. The 18th (Saturday) will be the day, & my wish to go was strengthened by Gov. Hammond offering to go with me, & spending a day or two more on the Ashley. The plan has been arranged Thus: His grand military review is to take place tomorrow, (which was the object of his visit to Charleston,) & I will go up the river tomorrow morning in my boat, & the next morning either come or send the boat to meet him at Mr. Brisbane's house, to which he will come in his carriage. In addition to my having the pleasure of the governors company, I think a beneficial influence will be introduced by this attention on his part to the survey & to the marl of the region. Already Gov. H. has applied marl to more than 600 acres, though beginning only in 1841, which is six times as much as has been marled in all South Carolina besides, so far as I am yet informed. I spent this day in analyzing the specimens of marl obtained from Ashley river, which were found to be very rich, as I had supposed, varying from 54 to 78 per cent of carbonate of lime.

17 FEB. Had appointed with my boatmen to set out immediately after breakfast, but found that one of them was missing, he preferring to attend the review today. After some delay, obtained another, & set out. The wind was pretty strong, & ahead as well as the tide; & when the latter had reached its full strength, we made scarcely any progress. It took 5½ hours of hard work to reach Mr. Brisbane's plantation, (between 9 & 10 miles from Charleston) when it was too late to think of going farther as intended, & as it would have been still against the tide.

The borders of the Ashley to this point are more generally of highland than of marsh, & of but little elevation. The mansion houses of different plantations are numerous, & evidently the situations were beautiful in past time. But now almost every place is deserted as a residence, & there is in all such places a melancholy appearance of abandonment & decay.—10 miles.

18 FEB. Examined the bottom of the river opposite Mr. Brisbane's landing, & found the marl at the depth of 26 feet below the then height of tide, and about as much below the surface of the marl. My instrument is a pole of 32 feet, pointed with a steeled gauge, which brings up enough of the bottom to furnish a specimen to analyze.

After waiting until past the time agreed upon, we proceeded to the club house across the river, where we met the governor, who had been induced to take the lower ferry road, & was accompanied by Dr. Bachman, who takes a strong interest in the survey, & whose influence I trust will not be exerted in vain. About 16 or 18 gentlemen were present. A few of them only seemed much interested about marling, & anxious to converse with me on the subject, & who I believe will make a beginning this spring. Not the smallest application has yet been made on the lands of this river or its vicinity. The ignorance on the subject is greater & more general than any idea I could have formed. Even Gov. Hammond, now so diligent & zealous a marler, told me that as late as three years ago he was totally ignorant in regard to marl, & both in the State Agricultural Convention, & in the Legislature, then used all his influence to oppose an agricultural survey, which others then proposed & advocated in part to disseminate information in regard to marl & its use, & that opposition, by of [sic] reasoning & of ridicule, was directed especially to that point.

After a pleasant dinner, I again embarked, with the governor, to go up the river to Cedar Grove Mr. Dwight's plantation. The distance is about 12 miles, which we finished about an hour after dark. A bright fire had been kindled on the landing, which was very necessary for our guidance in the dark. A good supper & a hearty welcome awaited us within doors.—15 miles.

19 FEB. The tide was at its lowest state soon after daybreak and I was anxious to have that opportunity to hunt for more fossils at Gillon's shore, which is about a mile below Mr. Dwight's house. Accordingly I was in the boat for that purpose soon after daybreak, but found nothing better than at my previous visit. I feared that this early exposure, as well as that of the night before would have still more increased a most distressing cold in the head which was very bad the night after I left Charleston, & was much worse last night. However it was better this morning, & continued to improve through the day.

Mr. Dwight's plantation is on the left (or east) bank of the river. It was the seat of one of the Izard family. The mansion is large, built in the old style & in excellent manner peculiar to past times. It was one of the highly improved seats (as to buildings & the grounds around,) & though all the ancient splendor of artificial decoration is more or less abated, still the place is in good repair for use, & such as is now found in few estates on this river. The magnificent live oaks, planted by the early proprietors, form here as in most other places the noblest & most striking ornaments.

Mr. Dwight had traced the marl as forming the entire beds of the creek near his land, but had not seen it elsewhere. In half an hour's walk before breakfast, I was enabled to find in three several places where it had been dug up in shallow wells, & another shallow digging. His mind was directed to these places by my remarks, & without my presence he would have come to the same result. But like almost every other person, though knowing well the hard earth or rock as it is called by the boatmen, which is exposed by the river, Mr. Dwight, though an intelligent & observant & well educated man, & a good planter, had taken no other notice of marl otherwise.

Several of the gentlemen who had dined at the club yesterday came by appointment to take breakfast with us; & afterwards we all embarked to examine the river to the head of ordinary navigation. Within a mile above Oak Forest, which was the limit of my previous excursion, (& as Mr. Brisbane thought the limit of the visible marl,) is the site of the former village of Dorchester,[1] once a place of some importance for its trade & its number of inhabitants, & of which every habitation has long since disappeared, & the whole place is grown up as waste land. The rights of property therein have been so little cared for, & so much forgotten, that a neighboring individual has recently taken up the whole as unappropriated land. The ruins of the church & the fort still remain, & present the most interesting objects of the kind that I have ever seen. The Church was of a superior style of construction, as is evident from the small portions remaining which are the steeple, with the exception of its spire which was of wood, & part of the opposite wall of the body of the building. The lower story of the steeple is square, & the next above octagonal. The mouldings executed in the brick-work are remarkably elaborate, & of good appearance. The upper story of the ruin, over topping the surrounding trees, is seen from the river, & offers a most picturesque object. There was formerly a chime of bells in the steeple, which unusual appendage would alone indicate a church of unusually high order for this country. The long deserted & silent spot is full of melancholy interest.

The old fort stands within 20 or 30 yards of the river, & the walls are mostly in good preservation, & seem likely to continue so for many centuries. The only part wanting is a small portion, which for some cause was built of brick, & which, like the greater part of the church, has been demolished to remove the materials for other uses. All the other parts of the walls are wholly of tapia work,[2] or formed of oyster shells & lime mortar, which have formed a mass as hard as can be conceived. The walls are about 3 feet through at bottom, becoming somewhat thinner as they rise. They are of a deep gray color, & partly covered with green lichens, so that as seen from the river the walls might be taken for granite rocks, but for the sharp angles & other artificial lines of fortification. It is filled & surrounded by trees & pines, so that I found it difficult to pass along the entire circuit, so as to see the whole. One of the gateways, fully 6 feet wide, had formerly been crossed over the top with strong & broad wooden lintels, and on the outside & inside, over & between which a cover of the tapia work had been laid, which is only 6 or 7 inches took even the places of the lintels, & a foot thick for a narrower space between them. The lintels have long been gone, & though since there is nothing to support the thin stretch of tapia above, it still stands firm, & without any indication of yielding. There have been sundry excavations made in the area, & which still remain open, which were the work of different persons, seeking in vain for hidden treasures.

Many of the bricks of the old church which had been taken down & cleared of their mortar, were still there, piled up for future transportation. The continuation of these abominable acts of thieving vandalism will soon leave nothing of this venerable ruin.

The entire disappearance of a town, its inhabitants & its interests, & of all value attached to the land on which it stood, was to me a novel state of things, which can scarcely be accounted for by all the reasons stated. There were the cessation of this being a military post, after the British forces left it—the change in the course of trade to Charleston—& above all the increase of malaria, which more recently has rendered this place, as most others on the Ashley, unfit to preserve health or even life during the summer & autumn. If this fell destroyer is not arrested in its progress by draining & other agricultural improvement, will not all the lower part of this state be hereafter an abandoned waste?

From Dorchester the river winds about 5 or 6 miles higher to Bacon's bridge, the head of sloop navigation. Still, if time had permitted, the boat might have gone much higher in the stream through the large cypress swamp which furnishes the head waters of the river. The river from Dorchester to Bacon's Bridge has an average width of not more than 50 yards, & below to Oak Forest of not much more. The marl shows at almost every washed bank, & forms the entire naked bed of the river, & of the upper creeks, except at some places where thin covering deposites of mud are left by the river. . . . I have no doubt that the stratum underlies the whole course of the Ashley & Cooper & all the adjacent lands. And on the Ashley, it would seem to be easily accessible on every farm, except within a few miles of its mouth, where the marl dips too deeply.

The lands and residences are kept in better condition on the western than the eastern side of the lower half of the Ashley, though still greatly fallen off from their famous state & appreciation. Mr. [J. A.] Ramsay, one of our company, bought the plantation on which he had resided, on the west side, since my first visit, & which he therefore knew well & deems valuable. He bought it at $1500 for 600 acres. He was one of the executors of the estate to which the plantation belonged, & stated in advance that he would give that price. But for this commendable delicacy, Mr. Brisbane says that Ramsay might certainly have bought the plantation for much less, as any one else might, if he had not been a bidder. This place is 15 miles from Charleston, & (as every other place thereabout) most abundantly supplied with marl. The blowing down of a large tree on Mr. Ramsay's land, a mile back from the river, had recently exposed the marl near the surface.

Mr. Dwight's land, 17 miles from Charleston, is evidently under good management compared to most on this river, & the land lies & appears well. He bought the plantation [E. R.] acres when prices were much higher than now, for $8000 dollars, which sum could scarcely pay for building the large & excellent mansion house alone.

By personal observation on Mr. Dwight's land, & by conversation with other planters at the club, I learn that there exists more reasons than I had supposed for the general objection to the plough, & especially as to the culture of the long staple cotton. Still I do not admit the force of the objections to be sufficient. It is supposed that the preserving the land unbroken & hard, except for the shallow surface is essential to the well-doing of the sea-island & long staple cotton crops. A more important ground of objection on the Ashley lands is that though the soil is generally light, the subsoil is a close clay impervious to water & near the surface. Of course, the lands, which moreover are very level, need much drainage, & are rarely enough drained. Still numerous ditches are kept, which would obstruct the passage of the plough, or be filled by ploughing if passed over.

The distance had been longer than was supposed, making it full 24 miles before our getting back to Mr. Dwight's to dinner. In the evening returned with Gov. Hammond & Mr. Brisbane to the home of the latter. We were met by the flood tide sooner than it was expected, which kept us again two hours in the night.—36 miles.

1. Dorchester was described by an English traveler in 1774 as "a pretty good sized town, upon Ashley River about 20 miles above Charles Town, and navigable all the way up to it . . . for vessels of above 100 tons burthen" (Merrens, ed., *Colonial South Carolina Scene,* p. 288).

2. Otherwise known as tabby, this substance is lime mixed with stones, gravel, and shells, which is very hard when dry.

20 FEB. To save the last hour of the ebb tide, it was arranged over-night to set off this morning by daybreak. And the governor & I were ready on the beach at that time, though the weather was made very unpleasant, by a thick fog, or mist. But the boat had been carelessly suffered to get aground, & more than an hour was lost, before it was afloat. This gave us a head tide all the way to Charleston. The fog continued the greater part of the time, so that we could only see the way by keeping close to the shore. Reached the city by 11 A.M.—10 miles.

10. Long-Staple Cotton on James Island (21 February–22 February)

21 FEB. Having appointed the 23rd to meet Mr. Whitmarsh [sic] B. Seabrook here, it was convenient to use the interval of time between to pay a visit to James Island, which I had before promised to Mr. John Rivers, a planter on that island, & senator from the parish of St. Andrews. While waiting for the arrival of Mr. Brisbane, who had agreed to accompany me, & who arrived by the appointed hour of 11, Mr. Rivers & Mr. Lawton came, & we returned with them before 1. in the boat of Mr. Lawton, & to his plantation, where he resides, which is

directly across the Ashley from the city, & where, after a passage of 15 or 20 minutes, we were truly in the country, with none of the usual appearances on the ground of the close neighborhood of a great city. The contrast is remarkable between the quiet & regular plantation culture, & the full view of the most dense & busy port of Charleston which is so near.

Our conversation while crossing, was in regard to marl, & my recent observations of the immense exposures on the Ashley; & from all said, then & previously, there was no indication that there was any such resource on James Island. The first thing that attracted my observation after landing, was a spot so thickly covered by an ancient heap of oyster shells, that it was manifestly an incumbrance & injury to the cultivated crops. "Here" said I to Mr. Lawton "is land marled in effect, & manifestly to excess." We proceeded to walk over the fields, & I soon saw, & heard that such was the case generally on the islands, that these old heaps, made around the huts of the aborigines, were so numerous, as to offer a very considerable & very cheap supply of calcareous manure. In some cases these spots were the most productive in the fields; but more generally, where the shells are most thick & abundant, they are nearly barren. Yet neither good nor injurious effects had yet induced any one planter to remove earth of the shelly spots to other land, either to use it as manure, or to get rid of it as a nuisance. One of them, who is the most abundantly supplied with this artificial marl, & who moreover is an educated & intelligent gentleman as well as a large & successful planter told me that he would have carted off these shells, to get rid of them, but that there was no convenient place of disposing of them, except on other parts of his fields, & as they injured the crops where they were, he thought it most probable that the injury would only be diffused by the scattering. The gentleman I trust is already so impressed with opposite & more correct views, that he will as well as others forthwith proceed to make large use of this abundant source of manure.

After dinner, we rode to Mr. Rivers' plantation, on Stono river, & had time to walk over part of it.

22 FEB. The next morning our party, joined afterwards by Mr. Lawton & Dr. Thomas Legare,[1] rode over the greater part of the island. There is so much of sameness in the surface & soil of the island, & of its cultivation, which differs but little otherwise in the degree of good or bad management, that my remarks will be more general than particular.

. . . The surface is nearly level, but very slightly undulating, depressions being of rather moister grounds, but still in dry culture. There is but little waste land of any kind & few creeks or swampy passages or receptacles of water. The soil is sandy & very light, about 4 to 6 inches deep usually lying on a sandy subsoil. Sometimes the subsoil is somewhat stiffer, & is called clay; but it is much mixed in with coarse silicious[2] sand, & no where deserves the name of clay. But little of the land (excepting around the houses) is very rich, but nearly all seems in a middling state as to productiveness. From the kinds of grasses on uncultivated fields, I would infer that the soil had been originally productive, though not highly so, & that it was naturally supplied with some calcareous earth; or in other words is not an acid soil. Like the other sea islands, this had been reduced to general poverty by exhausting cultivation; but recently has been much improved by the new system of manuring, which however has to be continually repeated.

The necessity for & profit of using manure has been made so evident, that great attention is paid generally to the collection & application. The resources used are unlimited; but it is believed by the planters that there is no use in increasing the quantity of each application beyond a certain & the usual quantity, & that moreover that quantity, of the principal ingredient, the salt mud of the marshes, ceases to do good after many applications. This mud, marsh grass, & leaves raked up in the wood-land, form successive layers of a large compost heap. Such a one, without any mixture of animal manure, I saw at Mr. Rivers' plantation. The carts were removing it to apply to the cotton land, & it seemed to both sight & scent, to be a rich manure.

The great & celebrated crop of this as of the other sea-islands is the long staple cotton, which grows only in perfection in the very limited space of the coast from Charleston to the Southern line of Ga. On no other lands is the fibre equally fine, silky & long; & even here, there is so much difference of quality, owing principally to selections of seeds of the finest cotton, that the price varies as three to one, between the best cotton of one planter & the best made by another, & even between different acres, or plantings, of the same person. This year, when prices of short staple (green-seed or upland) cotton in Charleston has varied from 6 to 8½ cents, the best going at the latter price, & when the long staple black seed from as near as the Santee canal sells for 18 cents, Dr. Th. Legare of this island has sold of this last crop some 6 or 7 bales at upwards of $1 the pound. Mr. Lawton sold part of his crop at 60 cents. But the crop is small in proportion to the fine quality of the fibre, though not in proportion to the advance of price. Of Mr. Lawton's 60 cents cotton, he did not make more than 30 lbs to the acre. Of the long staple Santee, on the main the crop varies on good lands from 90 to 150 lbs.

The cotton culture on this island is altogether by the hoe. Mr. Lawton is the only planter whom I heard of as ever assisting by the plough his hoes in the first breaking of the land after its usual year of rest. There is no such thing as a rotation of crops. The great crop cotton is by the best managers raised every other year, on the highest or dryest lands. The intermediate years, the field brings such meager growth of volunteer grass[3] & weeds as it can, without seeding, & under the grazing of cattle. I should guess that they cannot get much by grazing, & that their treading is beneficial, by consolidating the too loose soil. The cotton field is manured every year it is cultivated, which is essential to maintain its production. The corn crop is put on the lower & moister lands which would not suit so well for cotton. Sweet potatoes form a large & valuable part of the production of each farm. Dr. Legare chooses his poorest land, which would otherwise come under cotton, for his potatoes. The land is well manured for them, & they are tended in high ridges as is every other crop. After the potatoes are dug up & removed, hogs are turned in; & to this time, judging by the deep & recent rooting, they have not got out every small root left. This operation by the hogs is in effect a ploughing of 8 inches or more in depth of all the land after potatoes & this is the only deep working in the island. In the winter after the crop, the ridges are levelled by the broad hoes, covering all the potato vines etc. in the former alleys & the land is left under the grazing of cattle for a year when cotton comes again. This year of rest & trampling is deemed necessary to consolidate the soil for cotton. It is generally supposed that this plant requires a hard soil, & the soil to be tilled shallow. This is one of the several reasons given for discarding the use of the plough. The soil is not so sandy as I had supposed. Still it is enough so to be a great evil, in the moving of the surface by the winds, when dry & in tillage. Even this day, the dust was so driven by the wind as to be some annoyance to us; & in some parts of the roads the sand was placed by the wind in ripples as the waves on the sandy beach of a wide river.

On account of the prevalence of wind, & the injury thereby suffered by the cotton, a singular culture has obtained. This is to plant rows of corn in the cotton fields, one row of corn to two or three of cotton, for the purpose of protecting the latter, & opposing the drifting of the sands by the screen afforded by the taller & stronger corn. This end has not only been obtained to useful extent, but another which had not been expected. Corn requires much more moisture than cotton; indeed cotton here rarely has the earth too dry. The rows of corn, by absorbing more moisture, relieve the cotton from it; & some planters think that the three rows of cotton to one of corn produce as much cotton as four rows, if no corn had been made, thus making the product of the corn so much clear.

The few low bluffs on the Islands, where it was supposed marl would be visible if anywhere, were examined for it in vain. Indeed, I did not expect otherwise, being sure from previous observation that here & elsewhere as near the sea, the dip of the marl to the eastward makes the top much below the height of the water in the rivers & in the wells.

But the absence of marl is scarcely to be regretted, where other supplies of calcareous matter are so abundant & cheap. Besides the remaining Indian deposites of decayed oyster shells, which would manure hundreds of acres as a very small cost, lime may be bought at the kilns at 5 cents the bushel either slaked[4] or before being slaked. Mr. Lawton burns shells regularly, & both he & Dr. Legare were troubled by the great quantity of refuse lime & half burnt shells forming the remains of old lime-kilns, & were afraid to put them out as manure, lest they should be useless, if not absolutely hurtful. In addition to applying all this without fear, & as quickly as possible, I advised Mr. Lawton to become his own customer for lime, & apply it as manure instead of selling burnt shells for 5 cents the bushel. This lime is made by burning the living oysters as well as the shells of the dead intermixed, which are in all the neighboring shallow waters between tides & below in inexhaustible quantities. And these being mostly small & the shells thin, from their crowded state when growing, I think they might easily be crushed fine enough for manure, & thus counting also to manure all the animal matter. As to the cheapness of attaining them, I was informed by Capt. Bowman (commanding at Fort Johnson,[5]) who spent the first evening with us, that he has been taking an immense quantity of these shells from a bank about a mile distant; & that in the collecting, transportation by a lighter, landing & carrying a short distance in wheel-barrows,

& depositing them, 5 men remove & place 1200 bushels a day, which at $1 the day for wages, is less than half a cent the bushel.

After dining at Mr. Rivers', our party, increased by the ladies of his family rode to old Fort Johnson, & took tea with the family of Capt. Bowman. I found this gentleman to be well informed & his conversation very interesting. Though strictly confined to his military duties, he is in taste a devoted agriculturist. His opinions & his zeal have already been useful to his neighbors, who highly esteem him; & I am sure that if the islanders would make him a present of one of their best farms, they would be repaid doubly by the benefit they might derive from his example.

Old Fort Johnson two miles below Charleston, is no longer maintained for defence & indeed there is no fortification, except a ridiculous watch[?] Tower. This site of the fort is used by the planters of the island as their summer residence, & there is quite a village of small houses, of plain & unpretending appearance. I heard here some curious facts in regard to the local limits of the malaria from which this spot is exempt, though no person's life would be safe if sleeping one night but 100 yards back from the beach. The old hospital stood about half as far in the rear; & every physician who successively attended it was of opinion that one end was healthy & the other sickly, from being subject to malaria. A few of the houses of the summer residents are below & a little back from the water behind a narrow marsh. This situation is as healthy as the others on the beach; but it is supposed that directly between these houses & the others, though not 150 yards apart, there was an interval subject to malaria, & to avoid walking through which at night, a foot bridge was made across the narrow marsh to the beach.

From Mr. Lawton's house, returned to Charleston, after 9. at night. The passages of the day made about 20 miles.

1. Ruffin omits here, and below, the acute accent on the second "e."

2. Converted through geological time into dioxide of silicon, it becomes rigid.

3. Any grass growing from self-sown seed.

4. Burnt lime or quicklime treated with water, or exposed to air, to produce hydrated lime (calcium hydroxide), then dried and powdered.

5. Fort Johnson was a stone building, begun in 1704, to defend Charleston from the sea, but, according to Governor William Bull in 1770, it was of "very unfit construction, and therefore neglected" (Merrens, ed., *Colonial South Carolina Scene,* p. 261).

11. Letter Writing and Specimen Analysis in Charleston (22 February–25 February)

Found the Carolina [sic] Hotel so full that I could only get a cot mattress in a large room where some half dozen were placed. Mr. Whitemarsh B. Seabrook, president of the State Agr.l. Society, met me by our previous appointment, to confer upon my movements, upon which we conversed until 11.

23 FEB. This is the race week in Charleston, which with the Governor's review just concluded, & the encampment service, makes this time a sort of carnival or great annual meeting of planters from a large portion of the state, who even if caring nothing for races or military parade, or choosing to avoid both, as I do, come here on this occasion to transact their private business, & to meet many persons whom they could not so easily see otherwise. On this account I was advised to be here part of this week; & after having determined to set out tomorrow, with Mr. Seabrook for Edisto Island, the governor has so strongly urged a longer stay to meet and converse with planters now here, that I have so changed my purpose. I had forgot to say that upon my returning to Charleston from my first excursion after Gov. Hammond's arrival, he had again so warmly urged my joining his mess & as such occupying his sitting room, stating as the inducement his wish to make me acquainted with the many planters who would call on him, that I acceded, & have found the change altogether agreeable.

This day I used to attend to sundry small matters of necessary preparation, as well as to write letters & bring up my journal. Dined by invitation with a small & agreeable party at Mr. Roper's.

[One of these letters was to Whitemarsh Seabrook as President of the State Agricultural Society. It was published, and a cutting pasted into a later diary entry.]

> Sir:—For the fulfilment of my duties as agricultural surveyor of the State of South Carolina, it is essential that my efforts shall have the support and aid of the agricultural community whose interests the survey was designed especially to promote. Deeming it the most proper mode of seeking this needed and important aid, I take the liberty of thus addressing the planters of South Carolina, and more especially the many agricultural societies, through you, as the

president of the State Agricultural Society, which is the representative body of all the others and requesting your agency to induce the desired action of the different agricultural societies. The act of the constituted authorities of the State of South Carolina which invited me to her service, is the most honorable distinction which has yet been earned by my long continued efforts to advance agricultural interests; and I am impelled by every consideration to desire that the services sought in my appointment shall be as faithfully performed as zeal and industry and my imperfect ability and means may permit. But all these, however faithfully applied, will avail but little, unless aided and given effect to by the co-operation of some of the well informed agriculturists of every section of the State, and in every department of agricultural practice and inquiry. Such co-operation and action, properly applied by the agricultural societies, will leave to the surveyor the more humble, and yet the most useful function of his office, that of serving as the organ or channel of communication, to receive and bring together, for diffusion and general use, the vast amount of information already in each particular section, and which may easily be collected and reported by the agricultural societies, or by individuals. In this manner, the agricultural survey may lead to important and general benefit, by inducing the furnishing of full and general information. But without such aid, the surveyor's services must necessarily be confined to narrow limits, whether of space, or of subjects of investigation.

With these views, I respectfully submit for consideration the annexed list of general and particular heads for inquiry, merely as suggestions or memoranda, on any of which, or on any others more appropriate to the general object, information is requested to be furnished in reports from agricultural societies, with the design that such papers shall be annexed entire to the surveyor's more general report of the progress of the work confided to his charge. And besides such reports from societies, embracing many subjects, and the general practices and the received opinions of practices in every section, it is hoped that particular subjects of investigation and of instruction will be undertaken and reported upon by any individuals disposed to render such service to agricultural improvement. In regard to a few particular subjects of agricultural improvement, I am not so distrustful of my ability to offer useful and profitable instruction; and therefore to these subjects my own efforts and personal labors and researches will be first and more especially, though not exclusively directed. These are the seeking for, examining, and endeavoring to induce the use of the very extensive and as yet scarcely touched sources of calcareous manures in South Carolina. To the subject, my personal attention and particular advice and instructions will be given, and as early as possible, whenever they will be enough valued to be put to immediate use; and if I can thus induce some hundreds of cultivators to begin during this season to apply marl or lime, there will be no danger of the agricultural survey being hereafter deemed a useless or unprofitable measure, or that its cost will not be repaid ten fold in the results of the first year's operations. Millions of dollars in value of newly created agricultural wealth will accrue to the state within a few years after the commencement of the general use of calcareous manures. From my examination already made, I feel authorized to assert that the marl (or soft limestone) formation of South Carolina, is more widely extended and abundant, more rich in calcareous earth, and more generally accessible, and that the proper application will be more profitably compared to the necessary outlay, than of any other extensive region yet known. Believing as I do that a new era of improvement will soon be entered upon, by the bringing into proper use these calcareous manures, it is especially interesting to mark the earliest progress; and therefore it is desired that every case may be reported generally, of the use already made of marl or lime, and the results, whether of benefit or supposed failure, and also all experiments that may be made during this year, of which I trust many will be commenced even after this time. Very respectfully,

EDMUND RUFFIN,

Agricultural Surveyor of the State of S.C.
Charleston, Feb. 23, 1843.

[The questionnaire has 52 subdivisions. Only the "General Heads" are reproduced here.]

I. Geographical character of the particular Agricultural District or section of country under consideration.

II. General description and management of land.

III. General market prices of lands, past and present, and rate of rents—and products in usual crops compared to these prices.

IV. Drainage and Embankments.

v. Implements and machines for Agricultural operations
vi. Fencing and enclosing.
vii. Grass husbandry and grazing.
viii. Live stock.
ix. Dairy management.
x. Manures.
xi. Orchards, Vineyards, and Fruits.
xii. Wood land.
xiii. Waste lands.
xiv. New or recently introduced and valuable processes or improved practice in Agriculture.
xv. Notices and suggestions of any new resources for fertilization or for agricultural improvement or profit.
xvi. Obstacles to agricultural improvement and profit.
xvii. Diseases of residents caused by climate and condition of the country.

24 FEB. Prepared for and made beginning of analyzing my specimens of the marls of Cooper river, which had been brought to loan for me by Mr. Roper. Sunday visitors, called on me in my laboratory, which occupied much of my time, though not unprofitably for my business.

A boring to obtain water to the depth of 335 feet was made in Charleston in 1824, under the direction of Dr. P. Moser. I had read the account of this boring & the description of the earths passed through in Mills' Statistics of S.C.[1] but before I had seen any of the marl in place, or had entertained a thought of the great extent of the bed, both horizontal & perpendicular which my personal observations, & what I have learned of the soundings in the rivers & Charleston harbor have since taught me to believe. Hoping that the stratum below the marl might possibly be reached & its kind ascertained by digging a well of some 60 or 80 feet, or otherwise that the important fact would be learned of the marl being still deeper, I had tried to urge on some of my new disciples on Ashley river to make such a digging, & had offered to pay a share of the cost. Since I heard from Dr. Johnson that the specimens of the earth obtained at different depths in Dr. Moser's deep boring had been preserved, & were in the Medical College of Charleston. This was a most interesting communication; & today, at the first hour I could command the time, I went to seek them, & at the same time to call on Prof. C. U. Shepard, who has charge of the chemical department, & who I knew by correspondence, & having before employed him to analyze the Gypseous earth[2] of James River. I could not have access to the specimens until a later hour; & meanwhile Prof. Shepards report of the specimens was that they were so homogeneous in appearance that nothing was to be learned from inspecting them, or of their constitution, short of chemical analysis, neither which, nor any particular examination had been made. Indeed I inferred from what I heard, & the neglected state in which I found them, that they had attracted scarcely any notice. Prof. Shepard however said that he had shown them to Mr. Lyell, who had been satisfied by the inspection that they contained no organic remains, & that nothing was to be learned from them. I found that the specimens had been moulded by hand roughly, when first obtained & soft with moisture, in the form of small bricks, on each of which was marked the depth at which it was obtained. There were a great number of these bricks piled up on a high shelf without order, & I had all to handle separately, & only about half an hour to do so & to obtain specimens, for which I was furnished the means by Prof. Shepard, who accompanied me to the place & left me to go through my job. Though the general outside appearance was indeed very much alike, varying only in being more or less of dark gray color, (caused probably by the black mud of the higher strata mixing with & coloring all below, or by the bricks being moulded by muddy hands—) I soon formed the opinion that nearly all, if not all, were of the same rich marl which I had traced to the depth of some 28 feet, & had guessed was 60 or 80 feet. My heart leaped at the first indication of this new fact—which however there was no means at hand to confirm, (as the touch of a drop of acid would have served for—) & I hastened to cut off & mark some twenty specimens from the extremes of depth found, & of enough intermediate distances, so as to leave no question of the uniformity of character. I found no specimen obtained higher than 120 below the surface, & the lowest was from 309. Yet within these extremes, several specimens would be marked as from depths varying but a few inches, & some not a full inch apart. I hastened to my laboratory to test the specimens, still in fear of disappointment; but to my delight, on throwing a bit of the deepest boring into acid, its strong effervescence proved what the texture had before indicated, that it was rich marl, & precisely of the same stratum of which the upper surface is exposed on Cooper & Ashley rivers. . . .

Here was clearly made out the important fact that the stratum of marl is at least 189 feet thick under Charleston, & it may be presumed as thick wherever it rises to or near the surface, through the wide space where it has been found at various detached & distant spots. For this great depth is the surest guaranty of great superficial & unbroken extent. But this depth is not near all. From Dr. Moser's account of his boring this marl must have been reached at 67 feet, at which he says "the pipes settled firmly on a stratum of olive-colored clay marl, which when heated, became of a white color; & so well preserved as to render additional pipes unnecessary. *The auger penetrated this stratum with the greatest facility,* which did not vary from 67 feet to 233 feet 9 inches. Here it appeared very tenacious & on washing gave out one tenth part of very fine white sand. From 233 feet 9 inches to 253 feet the stratum the same, nearly with the one eight part of sand. From 253 to 254 feet, calcareous earth & small stones, so that the auger was bent in penetrating it. From 254 to 263, white clay marl; here the auger rested on a hard carbonate of lime which it could not penetrate."

1. Robert Mills, *Statistics of South Carolina, Including a View of Its Natural, Civil, and Military History, General and Particular,* published in 1826.

2. Earth with gypsum (hydrated calcium sulfate or $CaSO_4$) is found in irregular masses in a variety of geological formations. In differing forms it is used as plaster of paris and alabaster. It supplies calcium to soil.

25 FEB. Finished analyzing the specimens of marl on hand, & a sufficient number of those of the borings for 120 to 309 feet. All highly calcareous.

Spent about two hours at night at a small party at Dr. Ramsay's to meet & convene with some gentlemen interested in my pursuit. After returning, packed up my luggage for leaving the city & wrote my journal & letters until past midnight.

My water travel had been attended with so much trouble that I gave it up after the last trip, determining to try travelling by land. I have bought a neat & light buggy for $75, on four wheels, & a stout tolerably ugly & I trust serviceable horse for $70; or rather they were bought for me by a better judge than myself, & I have not even yet seen the horse which I am to drive tomorrow morning.

12. Sociable and Improving Planters on Edisto Island (26 February–28 February)

26 FEB. Set off immediately after breakfast for Edisto Island, in company with a gentleman travelling the same road, which was to me very serviceable. Crossing the ferry over the Ashley, a mile above the city, we passed upon a very level road along plantations lying on the Stono river to Rantoule's Bridge's where both parts of the Stono were crossed on bridges. Saw indications of calcareous earth as I thought in the growth on the bank of a ditch on high land about 4 miles from Charleston, & soon after, on a newly made bank in lower land, there were stony lumps, which on testing proved to be calcareous. I infer that marl is not far below. At the tavern at Rantoule's, where we stopped to dine & feed the horse, I learned that the bed of the upper branches of the Stono was rocky & which I doubt not is the marl.

The road from the Ashley to the Stono passes along good land, but too level, & drained very imperfectly. After leaving the Stono, Colleton district is entered. The land is mostly pine barren, some very sandy, some slashy,[1] & very mean, until the road diverges southward & reaches the water courses. Before reaching the ferry over the Dawho river, which separates the main land from Edisto Island, the road passes for a mile over a causeway through the tide marsh, of which there appeared to me many thousand acres in one view, & in one body, except as cut up by the river. In sight of the road, in the marsh, or probably on a low island in the marsh, there are many dead remains of large cedar trees,[2] some upright, some inclined, but all in place; & by there being no living tree or shrub left, furnishing one of the many proofs of the land being subsided, & being now so low that the salt tide passes over land formerly free from it. Another evidence of the same truth is seen at the neighboring creek, Toogoodoo, in large stumps, or roots in place, in the salt marsh, where no tree can live.

It was sunset when I entered Edisto Island. Mr. Whitemarsh B. Seabrook had sent a servant to guide me, to his house, which was 5 miles farther, & which I soon reached. 36 miles.

1. After logging, the land would be covered with branches, tops, logs, and stumps.

2. *Juniperus silicola,* southern red cedar, is restricted to the outer coastal plain.

27 FEB. After breakfast set out with Mr. Seabrook in his carriage to ride over the island. Stopping for a short time at Major Wm. Murray's, he & Mr. I. J. Mikell there joined our party by appointment, when we proceeded to the open ocean beach on Edding's island. Though several times before on the sea, & oftener in sight of it, this was the first time that I had been on the shore, of the broad ocean, where altogether exposed & unprotected by any farther jutting out of the land. The scene was to me delightful, & its beauty & sublimity beyond description.

The weather was clear & calm & the sun's warmth soon became greater than was desirable. The tide was low, & the beach of perhaps 100 yards width, left by the last tide, was smooth, firm & dry beyond my previous conception. The wheels of the carriage scarcely made an impression, as we drove along upon the sand; & in walking, scarcely the traces of the footsteps were left. The sea shells were more numerous & in better preservation than I had expected to find; & if time had permitted, I could with pleasure have spent a day in searching for them, & might have gathered bushels of such as I would have valued at home, & of which kinds, I already have obtained by purchase, & have in my cabinet. However, I contented myself with gathering for half an hour, to observe the kinds; & left the making for me a full supply of better specimens to the gentlemen present, who promised to do so in the autumn, when they will reside here, & when the shells are more abundant & fresh.

The spot of higher sand hills immediately adjoining this beach is the summer residence of all the planters of Edisto, & their residences make a considerable village, which is quite pretty, owing to the uniform neatness & regular appearance of the houses. All are plain & unpretending & small, compared to the splendid mansions of some of the same proprietors on their plantations. But every house is comfortable, & large enough for its design, & the customs of the time & place. Here sixty families assemble the first of June, & remain until the first good frost. They all come at once & go together; & the contrast must be most startling of the state of crowded population at one time, & of entire desertion the balance of the year. Not an inhabitant remains, except that latterly a few servants have been kept here to guard the houses from depredations. I saw none of these, & the place would have appeared as if utterly deserted, but for there being some workmen engaged in removing farther back from the beach some houses which the encroachment of the waves had reached, & would soon wash away. The result is not infrequent; & in time the whole of the site of this village, & the small island on which it is erected, will be washed away, & not a trace be left to mark the place.

The social pleasures of the inhabitants must be much increased, & also their social virtue cultivated by their being thus brought together part of every year. These benefits, though greatly inadequate indeed yet make a valuable compensation for the families being compelled to leave the plantations for half the year. But no where can that evil bear more lightly than on this island. For every planter is within a short distance of his fields, & may & generally they do visit them every day. For there is no danger in being upon the most sickly lands every day & nearly all day, provided the proprietor does not sleep there, (of which a single occurrence would generally cause disease & endanger life,) & is not there either very early or late in the day, & also avoids getting wet by rain, or other excessive exposure or fatigue. With these precautions, which are seldom neglected, the planters of lower SC & their families are as healthy as any people in the world. And on this island the negroes also are very healthy, though of course encountering all the dangers of the malaria season & region. And such is the general report of the rice lands elsewhere. But as to these accounts differ; & I was told that negroes suffered much by bilious disorders, which though usually slight & not at all dangerous at the time, affected the constitution, & ultimately produced dropsical affections, which often were the causes of death after a long time.

There was one evidence of healthiness among the slaves of Edisto which was brought to my notice in a manner which was ludicrous as well as gratifying on other grounds. It seems that my visit & its objects have been heard of by the slaves, & no doubt odd speculations thereupon are made by them. The old woman who (as is usual) has charge of all the young children during the day, on Mr. Seabrook's plantation, came to her master to request that I should see her children, in which I very willingly gratified her. As we were just going to ride, there was no time to wait for the children to make their toilet, & we came upon them before the old woman had induced the washing of more of the faces of half of the 42 children under her charge & all under the working age. However, they did credit to her care, & she exhibited them with as much pride & gratification as my friend Parson Turner would have done with his Berkshire pigs. I never saw healthier & better looking children; & their clothing was woolen, which I had not expected to see in this climate, for children, but which is no doubt a judicious economy, as a safeguard against accidents by fire, & exposure to disease, as well as an indulgence to the comfort of the wearers.

At Major Murray's fine plantation, situated just back of Eddings'ville & in full view of the ocean, I saw in the fields immense quantities of decayed oyster shells in old Indian banks, which here, as on James Island, are in greater or lesser quantity on every plantation. No use has been made of any for manure, or for other purposes, except by the laborious process of separating the larger shells & burning them for lime—in which way Mr. Seabrook has used lime in a small way for several years, & with good effect. But none had thought the whole mass worth removing & dispersing as manure, which is the proper mode, & in which way I am sure that on several plantations which I saw, fifty acres of land may be well manured from these banks, which are encumbering & injuring parts of the land, at a cost of less than $1 the acre.

In returning from the sea-shore, & riding over a bridge on the main public road about 3 miles from Eddingsville, my eye caught what appeared like blue marl just above the water-line of the small creek, then at low tide. Upon examination, it was found to be a moderately compact mass of silicious earth & shells, of which the kinds appeared the same as the recent sea shells just before seen on the sea shore. Still this cannot be a recent deposite of the ocean, as it is covered by several feet depth of alluvial marsh, & separated from the ocean in the nearest course by several miles of the ordinary high land of the island. . . . Is not this what geologists would designate as the newest pliocene[1] formation? . . . Two of the planters with whom we dined at Mr. Seabrook's said that there was plenty of this in the creeks on their lands, but to which none had attributed the least worth, & had scarcely noticed.

Five or six of the neighboring planters dined with Mr. Seabrook, & we conversed on agriculture until nearly night.

1. Pliocene, the latest subdivision of the Tertiary period, lasted from 13,000,000 to 1,000,000 B.C. and was contemporaneous with the early development of man.

28 FEB. After a walk over Mr. Seabrook's fields, a fine boat or canoe of Mr. I. J. Mikell's manned by six oarsmen (& which was large

enough for 10), came for us, & called by for Mr. M. at his landing. We proceeded down the South Edisto river to its outlet into the ocean, & then ascended a creek which runs back of Eddingsville to one of the objects of the voyage. This was what is called "The Mount", & which is in fact the highest point in the Island, except some of the highest sand-hills along the beach. It is a mound formed by the aborigines, & which is entirely of shells, except some considerable intermixture of ashes, & bits of their broken pottery, broken bones & charcoal. The shells are of various kinds, of the neighboring river waters & sea, but principally of oysters. The mound is eliptical [sic], & measured by stepping over, is 150 feet long, & 48 wide to a perpendicular break on the creek made by the inroads of the water, & which apparently has washed away about 18 feet more of the side. The perpendicular section of the shells where exposed by this loss, is 10 feet, & 12 feet in all to the summit (above the ground of ordinary height, on which they are placed). The surface, except at the perpendicular cliff, is covered over with rich soil, & a growth of small trees & shrubs.

It is difficult to believe that an Indian hut or successive huts could have been continued long enough on one spot to permit its inmates to construct this enormous mound of the shells of fish which had been eaten. And even if this were so, the same explanation would not account for some others on neighboring islands, which are still more artificially shaped, being regular, circular ridges, hollow in the middle. Such a one I saw on James Island, from 3 to 4 feet high, of oyster shells & periwinkles, in the centre of which stands Dr. Legare's mansion house. And there are two others, which have been described to me, one on John's Island, & the other on a small island in the marsh attached to Edisto.

Upon leaving the mound, & returning, I saw a bluish appearance just above the water (then at low tide,) in the section of marsh soil through which the creek passed. Upon examination it was found to be a collection of oyster shells & various sea shells, filled in with earth, & forming another variety of recent marl. It is less compact than that found yesterday, & the sea shells of larger size.

The next object was to examine the dead shells of oysters which stretch along on the water margin of the marsh in some places for great length & in enormous quantity, & especially on Mr. Mikell's land; & to give my opinion as to their value for manure, & the best mode of using them. The quantity is enormous; & on digging into them, I showed what had not been suspected, that the greater part of the mass is fine enough or easily enough broken, to require no preparation for manure. But besides this, the beds of living oysters on the shores, between tides, offer a cheap and endless supply, for the whole island. These oysters grow differently from any I have before seen, though not much acquainted with them in their beds. They are on deep marsh mud shores, & the hinge end of the lower oysters are as it were rooted in the earth, with the opening of the valves directed upwards. On the tops of these, as branches on a trunk, other oysters grow, still extending upward, & branching & spreading outward. Thus the mass grows, & in form might be taken for a vegetable production (much like prickly pear, in manner of growth,) & rising to two feet or more in height, & as close as the bunches can stand, forming large beds; & immense quantities. Still, the oysters stretching into separate branches, & connected but at points, are easily broken up & separated by a grubbing hoe; & the shells being slender & thin, for the greater part, could be easily crushed by a heavy stone or wooden wheel running on edges in a circular trough, like a tanner's bark or apple mill. The oysters would greatly enrich the mass, & the animal matter could combine with & be fixed & saved for manure by the calcareous.

Arrived to dinner at Mr. Mikell's beautiful residence, having the ocean in full view. Spent the afternoon in seeing his fields, & his steam cotton ginning establishment. A rotatory steam engine drives 20 of the roller gins, such as are generally used & moved by the foot for the sea-island cotton. A laborer operating alone on one of these, gins 25 lbs of cotton. But the same driven by steam, will each gin 50 lbs, on an average, & a prime hand can gin lbs 75. At Dr. Legare's I saw a steam engine, & a newly invented gin deemed at first of much greater value; but which are not approved. Another still later has been invented by Macarthy of Alabama, which is supposed to be the best, & an admirable invention for preserving the fibres unbroken, but which has not been yet fully tried.[1]

Edisto Island, in its natural features, agrees generally with James island before described, but was of better natural soil, & is much better cultivated, & indeed, except the rice lands, is said to exhibit the best agricultural condition & treatment in S.C. The original quality of the soil of this island was generally good, capable perhaps of producing 25 to 30 bushels of corn per acre on the better lands, & there was very little of pine barren, or very poor land. The natural growth, as seen in the little forest now remaining, was mostly pine, with some oak & other trees indicating richer soil than pine. The surface of the dry land is level & from 3 to 10 feet at highest above tide. Much is intermixed of firm but wetter soil, & which is not generally sufficiently drained, though regularly cultivated. The soil is sandy, & of very uniform character—though a little is of *blowing* sand, & a small part of Aiken's highland is an extremely stiff & intractable light colored clay. I have seen no other clay or even moderately stiff soil on Edisto.

The system of cultivation & improvement of land is as uniform as the character of the soil, there being scarcely any difference of opinion or of intended or desired practice among the planters, & only such difference of results as will everywhere exist between the more adventurous & careful individuals, & those more deficient in these qualities.

There is more manuring done here than I ever knew elsewhere. The supply of materials is inexhaustible, but still the labor necessary to apply them is very great. Most of the planters manure annually every acre of their main crop, cotton, of their potatoes, (which is a large & very important crop,) & more or less, & sometimes all of the corn crop also. And the same or like operation of manuring has to be repeated every year, & barely serves to maintain the recent & present rate of product, without increasing it, or offering promise of making a much better soil, or its highest productive power more enduring. This manuring system however has greatly increased the product of the island from its previous exhausted state. Probably it has now been brought up to the original state of fertility, & therefore can be raised no higher by putrescent manures alone.

The rotation of crops, if it deserve that name, is the two-shift of 1. cotton, 2. rest from tillage, but grazed by cattle. The potato crop occupies part of what would be otherwise in cotton, & to that extent only, there is a partial & irregular change of crops. But as the potato ground is not changed with any view to alternation of growths, probably the greater part of the land has no change, other than cotton & weeds or natural grass, in alternate years. And the same of corn, which is put on the lower & moister parts of the land, which are therefore less fit for cotton than corn, & the shifts are 1. corn, 2. weeds or natural grass grazed. Cattle are kept more for manuring than for any other use or profit. The supply of dairy products is small; & 15 years ago no oxen were worked, & some planters have not yet begun to work their oxen. The plan of manuring is by running pens,[2] through all the year generally, & only excepting part of the winter on some plantations, when the cattle are penned on vegetable matter, principally marsh grass, pine leaves & corn straw. Cotton seed[3] form an important article which is used separately. The plan of penning in summer, (always bad for the cattle,) is here very much better for the land than that of continuing a pen in one place until all vegetation is killed, & there is no speedy growth to take up the wasting excess of excrement as is done in Virginia. Here the pens are shifted every day, or at most every two days; & the plants soon spring again, & the manured space is soon covered with vegetation to absorb the soluble putrescent matter & to protect all in some measure from waste. In this manner, the best planters pass their cowpens over all the land in weeds designed to be planted next year in cotton or potatoes. Others cannot do as much, but extend this manuring as far as they can. Besides this, salt marsh mud is applied at 40 single horse or mule loads to the acre (350 to 400 bushels;) & where the raked cowpens do not extend, is dressed by the winter made farm yard & stable manures, at about 20 loads to the acre in addition to the 40 of salt mud. The mud is taken of the upper part of the tide marshes, or as deep as filled by the living or recent roots of the marsh grass, which is deemed much better than to dig it lower. The marsh grass, which makes tall & heavy growth on all the marshes, offers any quantity of litter for manure, & indeed good hay for stock; but little use is made of it for hay. The labor of manuring with muck is very great. Mr. I. Jenkins Mikell, one of the best improvers & cultivators, computes his manuring to require from 10 to 12 days work of one hand, which includes the daily moving of the cowpen of a quarter of an acre, which one hand performs.

Edisto exceeds all the other sea-islands in the general quality & character of the fine cotton which is its most important crop & only market crop. Much of this fineness of fibre depends on selection of seed; & many improvements have been so made, & are still made by new selections. The finer the staple, the smaller the products; but still the greater the profit, owing to the very small quantity & high price of the finest products. This fine cotton is produced nowhere in the world, except on the sea-islands of S.C. & Ga, & even of these, it becomes coarser as proceeding south along the coast of Ga.

Edisto is in fact composed of several different islands, separated from each other, by different narrow passages of tide water. There are sundry other such islands, besides the principal ones which compose the whole seaboard south of Ashley river. And in like manner are formed all the numerous islands from Charleston to the Savannah; the extensive low coast lands being merely intersected by numerous narrow passages of water.[4]

The general products of Edisto, for the greater part of the island & rejecting extreme cases, may be counted at from 10 to 25 bushels of corn per acre, for the land put in that crop, which is not usually the best—100 to 200 lbs cotton, before the *finest,* or selected varieties were recently introduced, & 50 to 130 since. Potatoes, 100 to 250 bushels & perhaps 300 for the best crops. Land sells for prices $40 to $80 the acre.

1. This machine was also known as a Florida gin. Cotton was drawn between a leather-bound roller and an iron plate, with the seed separated by a fast-moving steel bar. Reckoned to be equivalent of five old gins, it could be worked by horse or mule as well as by steam. It cost about $100. It was first used in the state by William Lawton and Whitemarsh Seabrook and subsequently gained popularity.

2. Pigs, cows, and other animals were confined in a fenced area for purposes of feeding and manuring.

3. After lint was removed, the boll was used for oil, cattle feed, and fertilizer.

4. Robert Mills remarked in his *Statistics* that these islands would already be mainland if not for human effort to keep channels open.

13. Jehossee Island and the Edisto Agricultural Society (1 March)

1 MAR. Rode with Mr. Seabrook to the opposite extremity of the island, on the South Edisto. The road passed through the plantation of Mr. Hennehan, who has a large extent of embanked marsh in cotton culture, & which produces well. The hands were "listing"[1] the land, which was of just such appearance as the most compact soil of fresh-water reclaimed marsh that I have seen in Va. This is in pretty good order, but the drains not scoured deep enough at present, & the earth hoes up in clods which must be troublesome to reduce. If this had been done in November or Decr. the earth would have been fine & light. This land is cultivated in cotton every year.

An artificial canal, which crosses the island & is used for navigation, separates this land from the celebrated rice plantation of Jehossa Island, which belongs to Mr. Aiken, &c, to see which was the main object of our ride. The vast extent of rice ground of embanked marsh in fine order, was an interesting object, but not varying except in extent of surface & the large size of the canals from other rice grounds. But a more interesting sight was the negro houses, which in number may be considered as a large village, & certainly the most regular & handsome village of its size that I have seen. For all villages, & even towns, though they may have some splended edifices, have also many more that are mean, wretched, & offensive to the eye. In this negro village, while every building is plain & humble, every one is also neat, comfortable & pleasing to the view, & still more to the imagination. The houses are of uniform size & construction, (except the church,) & are all neatly built of frame-work, & each containing two tenements, & having a brick chimney in the middle. They are all white-washed; & as seen when approaching the plantation, appear to stretch continuously for nearly a mile. In fact the village must be more than half a mile long, besides a vacant interval along the wide bank which forms part of the road. In the first part of the village there are four rows of houses. Across the causeway three rows, & at the farther extremity two rows. Each house has attached its garden ground, of uniform size, & well enclosed by stake poles. The negroes residing in them are about 700 in number.

The main drains of this great rice plantation are canals which seemed to be 20 feet wide, & furnished good navigation, & seeming to be designed to be so used. The rice is thrashed by steam, the engine driving three thrashing machines & a fanner for each. The pounding mill is separate, & worked by tide waters, part of the rice ground serving as the pond.

We saw neither the owner, who was absent or his overseer, who was engaged on another part of the plantation. Mr. Butler, the overseer, receives $1800 a year as wages, besides the support of himself & family.

After a hasty & limited view of this great property, we hastened back to meet the Agricultural Society of Edisto, which had been called together to confer with me, & to have submitted to it a proposition to aid the agricultural survey, by preparing a full report of the agriculture of the island. The society has only 15 or 16 members, of whom several were absent from the island. All the others assembled to meet me, & also as many other persons as amounted to about 40 which were nearly all the male proprietors of the Island. I had arranged with Mr. Seabrook, because of my never being accustomed to public speaking, & having no ability for it, that no *speech* was to be expected from me, & that I should merely answer fully such particular questions as would be asked by any of the members or auditors. Still, even with this help, it was to me a new business, & I expected to be much embarrassed. But the matter passed off more easily & better than I had counted on. When invited in & introduced to the meeting, questions were put to me by different persons, in regard to marl & lime, & matters connected therewith. In my answers, I aimed to confine myself to the particular question asked, but to be as full & as plain as was requisite, and as the occasion permitted. Of course the questions, many as they were, occupied but a small portion of the time; & when my examination closed, I was astonished to learn that it had continued nearly two hours. My auditors were very attentive, & seemed deeply interested; & if I may trust to appearances, the procedure will do more good in forwarding my object here, than any thing that I could have done. Every member present professed his intention of forthwith commencing experiments of calcareous manures & several of those who were not members. The proposal was then submitted, & resolutions conforming there-to, which were passed unanimously.

After all dining together, those feeling most interest in the discussion remained some time at table, & continued to ask opinions of me & to receive answers in a less formal manner than when in session. Here too, the opinions & objections of my auditors were freely stated; & I was astonished to find such utter ignorance of every thing relating to calcareous manures in gentlemen otherwise well informed, & experienced & successful agriculturists. And if this observation showed me much more clearly that my instructions & labors were greatly needed, & might be greatly beneficial, it also showed that I had before me uphill work to induce practice by teaching the theory & rationale of marling & liming. Altogether, I am much gratified with the results of the day's operations; & shall be willing to stand another such public examination before any other society that may desire it.

There were brought to me at the meeting specimens of the new marl I found here from another creek, & reports of its being plenty on still another—all the three creeks emptying into South Edisto near its outlet. I have no doubt of its being very general in this island, & regret that the time appointed for my departure does not permit my farther personal examination of the bed at this time. However, I hope to find it elsewhere.

The nights are still cold, & the days not warm, except in the sunshine & out of the wind. But vegetation is well advanced. The red plum trees are in full bloom; & the fragrant woodbine, which borders every road, has already many of its flowers fully open.

1. Furrows would be opened some inches deep with a lister (a plow throwing soil both ways) before depositing seed in raised beds in between.

14. West through Colleton (2 March–4 March)

2 MAR. Left Edisto, accompanied by Mr. Seabrook, for Beaufort & St. Helena. Mr. Seabrook & I took the driver's seat in his carriage, & his men drove my buggy. Our first day's journey extended only 10 miles, to the seat of Col. John Ashe, on Toogoodoo creek, emptying into North Edisto river. Four other gentlemen by invitation met us here, & a very pleasant day was spent. Mr. Wm. Elliott, president of the Beaufort Agricultural Society, was one of them, a gentleman of literary talents, & very interesting & amusing conversational powers. I should have been gratified to take him with us in our excursion, as a sufficient substitute for all the other members of his society, whom he informs us are so scattered abroad on their distant plantations at this time, that there will be no chance of their meeting at this time of year. Indeed, the society, like the town of Beaufort may be said to be in a state of suspended animation, during the winter season.

I heard here in our conversation, of two interesting facts, which if better known, ought to remove much of the existing incredulity in S.C. as to the effect & permanency of calcareous manures. Two of the gentlemen who met me were sons of an Englishman named King who was a very successful planter on Burden's island, close adjeacent [sic] to this place. He applied there a cargo of lime, on some of the worst land, about 20 years ago, and the benefit still is apparent. On a neighboring rice plantation, on Dawho, when sold some 12 or 13 years ago by Mr Linning to Morris, one part of the rice land, about 40 acres was of very inferior quality, & could not have produced more than 40 bushels of rough rice to the acre, when the balance made 60 as usual there. Besides, so uncertain was its product, that for 3 successive years after Morris bought & planted the swamp, the crop failed on this poorer part. He then applied lime to it, at from 70 to 80 bushels to the acre; & ever since, (about 9 or 10 years) the product has been equal to the better part, or supposed to be increased annually by 20 bushels per acre.—10 miles.

3 MAR. To Mr. Elliott's, by appointment to lunch, but which he would convert to an early, yet luxurious dinner. In the afternoon continued our journey, across the Edisto, by ferry, there a rapid fresh river, but still subject to tide; next over the Ashepoo ferry to Mr. Edward Linning's, to lodge, whose invitation had been sent to me at Charleston.

The ride of this day has been mostly through a poor country, almost abandoned. The Ashepoo, where crossed is fresh tide water. Its borders are very extensive swamps, connected to rice grounds. A singular stone in thin layers is abundant here, in the high-land & also is known in different parts of the bottom of the river for 8 miles down. It is in color like whetstone, such as fine *hones* are made of, & not at all calcareous. At the ferry landing picked up pieces of hard white rock, which on touching with acid was found to be limestone. This was a very interesting indication & I returned to the ferry to examine farther. But the only *fixed* rock (the whetstone) being *not* calcareous, & this being a place for sea vessels to receive cargoes at, induced the belief that the loose calcareous stones were of ballast, of foreign origin, & not (as the ferryman said) from the bottom of the river.

The upland soil here is very stiff clay. Mr. Linning cultivates more than 100 acres of highland annually, besides the rice, which is his main crop, & does not use a plough at all. Saw on his land a good hedge of Cherokee rose, only 4 years old, & an effectual fence, unless against hogs. This was on a ditch bank, & grows better for being on clay soil.

Among the evidences of decline seen on the journey today, the most striking was the town of Jacksonborough, which was once for a time (before [sic] the revolution) the seat of government, & a place of importance.[1] Now only 4 or 5 houses seem to be inhabited, & one only, the tavern, is in good condition. This is on the main stage route from Charleston to Beaufort.—20 miles.

1. In this town thirty miles from Charleston, the newly elected revolutionary assembly met in the first two months of 1782, legislating, among other things, to confiscate loyalist property.

4 MAR. The distance from the Ashepoo across the country to the ferry over the Combahee, 10 miles, except the plantations on both rivers, showed no sign of habitation, or of cultivation, except two or three inland rice swamps, the highland pinebarrens, & others of moist swampy appearance. At Combahee, the rice grounds of immense extent. Passed through them on a straight & excellent causeway or embankment of more than a mile long to the ferry. The negro houses of this great estate (Hayward's) formed another large village. The banks which secured this vast extent of rice ground from the highest tides seemed to my eye to be only from 2 feet in large part, to 3 feet high. Certainly the limit of the high tides in this state, must be much lower & more fixed, than in Va., & the early expense of embanking marshes is very far less than in Va., besides the work being very far more permanent with these much higher tides. The marshes must be much higher above ordinary tides.

15. Sea Islands of Beaufort (4 March–8 March)

The next ferry, over the Coosaw, (salt water here,) took us into Port Royal Island, & 10 miles thence, mostly over pine barrens, & some inferior cotton lands, brought us to the town of Beaufort, about 4 P.M. The roads today firm & excellent, except the last 6 miles, of which the sands are very deep.

At the Coosa ferry, the abutment was covered with stones like both kinds found at Ashepoo (one calcareous & the other not) & also two species of recent coral or madrepore[1] in large lumps! All these the ferryman assured us were brought from Huspa creek a few miles behind us. If this be true, even omitting the coral, there is reason to believe that the white limestone found at Ashepoo was from the river there, as well as in Huspa creek. The whetstone is certainly the same. Heard at Beaufort that these stones are imbedded in the mud, of Huspa creek, in great abundance, & are commonly supposed to be petrified live-oak roots.

Slight showers of rain had occurred at intervals during the last 10 miles which became steady as we entered & passed through Beaufort, to Mr. Moncock's private boarding house, where we found neat & excellent plain accommodation.—30 miles.

1. Madrepore is perforate coral.

5 MAR. Sunday. A steady & heavy rain all last night, which continued more lightly some hours this morning. Attended public worship at the Episcopal church, which I am fearful will increase my cold as, according to the time-honored usage of us southrons, the door was left open through all the service, & the damp wind was felt unpleasantly on my head.

There are two curious plants seen here, & one of which I am told is only known in Beaufort, & is probably an exotic. This is a cactus, very like the common prickly pear,[1] (of Va as well as of S.C.) but differing in being of much greater size, & in the color, size & flavor of the fruit. This plant has its leaves ¾ of an inch thick, & sometimes 5 of them, attached end to end, forming stems, which, if standing erect, would be three feet high or more. This is the peculiar manner in which all the numerous varieties of the cactus grow. But it is rare in Va, to see in the only native species, more than the lower leaves, & of but a few inches in length. The fruit of the Beaufort plant, as now seen, is as large as a middle-sized fig, & nearly of the same shape—of a dark purple color, & acid taste. These plants form large beds on the most sandy ground, & by the venomous thorns & prickles with which both the leaves & fruit are covered, forbid the touch of every animal.

1. This was probably *Cactus opuntia,* or common Indian fig, plentiful on sandy lands near salt water; it furnishes scarlet dye. The cochineal insect gains its brilliant color from feeding on the plant in spring.

6 MAR. After breakfast we crossed over the ferry (a mile wide) from Beaufort to Lady's Island, the nearest land. This is a long, crooked & narrow island, 15 miles from one extremity to the other, separated by a narrow creek from the longer island of St. Helena. At the landing the highland presents a precipitous bluff of 9 or 10 feet high, of nothing but yellow sand below the soil. The adjacent cultivated land is, in general hereabout, remarkably light, so as to be subject to much injury from being blown away by the high winds. A remarkable & un-

looked for feature was in sight, which a proprietor who accompanied us said was common throughout & peculiar to Lady's Island. There are numerous sinks, or basin shaped depressions of the land, of various sizes & shapes, but mostly circular & of no great extent; & which by their number, & sometimes by the steepness of their sides, are deemed serious impediments to tillage, & serve much to lessen the value of the lands. I have long ago supposed of such sinks in mountainous limestone regions, (where they are very common,) that they were formed at a very remote time by the falling in of the roofs of caverns, & the surface earth thereby sinking, & being gradually smoothed, & closed at bottom, by the action of rain &c. A few such sinks I have seen in lands underlying rich marl, & which I have supposed to be caused in like manner. And though there is no appearance of marl in Lady's Island, & no other indication heard of, I am confident that the sinks have been caused in that manner, & that below at some depth there is a stratum of marl, rich & compact enough for caverns to be found by the passage & solvent action of water.

Three miles of the main road took us to the bridge which connects it with St. Helena. The main object of my journey to the island was to examine a body of marl long reported to exist on the land of Mr. Gabriel Capers. This we proceeded immediately to see, accompanied by Mr. Joseph Pope, who had come to meet us at the ferry. The reported marl was at one of the extremities of the island, & the necessity of heading creeks made it 12 miles to reach it. Upon inquiring for the marl, Mr. Capers carried me to the spot at his landing, on a creek, of Port Royal river. The body was only about 2 feet thick, and this, added to its peculiar appearance, different from all marls known, added to the previous doubts in regard to marl being found in such locality, soon convinced me that the deposite was artificial & accidental, & that it was probably the refuse or damaged lime left at the bottom of an ancient lime kiln. The layer was indeed covered with some 18 inches thickness of over-lying earth, which might have been deemed an objection to my opinion. But this covering consists of oyster shells mixed with rich black mould & of course a deposite comparatively recent, such as is exhibited at every one of the thousands of heaps left by the Indians, & also at negro huts, & which still remain to encumber the fields of these islands. There was evidence that indigo vats had been formerly close by this supposed exposure of marl; and lime was necessary, & was always provided by burning, for preparing that crop for market, by the old cultivators when indigo was here the general & great crop of the country.[1] The kiln, if it was one, was bottomed on the sand, above the height of common tides. If a storm tide had reached the bottom of a burned kiln, the lower part would have been water-soaked & spoiled, & might well have been left, as useless for any purpose to which lime was then applied. Further—if there had been a surplus of shells left, more than was used for the first kiln, to remove them from the tilled land adjacent, there was no spot so likely for them to be placed, as on that already rendered worthless by the spoiled lime. These shells, thus mixed with such earth as tillage & the rains might add from the tilled land on one side, or what the dashing of waves in storms might throw up from the water side, might supply all the earth which fills the interstices of & covers over the shells. A few days digging for use of this body of "marl", (which is very good, so far as its quantity will go,) would soon decide whether it is natural & extensive, as heretofore supposed, or accidental & as limited as I infer. But there is little hope of proof being so afforded soon. For though this deposite was observed 20 years ago, & has ever since been recognised as marl,—though exhibited to an agricultural society of the neighborhood, & also analyzed by a distinguished chemist—& though the proprietor has been using it at different times & in a small way for 10 years, he thinks that all his experiments have not amounted to more than 4 acres. Yet this slow progress has not been caused by any doubt of the extent of the body, nor from disappointment in the effects. For, although applied so lightly, that I would not have expected any certainly apparent benefit, Mr. Capers is confident that his applications have produced manifest & great benefit in every case. This case is a remarkable example of how little reliance is to be placed upon current reports & received opinions in regard to calcareous deposites & their effect or non-effect as manures, without careful & personal examination, & sifting the evidence of the most veracious witnesses.

It required nearly all the day to examine this one thing, & all gained was to learn that a generally supposed fact has no foundation. This however is important. Before I found the new (pliocene?) marl on Edisto, I had supposed it impossible that marl, such as found on Ashley, could be visible above water here; & I set out in the pursuit under that impression. But after the unexpected discovery of a new & higher formation on the neighboring island, I had hoped to find the same here. I had heard in Beaufort reports of marl in two places in Port Royal Island. It remains to see, on my return, whether there is any value of the deposites or not. At 4. P.M. we reached Mr. Pope's house, after riding 20 miles.

The lands of St. Helena, as seen this day, & as reported, lie higher & are lighter than those of Edisto, & are in the general, less productive though there is much of good land. The effect of high winds on the light sandy soil is a great & general evil. The great crop of the island is long staple cotton, of which the quality of this & of Edisto, is finer & higher priced than any others. Still, the prices, as well as qualities, are very different, the best & the worst being often the products of adjacent fields, showing that differences of results are caused by differences of seed, & management, & not of soil & of climate altogether. Still, no other than one or other of fine black-seed cotton is worth cultivating on these islands. And it cannot be produced of as good quality elsewhere, & not at all to any profit, more than about 40 miles from the coast. The average product of this island, for four fifths of the land, is supposed to be 75 to 150 lbs the acre in suitable seasons. Of corn, 10 to 30 bushels. The prices of land, from 30 to 60$ for good, as low as $12 for inferior. The same general system of cropping, resting, grazing & manuring practised here as in Edisto Island.

The heaps of oyster-shells left on the former sites of Indian huts are to be seen here in every field as on James, Edisto, & as I learn on all the other islands. Very few proprietors, & they but recently, have done anything to spread these at all. Most of them merely submitted to the incumbrance, & some have actually removed them off the land, to get rid of them, & others have thrown them into the rivers & creeks. The beds of oysters are as abundant on St. Helena, & as ready for use as manure, as on the other islands. No use whatever made in any judicious & designed manner, with any calcareous manure. The marsh mud here is so intermixed with sand as to be inferior. The cause of this is the blowing of sand from the high-land to the marshes & creeks. I rode this evening at low tide, with Mr. Pope, nearly a mile along the channel of a tide creek bordered by marsh, all firm enough with sand. Any such channel elsewhere, would have mired horse and rider instantly.

The object of this ride was to seek for marl; but in vain. No such deposite has been heard of in this island except Mr. Capers. Still I think it will be found & that the few planters whom I have met with today will be induced to seek it & to use calcareous manures.

1. Plants were steeped, fermented, and beaten in water, indigo, with the assistance of lime, sinking to the bottom as a coagulated mass. Additionally, vats needed strong lime cement to help them resist fermentation pressures. The product was a dark blue dye. The industry was killed by East Indian and coal-tar competition and the new profitability of cotton growing.

7 MAR. Raining heavily last night, & all day, with some snow falling fast. High wind, & very cold. The contrast is great between the temperature & the flowers now open in this land of the orange & pomegranate. Confined to the house all the day, & was thankful to be in good quarters, & with such good company & Mr. Seabrook & our host Mr Pope who is a well-informed man as well as a very good planter. My constitutional infirmity of suffering from cold made this day & still more the night after disagreeable in that respect, though constantly by the fire until going to bed under a load of blankets. I have latterly had a slight pain in my breast, for the first time in my life, which makes me somewhat uneasy. The exposure undergone since my leaving home is great compared to usage & comforts for years previously.

8 MAR. The weather still raw, cold, & very disagreeable. Left Mr. Pope's after breakfast for Beaufort. Within a few miles, & about the centre of the island, passed by, & stopped to examine the "Indian Hill". This is a mound of earth, made by the aborigines, which notwithstanding its having been subjected to cultivation, & thereby necessarily impaired in its form, preserves much of its original shape & elevation. It is in form a section of a very flat cone—the upper part being cut off parallel with the base. . . . By stepping, the base, nearly circular (& no doubt quite so at first, as well as the top,) is 126 feet in one direction & 135 across. The flat surface at top is 57 by 60 feet. The perpendicular height 14 feet, by supposition. This mound stands on land of the usual level of the highest parts, & is the highest eminence in the island. There is a smaller mound much flatter & about 5 feet high, about 60 yards distant. I heard of several other of the circular rings of shells, of Indian construction in this & the adjacent island of Hilton Head.

16. In and around Beaufort Town (8 March–10 March)

Reached Beaufort at noon. Though unsuccessful in my personal search for marl, I found that my enquiries & instructions had been more operative. Mr. Joseph Hazel, who resides in Beaufort, & owns Distance Island, had called to see me, & crossed the ferry with us on Monday morning on his way to his plantation. I showed him the marl I had found on Edisto & told him how to look for it as it is there found; & the same day he found it, & a large sample awaited me on my arrival at my lodgings. It is precisely similar in appearance to that found on Edisto, in the creek near the Mound, the shells being whole & partly large—& like that, is between low & high tide levels. Other gentlemen who called on me today state also that something similar to the specimens brought from Edisto is to be found in several different places; & I shall postpone my departure a day longer than was before designed, to examine some or all of the places, if weather permits, or to have specimens procured by others, if unable myself to visit the localities.

The Charleston papers which arrived here this morning contained the report of the proceedings of the Edisto Society, & the statement of my examination—together with my correspondence with the President of the State Society acted upon in that meeting. This had awakened the good people here, of whom I had seen scarcely anything before going to St. Helena. Several had called on me afterwards, & more came today; & a request was made that I would meet the citizens at the Town House, for a free conference on marl & lime, after the manner of the meeting at Edisto. This I readily agreed to, & in two hours after, (4. P.M.) we met. I required that I should be questioned by all present who felt disposed; & some six or seven did so. After answering an hour & a half, fearing that the meeting might be too long for the pleasure of my auditors, I put an end to the sitting, & rose to withdraw. But the talk was still continued, for half an hour longer, by persons putting further questions to me. The attention of all present seemed to be well awakened, & I hope that some will be put to work successfully. I believe that my best service is done by talking to & instructing persons on these subjects. It is both astonishing & mortifying to me to find such general ignorance in regard to the use of calcareous manures, upon which I have so long been writing & publishing & have sent so many copies of my publications to this state. But on the other hand, the less informed they are, the better ground it furnishes to hope for arousing them from their state & producing sudden & great action; for though they have neglected calcareous manures entirely, it is because they are entirely ignorant of their value, & all that I tell them is almost as new to my hearers as if I had never written or published anything on the subject. Rain in the afternoon.—9 miles.

9 MAR. The morning, like the preceding afternoon & night, cloudy & misty, & drizzly at intervals. After waiting an hour after the early time fixed for our departure, there being some appearance of clearing off, Mr. Hazel & I crossed the ferry from his land, which is Distance Island, a small part of Lady's Island, & 5 miles from the ferry. The delay made us lose the time of low water, necessary to lay bare the exposed marl which is along the beach, at & a little above low tide mark. The rising tide had covered it all. But upon scraping off the wet sand of the adjoining part of the higher beach, not yet covered by the tide, the solid body of marl was found within 8 inches of its surface, & of much better appearance & more compact than what was brought to me yesterday. . . . The marl approaches in general appearance the kind first found in Edisto. . . . On digging into it, the stratum at that place was not 2 feet thick; but it must be thicker elsewhere, or the marl would not be as low as low water mark, where it is visible, & was first observed. . . . The continued drizzle compelled the examination to be hasty. I left directions with Mr. H. to observe further & inform me.

Returning nearly to the ferry to head a creek, we visited Mr. McKee's plantation, where I had before heard of indications of marl. None was visible; but it had formerly been torn up in masses, from below a water trunk in a dam which had been blown up, & many shells still remained on the dam to show that they were similar to those of the marl on Distance Island. The place was a narrow tide swamp, of high level, & far up in the main land of the Island.

Upon seeing the marl found at Distance Island, some gentlemen said that they were confident that they had seen the like formation in the marshes of the Hunting Islands, about 14 miles distant.

Returned to Beaufort at 3 P.M. The weather still continuing misty & drizzly, prevented my leaving this evening, to proceed on my journey to the next reported appearances of marl on the main, & to see in place the stones & reported coral of neighboring rivers. If coral is indeed found there, it must be fossil, & part of the new marl formation.

Mr. Seabrook left me this morning, to return home. He had devoted two weeks to my aid, & has essentially promoted my objects.

The town of Beaufort is small & old, & yet beautiful in its situation & the appearance of the houses, of which an unusually large proportion are private residences, & many of them of men of wealth. It is however now very thinly inhabited, though full during the summers & autumns. It is then sought as a healthy place, & much of it has been found entirely so, except when visited by yellow fever. This would be deemed by most persons as an exception so fearful, as to be incompatible with a general healthy character. However the visitations of yellow fever are so rare, as scarcely to be dreaded. Yet in 1817, the last time of its appearance, that terrible disease swept off one fourth of the inhabitants of Beaufort.

The exemption from ordinary diseases of malaria does not belong to the whole site of the town. The point, extending most into the water is most healthy; & the opposite end of the town, although also on the water, is nearer to the main land of the island, & is sickly. Some of the finest houses there are now deserted, on this account; & a college which stood still farther, became unfit to use, & has been demolished. This town has no trade worth consideration. Strange as it may appear, it is sustained by the operation of malaria, which drives residents to it; & but for that cause, it is not unlikely that Beaufort would decline rapidly, & be soon almost deserted. As it is, its permanent population, nearly all proprietors of neighboring plantations, are of an order uncommonly elevated, & in the season when malaria abroad fills the warm tennantless [sic] houses, the society here must be excellent.

10 MAR. Though a most threatening morning, left Beaufort after very early breakfast, by previous appointment. The first designed calling place was close at hand, which induced me to risk such appearance of the weather. Before driving more than 5 miles the haze changed to steady rain, through which two miles more brought me to Mr. Milne's plantation the first object of my present trip. Here a new mishap befel me. My newly purchased horse, which had shown himself throughout skittish & unsafe, took fright at the attempt to take him out of the harness, owing to the carelessness of the boy who had him in charge, & he

ran off *very luckily* at such time as to do but little injury to the buggy & harness. It was obvious that it would not answer to trust him again; & in this place, where so few horses are employed, it is particularly difficult to obtain a suitable one on a sudden emergency, even at any pecuniary sacrifice. Fortunately, Mr. Robert Barnwell (formerly president of the S.C. College,) & Mr. Henry Stewart had kindly offered to accompany me a day's journey, & their carriage came up soon after my mishap. Temporary relief was afforded by their servant who rode on a gentle harness horse exchanging him for mine.

The object of the visit to Mr. Milne's plantation, & the issue, require full description, to show some of the difficulties to which I am subjected in my search for marl. As soon as I arrived at Beaufort, I was told that Mr. Milne, on Port Royal Island, had discovered marl two years ago, in his cultivated land, & rising almost to the surface of the ground. This was a strange & interesting fact, to which I attached the more importance because Mr. Milne had been for some years a subscriber to the Farmers' Register, & of course must know what is marl. I wrote to him by messenger despatched next morning (Sunday) before sunrise, requesting him to send a specimen, & stating that I intended as returning to call & see his marl bed. The messenger brought back no specimen, but a polite note inviting me to come & visit him, adding that he could not come to see me that day, but would be in Beaufort next day by 11 o'clock. My engagements took me early to St. Helena—& I left a message for him with my host, Mr. Moncock, requesting him to dig & make examination in advance of my visit. This was told him on his calling; but he said he had a borer, which required but little fixing to be put to work, & that when I came to his house, every examination desired could be made forthwith. When hearing this on my return from St. Helena, & fearing delay in this fixing for boring, on the operation itself when first attempted, I made another communication to Mr. Milne, again requesting that he would put his borer to work before my arrival, & preserve specimens from the different depths properly distinguished. The mention of a borer, involving the need of using one, greatly exalted my expectations. After still another day, I went this morning to see the result. No examination had been made, by boring or digging; & as I proposed that it should be commenced forthwith, we proceeded to the place, accompanied by a man with a spade. The spot was by the side of a small & shallow ditch through land regularly cultivated, & then ready prepared for planting. The digging the ditch had discovered the deposite, two years ago, & it was within less than a foot of the surface of the land. When the digging was beginning, I looked down the descent of the ditch & observed that at 8 or 10 feet distance its bottom was considerably below the level of the marl, & yet that it showed nothing of it. This was enough to prove a very sudden termination on that side. The digging was soon sunk to about 18 inches & there the whitish calcareous deposite seemed to change to the clay subsoil of gray or bluish color. It seemed before this almost certain that the deposite was not only artificial but very limited in extent, at least on the one side & probably the bottom; & a little consideration satisfied me that here had been the site of an old indigo vat, & that the "marl" was the lime used, as usual, for making the indigo. If it had not been still raining steadily, I would have set to work with my own hands to dig out & spread Mr. Milne's marl; & believe that in half an hour at most, I could have exhausted the supply, & finished the job, with much less trouble than our mere correspondence about it had cost. It would not have been surprising that some people should have left this deposite thus without digging & using it, or even examining into it; but it was indeed surprising as to Mr. M. who is not a native of this creep-easy region, but a canny Scot, & a zealous experimenter as a planter & moreover a man of inquiring mind, of extensive reading & even some scientific attainment. And thus have ended the investigations of the only two *marl beds* of which any information existed, & one of which was an important object to carry me to St. Helena. *That* proved to be the bottom of an old lime kiln, & *this* the much smaller remains of an old indigo vat. The rain soon ceased, though still threatening, & afterwards returning at intervals, & we proceeded immediately after the marl examination on our journey, across Port Royal ferry to Mr. Fraser's, on the south side of Huspa Creek. In crossing this creek I noticed the abundant & peculiar stone which is generally believed to be petrified wood, & which I found to be precisely the same as I had found on & near the Ashepoo, & which is certainly not petrified wood. It is not only abundant in all the bed of Huspa creek, but in the underlying fields adjacent, & indeed in all the neighboring waters, & is seen in many parts of the land. It is certainly a singular formation, & especially for a region which otherwise has no stone.

The snow had fallen in this neighborhood on the [ER] so heavily that it lay 4 inches deep; & in Mr. Fraser's yard we found a snow-ball, made on that day by rolling, which still stood waist high, notwithstanding all the rain that has fallen. 18 miles from Beaufort, besides two variations from the course making 8 more.

17. Broad River to Savannah River by Grahamville and Purysburg (11 March–14 March)

11 MAR. Proceeded to Capt. Huguenin's plantation south of Coosawhatchie, accompanied by Mr. Fraser, through the Pocotaligo & Coosawhatchie, the latter a very small village, & the former scarcely the pretence of one—but the heads of navigation of the creeks of the same names.—18 miles.

The whole journey from Port Royal ferry to Huguenin's Neck, is a curve around the head waters of Broad River, the road crossing them about the head of tide & of sloop navigation. These were in succession Huspa creek, Stony creek, Pocotaligo, Tulifiuny river all small & salt tide & where crossed; & Coosawhatchie river, fresh water, & bold rapid stream of black water, such as are all the fresh-water rivers & streams yet seen in S.C. This river is only swelled here regularly by the flood tide, without the downward course of the current being reversed. All the lands between the different streams, & which are distinguished as "necks", under different names, are of good land; & even where the road crossed them, at their wider & faster parts, & through wood lands principally, the quality of the land is generally pretty good, & much better than I have seen elsewhere for any considerable distance on the main land. Though such as I would elsewhere call level, there is some undulation. This especially observed at Mr. Fraser's, which is a very good body of land, & showing much better management than usual. There I began to find cultivation for which the plough is used & appreciated; though still less than it should be used. Mr. Fraser uses 10 mules & 60 hoe hands: cultivates 300 acres of black seed cotton, about 150 of corn (& is the first corn selling planter whose land I have yet seen,) and some rice on inland swamp. All this land to Huguenin's Neck is underlaid with, & the rivers & creeks showing abundantly, the same peculiar thin stone which attracted me to the Huspa, the great place of deposite, & which is generally supposed to be petrified wood; & which

I had sent me today by one of the best planters hereabout taken from his land, as *marl.* . . .

Capt. Huguenin, whose house was our destination this day, is reputed & acknowledged by all his neighbors to be the best & most successful planter & manager in this region. By his own industry & judicious cultivation only, he has made a very large fortune. His mansion & ornamental garden show that he attends to judicious & appropriate decoration as well as to the main objects of cultivation & a country life. Though not ever having to him a letter of introduction, & indeed visiting him, (as I told him at once,) on this occasion as a matter of convenience to myself & to my further progress, I could not have been more hospitably & kindly received & entertained. The rain (which had never ceased to threaten,) soon recommenced falling slowly, which, with the before water-soaked condition of the ground almost confined us to the house during my stay, & prevented either walking or riding upon any of the fields. This I much regretted. For though my object is not to examine planting operations at present with any minuteness, nor with a view to noting them & reporting thereon, still the superiority of Mr. Huguenin's practice would have made it very interesting & desirable for me to take a general view of this plantation & operations. And this I can scarcely expect, unless my survey should extend beyond the single year, to which I have determined from the first to confine my service.

By night, the rain became heavy, & continued falling steadily until morning, (Sunday, 12th), when it still continued at intervals until in the night.

I was much gratified in hearing Mr. Huguenin's opinions & observations, & if I take literally his expressions of approval, he was no less gratified & *instructed* by mine. The latter effect I had no right to expect, & certainly did not seek, with design, to produce. When a planter has through a long & secluded life, been highly & regularly successful in his labors & the improvement of his fortune, & when his merit in these & other respects as a cultivator is universally acknowledged, it is scarcely to be expected in any case that he will readily yield any of his established opinions, or receive others totally new to him. My arrival of course introduced the subject of marl & lime. I found it totally new to Mr. Huguenin, & that he was as little informed on the subject as almost all others whom I have conversed with—that is, entirely ignorant of its action, & its difference from other manures. However, so far from being bigoted, he was ready to inquire & anxious to be informed; & in a short time he was made zealous to begin experimental use. He has on his land, like all other plantations seen along the coast, many spots in his fields covered & encumbered by heaps of oyster shells left at the huts of the Indians; & seeing that these were generally injurious to the crop, & always in the way of hoe tillage, he had never thought that it could be advantageous to spread them over a much broader space. This now, he as well as Mr. Fraser, will begin forthwith.

Capt. Huguenin makes more use of ploughs than any planter of the low country yet seen. I think he said that he had about one plough (single mule) to every 5 hoes. He has been gradually increasing his use of ploughs, from the time when, like many others even now, he used none; & he thinks his net products, as well as gross, have been increased with every enlargement of his ploughing.

The strongest example of aversion to any substitution of teams for hand labor was presented in the circumstances of a planter of this neighborhood named Dawson, who owned about 100 slaves, & never kept a single horse, mule or ox, for plantation labor, or even for taking his rice to the vessels at Coosawhatchie. He lived 5 miles from the river, & all his rice, after being prepared by hand grinding & pounding, was *rolled* in the casks along the road by his negroes the 5 miles. Every year he lost something in price by the dirty outsides, & consequent bad character of his rice in casks—but neither this nor any other consideration produced any change in his system before his death, which occurred not many years ago.

Upon seeing my specimens of the island marl, several persons have stated that they had seen the like formation at different places on the shores of the marshes. This was said of the Hunting Islands, lying next to the ocean from St. Helena, & Mr. Julius Huguenin was certain that he had seen the like on Daw's Island on the Broad River margin.

Capt. Huguenin relieved me at once from my difficulty about my horse, by making with me an exchange in such manner as rarely can occur; for it was agreed upon by us before either had seen or cared to see the other's horse. He had a very old horse, which had been once remarkably able & highly valued & highly priced, but which he kept & used as his saddle horse, now merely in remembrance of his former services. Still, he thought him now fully equal to the service which I require, & perfectly gentle, which were all I wanted. I offered mine with all his imperfections known or feared, stating further that if he should fall short in value of the old steed, that I would make good any such deficiency.

Cloudy all day, & some drizzle. Before night, heavy rain began & continued.—18 miles.

12 MAR. The earth entirely water-soaked or sodden with the quantity of rain, which still continued, though more slowly, until near noon. In the afternoon set out, in company with Gen. Howard, who came yesterday evening to meet me, to his house in Grahamville, on our way to Pury'sburg.[1] Began to rain slowly again before we had finished half our short ride of 6 miles.

As soon as leaving the *necks,* on lands lying between the lower or navigable parts of the rivers, & travelling up the country, the land changes to level pine barren, which now shows extensive ponds, & much of what the eye cannot distinguish from the highest, is now overflowed by the heavy fall of rain.

Grahamville is one of the many villages in the pine barrens settled solely for health. From 30 to 40 families reside here in the sickly season; & only 5 or 6 through the balance of the year. The buildings here are quite neat & comfortable, & larger & of superior order of construction to any others seen. The plan too is pleasing to the eye, & no doubt desirable on other grounds. Instead of being crowded together, the dwelling houses are far apart, & stretch along the straight public road, (the main road from Charleston to Savannah,) for two miles. Each dwelling house is back from the road some 20 or 30 yards; & as the woods are but partially cleared, the whole village is concealed in the forest, except a small space around the observer. But it is much to be regretted, & also passing strange, that waste is constantly making of the trees, by cutting them down either for building timber or fuel; & if this work of destruction is not stopped, Grahamville will not only lose what constitutes its peculiar & great beauty, the cover of noble pines, but probably, as the consequence of this destruction, the very healthfulness of the locality, which was the sole reason for its being sought & built upon by its residents & proprietors.

1. Purysburg, on the east-bank bluff above the river, two dozen miles up from the coast, was established by the Swiss wine merchant Jean Pierre Purry (1675–

1736) and poor Swiss settlers in the early eighteenth century. They engaged in farming and silk production from wild mulberries. The town was later abandoned.

13 MAR. This morning cloudy & raw & cold, & no promise of fair weather, save in the N.W. wind & coldness of the weather. We started for Purysburg after breakfast; & after a few gleams of sunshine, it became clear by noon, after 9 successive rainy & some of them very stormy days. The road is straight, & excellent even after all this wet weather; & it must be one of the best in the world generally. Still through pine barrens, with scarcely any appearance of cultivation to the small & decaying village of Purysburg on the Savannah.

Here awaited me the disappointment of the sole object of my visit, & the whole journey this side of Beaufort. But previous like disappointments had so well prepared me for this, that I expected nothing better, even before the result was finally known. In Charleston I had been informed that marl was to be seen in the river bluff at Purysburg; & relying upon the information (as I shall not do in any such case hereafter when opposed to my general views of the position of marl,) I determined to examine & verify so interesting & valuable a fact, & at the same time so strong an exception to the general rate & inclination of dip of the marl as appearing elsewhere. The first glance satisfied me that there is no marl visible at Purysburg. There is part of an Indian mound left on the bluff, of which the greater part has been washed away by the river. The mound was made of earth of different kinds, & some of it is evidently such as was left about the cabins, containing ashes, & a few broken shells of oysters & freshwater muscles [sic]. . . . The whole Ga. side seen is an extreme cypress swamp, such as borders both sides of the Savannah generally for a long way up. The general height of the bluff extending along Purysburg was then about 10 feet above the water, & the Indian mound rose some 8 or 9 feet still higher. As the river washed along the face of the bluff, it left no place for walking at the foot, & I reached it only below the mound & a little farther. But that slight view was sufficient to dispel all of the very faint hope of finding marl. Immediately after being satisfied on this point, I turned my course for Eutaw & the Santee, to reach which I had designed to take a new route across the upper parts of Beaufort, Colleton & Charleston districts. Our stopping place in the evening was at Col. I. Lartigue's residence & plantation bordering on the then flooded swamps of the Savannah, five miles below Robertville. The latter village would have been my destination, but for a kind & urgent request for me to call sent to me by Col. Lartigue; & which for my own gratification as well as his, I am glad was done.—36 miles.

The tops of the tall cypress trees on the swamps along the Savannah river were frequently in sight so as to show the close neighborhood of its course to that of our road. Though so close to a great river, the land is still quite poor, though better than the pine barren, & very sandy. The plantations show that I have reached the beginning of another region of culture, much more like that of lower Va. & yet not more judicious than that of the seaboard. Though the land is but slightly undulating, & much of it quite level, flat culture & crop ploughing is becoming frequent, though the level surface manifestly needs ridge culture.[1] The green seed cotton is here raised entirely. Heaps of the seeds were seen in the cornfields ready to be applied as manure. This highly prized manure must lose much of its richness by the seeds not being first crushed or heated, so that they germinate in the field, & their oil is thereby converted to sugar or starch, much less nutritious as manure. Still, even as used, cotton seed is highly beneficial as manure, though giving nearly all its effect to the first crop, & its effect never seen beyond the second. Before I explained to Capt. Huguenin the different & permanent action of marl, & made him a thorough convert to my doctrine, he observed that if he could only apply 60 bushels of cotton seed for each acre of his crop, he would not care for any better manuring.

After our arrival, there was time to walk with Col. Lartigue through his field to the border of the swamp, & though the nearest distance is 3 miles across the swamp to the river, the rising fresh had already approached nearly to the high land. There were other evidences of the height of flood expected, in seeing some of the neighbors on horseback driving their cattle home from the swamps, lest they should be drowned by the rise of the river.

Wet as was the field, & in some places the mules ready to mire in the sandy subsoil, the ploughs were running. Indeed, it has not appeared to me that they have stopped, where used at all, in all this wet weather. Even the best planters seem to have no adequate idea of the extent of injury to land caused by ploughing it wet. Col. Lartigue's fields are slightly undulating, & very light, with sandy subsoil—& like most other high-land hereabout is very like the ordinary light land of Sussex in Va.

The first water mill seen in S.C. (other than tide mills,) is in this neighborhood. But there is little custom or demand for those which are established, so general & inveterate is the habit of grinding by the ancient & unimproved hand-mill, which is universally used in all the lower country. Even there, laborious as it is, & much worse than being merely laborious in other respects, I would welcome it for Va, if thereby we could get rid of the much greater curse of mill-ponds. But a far better mode than either would be either steam mills or horse mills, if horse-power can be thus well applied, which has not yet been done conclusively. There was seen the most promising plan in operation at Col. Lartigue's that I have before met with. It is a mill worked by two mules (& evidently light work,) by means of the horse-power of his cotton gin. When the stones are newly picked, it will grind 6 bushels of corn an hour—commonly 5—to the state of mixture of meal & small homony [sic] which is desirable here, where this homony is so large & favorite a portion of the food of all classes—serving with rice, to substitute bread almost entirely. Of 5 bushels of what the mill furnishes, about 2 bushels of meal is separated by the same, for bread, & the balance is fine homony & bran. But this was the report of the cook—& as to the work performed by the mill, it was only used for a few hours at a time, & therefore no fair estimate could be made of continued & daily operations. I like it however well enough to design to have one put up in Va. It is made by Mr. Right & Son, Winnsboro, S.C. & sold there for $80.

1. Ridge culture was throwing two furrows together and planting in the ridge between. This was common on flat, wet lands.

14 MAR. Col. Lartigue's carpenters began this morning to build a flat boat (or bateau) which is to be ready tomorrow morning to convey him & others to an island in the river for a week's hunting. As may be supposed, the building of such a boat is a slight affair, & it rarely is kept for another such use. This island is about 2 miles in length; & when a fresh covers the extensive adjacent low lands 6 feet deep, the deer & other animals resort in great numbers to this island; & when

attacked there by hunters, can scarcely reach any other high land as the inundation covers all but the trees for miles in every direction. The harassed animals when pursued by the dogs, or escaping from shots, take to the water, & conceal themselves for a while among the thickly standing trees; but soon are compelled to return again for rest to the island. I was much pressed to join the expedition; & as little disposed as I am for the tamer sports of hunting & shooting, I would have been glad to join in this new & exciting scene, if I had only my own time to spend & my own pleasure to seek.

My proper course was Braxton's bridge over the Big Salkehatchie near the upper part of the districts of Beaufort & Colleton. But the certain impossibility of crossing an intervening ford (McBrides) over the Coosawhatchie compelled me to aim for a higher route through Barnwell. Stopped for the night, near the Barnwell line because afraid that I would be losing distance by going farther, & not knowing where to stop if proceeding farther. This is the first day, since leaving Charleston, that I have not been accompanied by some one acquainted with the road & neighboring country.

Soon after passing the village of Robertville, I entered the main road from the Savannah R. to Orangeburgh. It passed for some 10 or 15 miles through pine barren; & no where else have I seen such great destruction of large pines by the general & abominable custom of all this country of annually burning the leaves lying on the wood-land. This is done, often by design, to make the scanty grazing better & more forward by removing the cover of leaves & dead grass. And where not designed, or even wished to be avoided, it seldom happens that accident does not cause at some time of the year almost every large piece of wood-land to be burnt over—always endangering & frequently destroying the fences & deserted buildings of adjacent plantations. For the broom grass grows wherever land is left two years uncultivated, & furnishes fuel to spread the flames, aided by dry weather & high winds, over every acre not recently tilled, or not defended by water courses & swamps, or by great & unusual care of the proprietors.

But besides the accidental & irregular damage caused to fences & buildings, there is enough of injury & destruction to the woods alone. The forests of gigantic long leaved pine, such as may be seen in a few places, or of which the appearance may be inferred by the few scattered trees yet spared by the fires, have a degree of beauty & even sublimity which is rarely surpassed. And even if these noble trees were not highly valuable, as they are, for furnishing the best of timber—& if their cover were not conducive to health, still, covering as they usually do lands not deemed worth cultivation, they ought to be preserved from useless & unprofitable destruction, if only to make the most splendid & extensive forest scenery in the world. But instead of this, the annual fires have already destroyed the far greater number of these once magnificent trees, & have left what now remains to stand as huge monuments, blackened & deformed, of the former grandeur of the entire forest. The fires for a long time do not attack the vital parts of the pines, their very thick & dead bark serving as a sufficient shield. But every drop of the exuding sap, forming turpentine, which reaches the surface, strengthens the flame, & causes it to enter more deeply. This brings out more turpentine before the next year's fire, to still more increase its fuel & destructive action. This effect is made much greater by a custom of all wood & timber-cutters, of trying many trees, to find such as will split most easily, or having straight grain, by cutting out a broad & deep chip. This of course becomes an outlet for sap, & a receptacle of hardened turpentine, the overflowings of which drip over & plaster that side of the tree to the ground. In this body of turpentine, the flame being so well fed, has ten-fold power, & every year the fire eats in deeper & deeper, until the tree falls by its own weight. And every year's repetition of the burning increases the power of the flames; so that the trees would be entirely destroyed, & the lands be converted to the prairie state, but for their being too poor & the soil too unfriendly to grass (for want of plenty of lime) to encourage its growth more than that of trees.

18. Getting through Barnwell and Orangeburgh (15 March–18 March)

15 MAR. My route, to avoid the most flooded fords & bridges, had been aimed for Buford's bridge over the Big Salkehatchie—but I heard it was impassible [sic]. Then I directed my course for the next below; but upon approaching it, learned that it then would require a horse to swim. I then had to go still lower to my original destination, Braxton's Bridge, having travelled to it 46 miles, from Robertville, whereas less than 30 would have reached it if the Coosawhatchie could

have been crossed in the more direct & usual route. This swamp was crossed by my higher route so near its head waters, that its volume was less than many small swamps not worth mentioning, but some of which rose over the front wheels of my buggy. After crossing the bridge over the Big Salkehatchee, my proper route was across the Little Salkehatchie at Carter's ford. But meeting a man who crossed there this morning, he told me that the water was over the back of his horse. I then had to turn upwards again to a higher crossing place, Ervin's bridge, about a mile past which I was glad to find the neat & comfortable accommodation, though of plain & simple fare, which I am now enjoying. 36 miles.

The Big and Little Salkehatchie (corrupted to Salketcher) which uniting below form the Combahee, are here, as all other rivers, but the middle channels of wide cypress swamps, which are always miry bogs, & overflowed by every heavy rain. The recent cutting away of dead & fallen trees in the main channel has made the Big Salkehatchie navigable as high as the bridge where I crossed; but it is not used for cotton, which is carried more cheaply (as thought) by the railroad to Charleston. I saw collected at the bridge several of the enormous canoes, each dug entire out of the trunk of a single cypress tree, & which are carried in a rough state to the lower country to be completed. These rough canoes are sawed in two lengthwise, & a central bottom or keel part from another tree used to widen & deepen the vessel. The three parts are secured by inside cross timbers, & the whole dressed over; & thus are made the excellent & beautiful sea boats, still termed canoes, common in Edisto Island which 10 rowers may work to advantage & as many passengers be safely conveyed. On the morning of my return from St. Helena, after the previous very stormy day, as we crossed the bridge to Lady's Island, one of these boats passed through the strait from Charleston, propelled by 10 oars & a foresail, & of which the crew of island negroes had been exposed to the previous storm on their voyage, without it being deemed dangerous. Of course they had not kept at sea, but found a safe harbor in some of the numerous creeks or inlets.

The lands seen along my tortuous journey for the last 50 miles, in the neighboring parts of Beaufort, Barnwell & Colleton districts, are generally of much better quality than the high lands farther down. My route has been on & near the lands bordering on the many swamps, which is a little undulating & always of better quality than the higher ridge land. It is still poor however & very light. Much oak mixes with the pine growth. This region reminds me much of the cotton counties of Va. back from any navigable water. The population however is more sparse, more poor, & the country throughout sickly. The log houses, the universal worn fences, & latterly the corn bread, all remind me of home. There are no large plantations, or expensive buildings. Most of the clearings seem to belong to poor men; & even the more wealthy probably live much in the same manner as the poor. In this region, almost every proprietor near the roads is accustomed to entertain travellers for their money. Before I knew this, & being without any invitation or letter of introduction, I felt awkwardly situated, & I barely escaped making as great a mistake as Tony Lumpkin produced in "She stoops to conquer",[1] though in a reverse manner. For instead of mistaking a private gentleman's mansion for a tavern, as in the case referred to, I mistook a house of public entertainment for one altogether private, in which circumstances compelled me to intrude on the owner's hospitality. The accompaniments of such an error may be readily inferred. The very proper conduct of the gentleman of the house, which was altogether such as would have been most agreeable to me if I had known our true positions, appeared to me so cold & careless, & so unlike the warm & often *oppressive* attentions to which I had been accustomed in the lower country, that I feared I was regarded as an unwelcome intruder. This suspicion kept me uneasy. The high political station of "mine host" (he is senator of his parish,) did not permit me to suppose for a moment that he would accept payment, or that my offering it would not be considered as adding insult to my unwelcome intrusion. However, in consequence of my making allusion to such intrusions as mine, & guessing that they were frequent, owing to the entire absence of taverns, I learned that the continued applications for entertainment had compelled him to the demand for compensation. This information relieved me at once, & was barely in time to prevent my taking leave without offering to pay my bill. The scales now fell at once from my eyes, & every previous misconception was cleared up. I saw that the manners of my host had been entirely suited to the circumstances unknown then to me, & which would have been most agreeable, but for my mistake. Upon asking to be permitted to pay, a very moderate charge was made, which I had no difficulty in persuading my host to permit my payment to exceed. My only difficulty thereafter will be in knowing where I may offer to pay, & where not.

This day clear, the first of eleven, & in the afternoon the sun oppressive for the first time lately; though again cold before dark. The earth slightly frozen this morning, & still more yesterday. Many persons are planting potatoes, & some have finished planting their corn, though others deterred by the cold & wet weather have not begun. A few acres of wheat on many of the farms in the latter part of my route.

1. The celebrated play by the Anglo-Irish writer Oliver Goldsmith (1773).

16 MAR. Before day break, rain recommenced, & it has continued to fall steadily, as if for a continuance. Though extremely anxious to reach my destination, I have two good grounds for congratulation; one of them is that I have passed over the Little Salkehatchie, & the other, that being thus delayed, I am in such a comfortable & agreeable house of private entertainment—& so private that perhaps very few other travellers have called. I was about crossing the bridge, when coming upon two men of respectable appearance, I inquired as to what house I might put up at for the night, stating that I wished to go to one in the custom of receiving travellers for their money. Several were mentioned, but the nearest was farther off than I wished to go. The answers were given in the kindest manner—& it was said, as had been told me several times before, that I might go to *any* man's house, & not fear being unwelcome—whether they would or would not take payment. Though both the men lived in the neighborhood, neither of them in any way hinted a wish for my custom. At last, at a venture, I asked if a Mr. Harwell did not live near & on my road—as his house had been named to me among others previously in the day. The older of the men answered that the other was the son of the Mr. Harwell, & could entertain me, to which the other assented, provided, as he modestly said, I could be content with such fare as he had. That, I said, it was my duty to be, at any rate, & there would be no difficulty on that head. And there was none. I found the house of a poor man—with less *pretension* of luxury or even common conveniences, perhaps than any house I have been a lodger in since my boyhood. And yet every thing is so neat, so good of its kind, & the disposition & manners of my host & his young wife are so obliging, & yet leaving me free from all troublesome attentions, that I have been better pleased here than with the accommodations of the best public houses in Charleston. There was no butter at table, &

no reference made to it as an unusual omission. But the nice fried ham & its gravy, with good corn-bread & good biscuits, left nothing to regret. Last night there was no milk with the coffee. But the coffee was better even without that important accompaniment, than is usual with it. Altogether my supper was an excellent meal; & the early breakfast still better, with the addition of delicious fresh trout fish, caught last night at a neighboring mill-pond where fish are very plenty—& with milk for my coffee, which I doubt not my hostess must have sent abroad to procure for my better accommodation.

Though no marl is visible, or has been ever suspected to exist from the lower Savannah to the Edisto river, there are signs of its being below, in addition to the general views which induce my belief of the great stratum seen elsewhere extending under this whole region. There are sink-holes, some full of water & others dry, in various places, & these I think indicate a stratum of thick marl at some depth beneath. A large & noted sink, marked in Mills' map as the "Punch Bowl" is near the road 6 or 7 miles north of Robertville. I heard from a gentleman whom I met, & who is a land surveyor, that he saw other sinks near the "Dry Gall Creek", which enters the Savannah river not much below the Barnwell line. I hear from my present host, that these sinks are scattered near to Lemon's swamp, which lies between this place & the Edisto besides the two marked "Clear Ponds" in the map; & that there are many limestone springs on the land of Mr. Pinckney on Little Salkehatchie, a few miles above this place, to which visitors for health used to resort. A great mistake, both as to the waters of "rotten lime-stone", as well as to the air of the swamps of this neighborhood.

17 MAR. The rain continued, last evening at intervals, & in addition the wind blew hard. After dark, the wind changed to the north & continued with increased fury nearly all night. The great change to cold, which I was not aware of until too late caused me to pass a night of suffering with cold, which scarcely permitted me to have two or three hours of disturbed & uneasy sleep. The first cock crow was heard with great pleasure, & still more welcome was the first sign of day light, & the making of the fire in the eating room, to which I hastened immediately to warm myself. I would have called for more covering, but I was not sensible of the want until I had been to sleep & all the family were quiet; & moreover I had been so largely supplied, that I thought

it doubtful whether there was more to spare. Besides, when I get thus thoroughly chilled in bed, no additional covering alone will serve to warm me, until I get to fire.

Clear ice in the bottom of a tin wash-basin over a quarter of an inch thick. The thin coat on the pools of water not entirely melted before 9. Set out at 7, immediately after early breakfast.

My host accompanied me about 2 miles to show me part of the way, which only he thought difficult to keep (in which he was much mistaken,) & also to show me a very abundant boiling spring, said to be of limestone. It is on the side of Lemon-Swamp, just after crossing Hamell's (formerly Lighter's) mill. . . . Its water was 5 feet deep, through which the ascending current came up strong enough to agitate the surface, & to throw up fine sand so as to cover a plank which crossed it, barely below the surface of the water. The water was perfectly transparent. . . . There were three trouts in the basin, which is about 8 feet across, seeming to enjoy the temperature of the water, the largest of them nearly or quite a foot in length. Though they darted off when I measured the depth, they almost immediately returned.

The roads travelled today, nearly throughout, were private or mere neighborhood roads, not wider or better than mere cart paths. They are altered continually by the extension of clearings, which & other causes left the road scarcely distinguishable. Of course I lost the road—& the 12 miles to the Edisto bridge grew to 17. The crossing is at Walker's bridge about 2 miles above the Colleton line, into Orangeburgh. Part of the logs of the abutment had been swept off by the fresh (which has now lowered here considerably,) & I should have certainly had my horse to sink in, & probably injure himself & the buggy, but that another one had done so just before me. Thus warned, I used special care, & with assistance, crossed safely.

Here, though my direction was rather downward, I had to turn up, both for want of any road, & because necessary to ascend to the bridge over the Four Holes Swamp, the longest tributary of the Edisto. Five miles after crossing the river, I arrived at Branchville at the junction of the Columbia railway with the Charleston & Hamburg. Then continued across 14 miles more, so called, but which seemed as long as 24, to the end of my disagreeable day's journey. I had hoped, by using so much caution, & going so circuitous a route in search of bridges or ponds not requiring swimming, that I had escaped. But just before closing my journey, I entered the ford of Cow Castle swamp, without having been warned of it, or knowing it, & therefore supposing it, notwithstanding its width, to be but another of the 50 or 100 shallow swamps & ponds crossed in the day. But I found it quite another thing, when it was too late & impossible to return, & useless to halt. Still I expected nothing more than had happened several times before, a slight dipping of water by the top of the wagon body sinking just below the surface. But this time the water was entirely over, & even after getting over that deep part, stood 6 inches under my feet & upon my luggage. I urged on the horse, to get over the deep part, & at the same time was picking up my book of maps, & smaller floating baggage. This prevented my guiding straight, & the front wheel was violently stopped by a large root standing out from the side, & which was covered by a few inches of water. Lucky it was that it was so high; for if the wheel could have risen over it, upsetting in deep water would have been certain. I got out on the root, & after much trouble succeeded in disengaging & backing the wheel, & then driving by. Still there was another deep channel to pass over, but this second wetting could do no harm. If my wagon body did not keep water *out,* it kept it in admirably. Not a drop was lost by leakage; & with all the waste of splashing over, I brought it several inches deep to the next house I saw, about a mile ahead, & where I was glad to take quarters, & to have my wet clothing dried. My larger box, fitting under the seat, received much leaking water among my clothes. A small leather trunk, though floating, received none.—36 miles.

The days ride through Barnwell, continued to show such land & tillage as the day before. Upon crossing the Edisto, the soil, though low, & evidently subject to freshes, & washed & channelled by them is a miserable poor sand. And the highland between the river & the neighboring [ER] swamp, is a naked white sand, which at a small distance looks like snow among the scanty & dwarfish trees & shrubs. It is too poor for pines to cover the land. There is however but little extent of such land here. Immediately after crossing the small swamp, the better quality of land is again seen, & continued generally through the balance of the days journey. The land seen in Orangeburgh, even when covered by pine exclusively, is not the pine barren of the lower country, & seems to be very improvable soil. Much of it is considerably mixed with oaks, showing better quality. But the whole is thickly set with high-land ponds & slashes, to a degree I have never seen elsewhere.

Upon inquiring for marl at the Edisto, I was told by Tucker, the proprietor of the land at the bridge that it shows there abundantly when the water is lower; but we sought in vain to see it today, it being entirely covered. It was dug up by the former proprietor, in one of his fields, & a kiln of lime burnt of it. No use ever made or thought of as manure. It is said also to be plenty on the Four holes Swamp. Yet the man at whose house I now am, a planter seemingly in good circumstances had never heard of marl, when I asked him if it were known to be in his neighborhood.

Tonight, & also on Monday night, there was the like strange luminous appearance in the same quarter of the sky. It was observed a short time after dark, & disappeared tonight before 9. It presented a straight stream or broad ray of regular faint light, but distinct & well defined, rising from not very far above the horizon, south of west, at an angle of about 45° to the south. Its length could not be estimated. Perhaps it might be equal to one tenth of the visible arch of the sky. It was straight, & widening slightly & gradually as extending upward. It was brighter tonight than when I saw it on Monday.

The family here say it was also seen on Tuesday. They are evidently alarmed at the appearance, & consulted me as to what it was a *sign* for.

18 MAR. Continued through similar private roads, like narrow, & often stumpy cart-paths, to the bridges over the Four-Holes Swamp, where I expected, from report another wetting, & made the best arrangements I could, to meet it with the least damage to my baggage. But the water did not rise quite to the wetting mark. The main body of water passes under three bridges, which are high enough & in good order. The danger was on the long causeways separating the bridges, which were covered deeply & entirely by the excess of water which the three main & wide channels could not vent. Upon getting over, I entered to my great satisfaction the public road from the village of Orangeburgh to Vance's Ferry—& found to my surprise that I was only 14 miles from Orangeburgh. Thus I should have actually saved distance, as well as had good roads, if I had at once made the one great curve of passing as high as that village. As it is, my route across from the Savannah to the Santee has been such as I think no man ever travelled before.

19. Springs, Sinks, and Marl Exposures at Eutaw (18 March–20 March)

When approaching the first stream that falls into the Santee, the road descends a considerable hill, & the first that I have seen in S. Ca. The remainder of my ride was within from one to two miles of the river, & some times within sight of the tall cypresses on the swamps which border it on the other side. If I had not known my position, or had never heard of this region being underlaid with marl rising near to the surface, I should have inferred as much from the face of the country. For the whole distance from the stream referred to, the land is undulating, & presents a succession of basin-shaped depressions such as I suppose to have been formed by the former sinking of the earth into caverns formed by the passage of water in calcareous earth or rock below. These basins are sometimes not more than 20 or 30 yards across, & deep in proportion, forming what are known as "sink-holes". But generally they are broad & comparatively shallow basins, the rim elevated nearly equally, but at any rate much above the centre. These broad & shallow basins are cultivated, & but a few had a little water remaining in the middles. Two deep ditches were crossed, which had been dug to drain such places. I have no doubt but a well sunk in the middle deep enough would reach some opening in the calcareous bed below, & discharge the water downward more cheaply. Indeed the same end is generally & evidently reached by mere filtration through the upper earth.

Before 3 P.M. reached Eutaw, the plantation of Mr. William Sinkler, to which he had invited me, & where I had appointed to come.

After dinner walked to see the great exposure of marl, on Eutaw creek, just by the house, & to see the river. It is here from 150 to 200 yards wide. . . .

19 MAR. Continued my examination of the Eutaw creek & the celebrated spring which supplies it. The valley of the creek extends back from the river about two miles, through the lands of Mr. Sinkler. The field on which the battle of Eutaw[1] was fought lies on the west side of the upper part. The now high waters of the creek (raised 6 feet by the fresh on the river) is from 70 to 100 yards wide generally, & as clear as limestone spring water always is. The slack-water[2] reaches to the head, & is there still several feet deep. The margins are narrow hillsides, of

more or less rapid slope, still covered with their native growth of trees, & dry to the water's edge. The bottom, which except the bed of the running stream, is left dry when the river is low, also has its original growth of tall cypresses and other low ground trees. . . . The volume of the principal spring appears to be equal to a very large mill stream—filling a channel, through which it flows rapidly of 5 feet wide, & about 10 inches average depth. Its temperature, which I tried carefully, is 64°. . . .

Through nearly all Mr. Sinkler's land, Mr. James Gaillard's, & several other tracts of land lying along the river & from Eutaw creek to Rocks Creek, entering the river 3 miles below, in a straight line, the lime rock is generally within 10 feet of the surface of the highest land, & frequently the stony masses show at the surface. The back line of this body of land is distinctly marked by a connected line of shallow sinks or basins, as well as by the great difference of the quality of the land. Without this line, though the surface is not much different in level, the marl is not within 30 feet of the surface, as known by the digging of wells. The body of land above designated, is throughout a succession of basin-shaped depressions, generally shallow, & forming a surface but slightly undulating in the general. This shape of surface, on any other foundation than one having numerous fissures, as the calcareous bed must have, would make numerous ponds, either permanent or temporary, in every field. But there are very few, in which water often stands so as to be of much detriment to culture. . . . In the remarkably rainy spring of 1841, Mr. James Gaillard had some 40 or 50 acres of the basins of one of his fields under water, & so long that no crop thereon could be made. . . . There are abundant evidences to me of there being plenty . . . fissures, & even longer caverns below. The numerous basins themselves must have been formed at a remote period by the breaking in of the roofs of caverns in the lime rock. Similar operations are every year still produced in the sudden sinking down, to a depth of 10 or 15 feet, of small spots of ground, rarely more than a few yards & generally but a few yards & generally but a few feet across. Such a one I saw in this field of Mr. Gaillard's, which had been suddenly formed within the last few weeks. There are many perpendicular holes, like roughly made wells, which retain their shape because in the solid lime rock. Near the side of Eutaw creek, & at the edge of the battle field, there are 13 such holes very near together, from 2 to 4 feet wide. I went down in one of these, after having the accumulation of leaves in part taken out. There was a low lateral opening descending towards the creek, into which I crept backwards as far as my own length when my feet reached water though still far above the level of the creek. On the opposite side was another opening but not more than 8 inches wide.

1. At the Battle of Eutaw Springs, 8 September 1791, Nathanael Greene and 2,300 men attacked and seriously weakened 1,200 British troops under Alexander Stewart. It was the last major engagement of the Revolution in South Carolina.

2. Part of river lying beyond currents.

20 MAR. Rode to the Rocks Creek & over the intervening lands. This Creek is, like Eutaw, supplied by lime-stone springs, so abundantly as to furnish water-power for a mill, which however can grind only when the river is lower. The pond is very small. On its bank, on the land of Saml. Gaillard decd. there are seen on the surface masses of the hardest lime-rock, & close by, the upper part of the general mass, very compact, but yet soft enough to be dug with the grubbing hoe, & which mostly crumbles as dug. It is as white as chalk, and almost pure calcareous matter. When observed closely, the compact mass seems composed of small fragments of shells and of exceedingly small corals. This is the general appearance of Mr. Sinkler's marl along the creek, and of which, I analysed a sample sent me some years ago, and without then knowing its precise origin, & found it to be 98 per cent. This harder part is but thin on Rocks creek, & below the marl seems composed of precisely the same materials, & as rich, but is quite soft. This kind has been dug for manure at a low place near the mill, where I found granules of green sand[1] intermixed with some layers in considerable quantity. Here I attempted to penetrate deeper, both by digging and boring; but could not owing to the water from springs which came in everywhere.

Visited Dawshee, another tract of Mr. Sinkler's on the river, and forming part of the body of land between Eutaw & Rocks Creek. One object was to trace the marl in the bluff. The water however was too high—and nothing of it was visible except a detached mass of the stone which is found higher than the general body. Part of this land was newly cleared of its growth of oak & hickory mixed with pine—and presented to my eye a soil of rich light loam, which, with my knowledge of the close neighborhood below the lime rock, and the soil being certainly affected beneficially by it, would have induced the belief that it would yield 30 or 35 bushels of corn. But I was told that 20 bushels was a fair crop. It is a prevailing opinion of the planters that the climate of lower S.C. is unfavorable to the growth of corn; & that the land cannot produce it as in the middle districts, or in Va. And this must be so, if this land cannot produce but 20 bushels.

The land of this neighborhood between & a little without the two creeks was of very good quality when new & still is good generally. Some of it has been cultivated in cotton almost every year, & has long sustained such barbarous usage. One piece, so treated, was mentioned as having been worn out & therefore thrown out of tillage three different times within the memory of the oldest inhabitants; and which had recovered with surprising quickness under the rest thus afforded. The valuable qualities of this land are caused by its being partly supplied with calcareous matter by the lime-rock below; but still it is not enough calcareous, & requires a further addition.

This fine body of land has undergone remarkable fluctuations of price. When indigo, the former universal sale crop of lower S.C. was abandoned, because of the reduction of price, & cotton had not been yet introduced, agricultural profits were extremely depressed, & consequently the prices of lands. The whole of this body of land with as much adjoining as made about 10,000 acres, was in one property; and could not be sold at that time for $1 the acre. Afterwards, being divided into smaller tracts, the land was sold at prices varying from $2 to $10, & generally from $4 to $7. Subsequently, at the time of mad speculation & advanced prices caused by paper money expansion at the close of the last war with G.B., the old Eutaw plantation, of 200 acres embracing the battle field & Eutaw spring sold for $50 the acre; & since was bought by Mr. Sinkler for $20. This year, his son bought Belmont, a tract embracing the upper part of Rocks creek, but which is of inferior quality, much worn, but has very good buildings, for only $5 the acre. This plantation has no marl visible, except the upper stony masses; but probably even in James its price would not have been varied by the presence or absence of marl. Such however is not the case even now.

On the Rocks plantation, belonging to the estate of S. Gaillard decd. & under the direction of his brother Mr. James Gaillard, there have been marled 50 acres of land, begun about 4 years ago; and 50 more just covered & which is to be cultivated this year for the first time after.

The rates, 90 to 120 bushels. This extent & continuance of operations would indicate that satisfactory benefit had been seen. But Mr. G. is not *sure* that any good effect has been produced. In this I am confident he must be mistaken—though there are reasons why the benefit should have been small, & not perceived. In the first place, the land being partly though not sufficiently marled by nature, could not show *immediate* benefit, as would an acid & bad soil. Secondly, the land had been in cotton continually for 12 or more years before the marling, & since to this time, which is the most unfavorable condition of land for marl to act. It is true that the land has been manured for every crop; but such manuring, from such materials as leaves, & the small resources of dung & of labor, and the large quantity of land so manured must have been very light, & not equal to the annual exhaustion of the field. Thirdly, the whole field was dressed, and no part left without, for comparison, to show the effect, if any, of the marl; and if any effect was produced, it would probably be ascribed to the other manure annually supplied. Still, on the land so treated the crops have amounted to lbs 150 of black-seed cotton—which as some other persons think could not have been possible without the marling. My views, when explained to Mr. Gaillard, were satisfactory to him; and he is going to continue the operation, on more correct principles. Mr. Sinkler had begun to follow my advice in applying marl before my arrival (in consequence of our conversation in Charleston,) & seems resolved to proceed rapidly & extensively. He is the first person whom I have yet seen actually engaged in the operation.

Although of the very few trials heretofore made of calcareous manures in S. Ca, nearly all have produced but little effect, & certainly but little immediate effect, it is gratifying that I find generally those very individuals who have made these fruitless trials, among the most ready to understand & acknowledge the causes of failures & anxious to renew their operations upon proper grounds, & with due regard to the true actions of such manures. The general & sufficient cause of failure to find the benefit expected, has been from considering marl or lime as *alimentary*[2] manure, and so applying it. It has been most generally laid along in the alleys between the old beds, over which the new ones are to be made, and these carefully covered over by all the successive drawings of the hoes, and without the least mixture of the manure with the soil. Of course it had as well not been applied. The next year, if the land remaining uncultivated (as is usual) the manure still remains separate & unmixed, & therefore necessarily inoperative. The third year the bed is reversed & cultivated, & the manure for the first time is partially mixed with the soil, or is for the first time really applied as manure. Then some benefit might be apparent. But by that time the proprietor has ceased to expect any, & therefore does not look for it; and moreover, the land is sprinkled over with putrescent manure, which serves to disguise the effects of the calcareous. Add to all this, that the proprietor is usually absent when the crop is well grown & ripe, and the observation of effects left to be made by an ignorant overseer, probably prejudiced against what he deems the nonsensical experiment.

In the battle field of Eutaw, & in other neighboring fields, the masses of hardest lime rock sometimes jut out of the surface of the cultivated land. These present impediments to tillage (which is by the hoe entirely, for cotton) & of course the spot is passed over, and as much of the surrounding land as the laborer fears his hoe might reach the rock below the earth. The existence of these few exposed rocks have caused the belief to prevail that the entire fields are abundantly calcareous, & their reduced condition, or want of superiority, is alleged as a proof that the lime has done but little for its fertilization. But, in truth, these lands are of the best quality of upland yet seen, for excellent texture soft & dry, for productiveness when first cleared, for long resistance of continual exhausting tillage—& when worn out, for rapid recovery by a few years of rest. These qualities sufficiently prove that Nature had limed these lands; but there is enough other proof that more lime is still needed. The hardness of the protruding masses of lime rock, the disuse of the plough, and the careful avoidance of the rock by the hoes, all serve to make the effect of these rocks on the soil in general, or even close adjacent, very slight. Still I have seen, on Cooper river, around such masses of fixed stone, a patch of soil so manifestly better (as apparent from the grass even in mid-winter) as to leave no question as to the influence of the calcareous matter. Sheep sorrel (rumex acetosa,) which in Va. is the most abundant and unerring test I know of acid soil & the great want of lime, is in S.C. a rare weed—probably because of the warmer climate being unfavorable. But I have no where seen it more abundant than on some of the outer & inferior land of the Eutaw neighborhood; & sometimes plants were seen on sandy hillsides on slopes of the fields generally of better soil; and within 20 yards of protruding masses of hard lime rock. For these reasons, I believe that the appearance of such masses at the surface is no evidence that the land in the field generally is sufficiently supplied with lime.

Besides what has been already mentioned in regard to sinks & basins, I heard sundry other facts serving to confirm my opinion of the existence of internal & deeply placed caverns. A well had been dug on a plantation back from the river (where the marl or lime rock lies deeper, & the land is poor,) & water obtained before reaching the rock, which is not always the case, & rarely on the river plantations. This well was used for some length of time. In a very wet spell, the water rose in it & stood very high. One morning it was found that no water was in the well, the curbing at bottom was not visible, & the negroes reported that "the bottom of the well had dropped out", as of an old cask. And no water has since returned in it. Another well on a river plantation was dug, & into the calcareous bed; & contrary to what is usual, water obtained in abundance. This was in the winter or spring when the river was high; but as soon as the level of the river sunk with dry weather, the water in the well sunk with it, & the well was left dry. It was evident that it had been supplied from the river, or by weirs in which the water was backed up by & kept as high as the water in the river. At a place lower down, belonging to Mr. S. G. Deveaux, a mill pond had been raised by a dam, which for a time served well. But one morning the pond was empty & yet the dam unbroken. A perpendicular leak had taken place, & a channel made from it under the dam, letting out the water in a few hours. The current washed out of its new subterranean passage numerous fossil shells, with sharks' teeth and vertebrae of some animals.

1. Greensand was Cretaceous sandstone with green specks of glauconite. In its purest form, without calcareous content but with notable potassic, and sometimes phosphatic, elements, it is valuable as fertilizer.

2. This form of manure feeds plants directly, rather than assisting through liberation of nutrients in other soil constituents.

20. Embanking and Marling on the Santee (21 March–24 March)

21 MAR. My time for researches has been contracted by the hospitable attentions of Mr. Sinkler & Mr. Gaillard, who successively had dinner parties of neighboring gentlemen to meet me. It is very agreeable

to me to meet with any persons with whom I can confer profitably—but would prefer that the form and show of regular parties could be avoided. This day, when I had appointed with Mr. Gaillard to join him on an excursion immediately after breakfast, or at 9 oclock, which Mr. Sinkler's breakfast hour would have permitted, we did not commence that meal until 10, because some *country* ladies who arrived yesterday from Charleston did not leave their bed-chamber until that hour. Then being engaged to dine with company at Mr. Gaillard's, left us scarcely any time for labor before dinner; & it was half past 5, before the close of dinner permitted me to pursue my journey. I had been invited by Mr. Mazyck Porcher repeatedly & had yesterday appointed with him to go this evening to his father's plantation on the Santee canal & river;—and he was also the bearer of an invitation from the Agricultural Society of Black Oak to join them in a conversation meeting next Thursday (23rd). I could not at so late an hour have ventured alone on a new route; but Major Porcher sent up a servant with an apology for not joining me at the dinner, & to accompany me to his house which is 9 miles distant; & which I reached some time after dark.

The marl shows in the public road I travelled at the outlet of the Rocks creek. It is also exposed similarly to lands above, on the plantation of Mr. Kirke, a mile below & back from the river. Below this & along the river side to Mexico (Major Porcher's) & some distance below, for a space of some miles in width there is no marl known. It must have an irregular & considerable dip at particular places of which this place is one. Also, in the Eutaw neighborhood, the bed of marl is manifestly much lower under the back lands than near the river.—17 miles.

22 MAR. Major Samuel Porcher is now 75 years old, & though suffering under some infirmities, is remarkable for his general health & retention of his ability to bear exposure & fatigue. He has by many years of labor executed a work of which he has a right to be proud, & of which the successful completion has doubtless added so much to his happiness as to compensate most amply for all his previous care, toil & expense. This is his embankment of swamp lands on the Santee, which I rode with him & his son to examine, through its whole extent, after breakfast, & which I shall proceed to describe.

This embankment extends from the northern outlet of the canal, 4¼ miles, on the swamp land of Major P. & 300 yards farther through his next neighbor's land, for the greater safety of his own. Its height varies from 8 to 14 feet of perpendicular height, & averages 10. The base varies from 35 to 60 feet, & most of it is over 50 feet. It protects fully, for dry culture 800 acres, & embraces about 1400. This great work was begun in 1817, & was not deemed complete until 1841; & even now, it is designed to add to its height considerably through its entire length. Besides the first construction, there have been at different times breaches made, & parts swept away, so as to cause about a mile of bank to be re-constructed. This is one of the greatest private works known; & is the more remarkable as being the work of one individual, & who was sustained by no encouragement of concurrent opinion or hope of other persons. Very few had any expectation of his ultimate success or profit in the undertaking; & even 4 or 5 years after the commencement, when the land had been embanked, & under tillage, & the worst breach & overflow occurred, Major P. was urged by his overseer Mr. S. Hawksworth, to abandon the work of embankment as hopeless. This overseer has been in his present employment for more than 20 years, which alone is sufficient evidence of his great merit as a manager, as well as his regard for his employer & faithful support of his interest. But in spite of the early disasters, & of all discouragement, the proprietor persisted in his labors, until their success was as complete as he had hoped for, & with the profit presented a full reward for all past sacrifices.

The plan of this great work was conceived in the mind of the undertaker before he was the proprietor of but a comparatively small part of the swamp land adjoining his high land. And after deciding in his own mind on the feasibility & propriety of the undertaking, he had still to effect, as he did successively, the purchase of other rights of property in parts of the area in question. This was done at various rates of price, & of disadvantage increasing as his object & plan necessarily became developed to his neighbors. The first & main purchase of swamp was of 900 acres of the higher & therefore the better part, at $4 the acre. Another & as good part, bought since it was left without his line of embankment, cost him but $200 for 250 acres. This, if owned at first, might have been embraced in the circuit with very little additional length of bank; but for which now it would not be worth making a separate bank. Another small lot of only 10 acres had to be bought at $50 the acre. Even now, however, if similar land were in market, it would sell very low—perhaps not for $2 the acre.

The swamp land was not, as I had previously inferred, incapable of

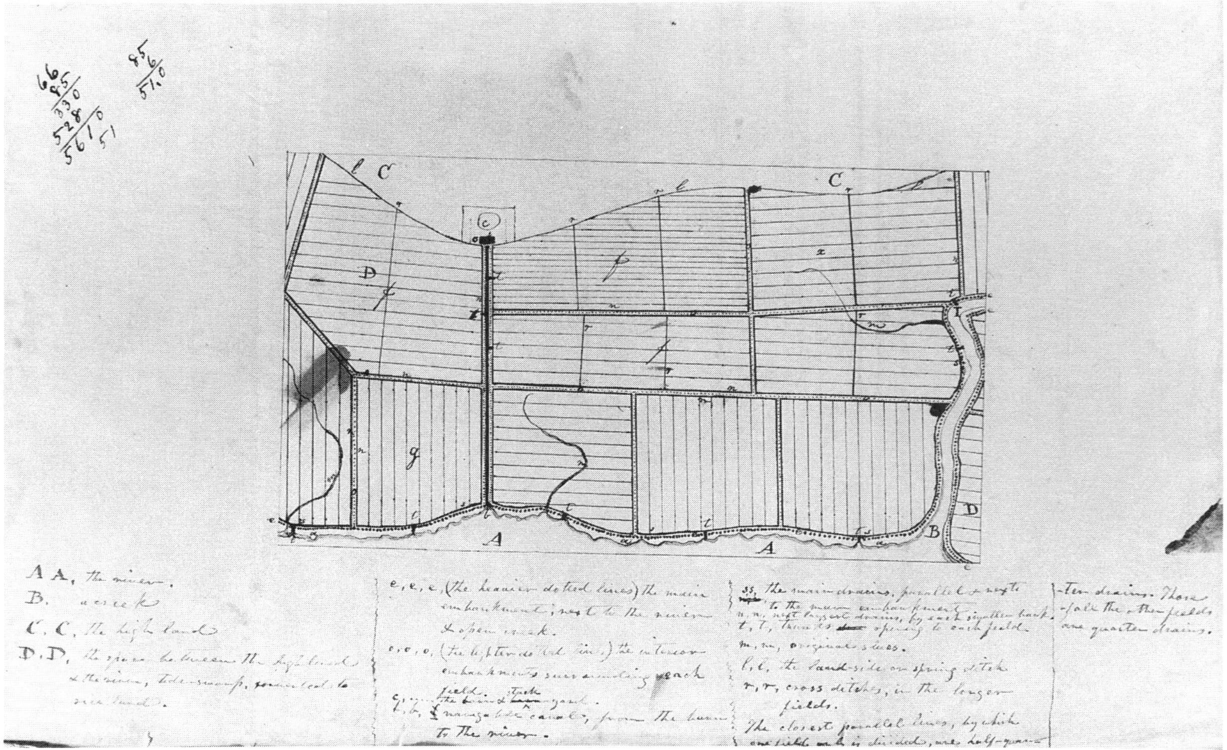

Drawing by Ruffin of embankments constructed by Samuel Porcher along his Santee River property. Such embankments, along with corresponding ditches, were employed to prevent flooding and to improve the drainage of adjacent fields (courtesy of the Virginia Historical Society).

being tilled before being embanked. When the river was at ordinary heights, the higher parts of the land was enough above the water to be worked; & hazardous as was the business, because of liability to freshes of frequent, but uncertain occurrence, & of irregular height & duration, still this land was annually cropped, & in indigo while that was the great staple & general crop of this country. And often as the crops were damaged or totally lost, the estates of the former proprietors of the adjacent high lands, were accumulated by making indigo in these swamp lands. But as time passed on, the freshes became more often & injurious. This no doubt was caused by the extension of clearing of the lands on the far distant upper waters of the Santee, which under the name of Catawba, receives tributaries from almost as far as the border of south-western Va. Major Porcher had lost by freshes 8 crops in succession on his then swamp fields before he began his embankment.

It was commenced on a small scale compared to the size it has since attained. The work was progressive, & necessarily slow; but it was not necessary to wait for its completion to derive much of its expected benefit. The first year's work served to defend the land from much of the flooding to which it would otherwise have been exposed. Each successive year increased the security, except when breaches were made by freshes which the embankment could not withstand. All these breaches, & some of which have been very disastrous, together have required as much reconstruction of entire sizes of bank, as amounted to a mile of length.

The worst breach was made in or about 1822. Part of the soil & subsoil, (which is of a good stiff brownish yellow clay, the alluvium of the Santee,) in the line, rests upon quicksand, over which the embankment was necessarily constructed. For subsequent heightening, pits were improperly dug on both sides, though far from the embankment, & so deep as to reach to the sand. The pressure of the water in the next fresh caused an oozing through the sand beneath, which increased until the bank was there blown up, as if by the force of gunpowder. The breach thus formed was 180 feet long, & 24 feet deep. It being impossible to replace this enormous quantity of earth thus lost, a new embankment was made within the breach, which was necessarily so much curved as to require 260 yards of the new to unite the separated ends of the old embankment, though the breach was only 60 yards long. It was after this terrible disaster that Mr. P.s intelligent & zealous overseer advised the entire abandonment of the work. It occurred in May, after the corn crop was well advanced, & which was totally destroyed. The passage by the breach was partially shut out as soon as possible, the land replanted as soon as it was dry enough, & that year the product was 20 bushels the acre. Ordinary crops, considered as good yield, make 30 bushels; & the best yet made was 47 bushels. Cotton is a precareous [sic] crop on this land, & is not now planted on it. Of oats, the best crop made was 70 bushels on one field, & 50 on another. The oats of this country are light & inferior. They are sowed here but one bushel to the acre, & as generally elsewhere sowed in narrow drills, & hoed. It is one of the universal opinions that broad-cast oats will not be near as productive or profitable as drilled.

The embankment does not exclude the water from the lower part, nor is it desired. There is a fall, supposed to be 18 inches the mile along the course of the river; & if the embankment secure, as it does, the land from the first approach of the river & along its course, the mere back water which will enter below is not regarded. Of course the land subject to this back water is more or less injured, & is not cultivated. The general slope of the swamp serves to drain all the upper & cultivated part into this lower part subject to back water, even during the continuance of freshes.

It is far from my wish to detract from the value of this truly great work, or from the merit of its zealous & indefatigable proprietor & constructor. But while I looked with admiration at the work, & the high degree of perfection of its results in keeping the land as dry as it was, I still thought that the drainage was deficient, & that it would be much improved by deeper ditches. The surface of the soil was indeed astonishingly dry, considering this defect. But the water in the drains & low natural sinks, acting as "leads" to carry off the surplus water, often stood within a foot of the level of adjacent tilled land; & under such circumstances the soil or the subsoil must be unfavorably affected. If this opinion be correct, it may throw some light on the ground for the strong objection of Major Porcher to the general use of the plough, on this land especially, & to any use of it here exceeding 1½ or 2 inches in depth. The soil & subsoil would indicate to me the opposite opinion—that ploughing & deep ploughing, if on well drained land, would be highly beneficial. Both soil & subsoil have evidently been deposited by the floods of the always muddy & yellow Santee. The soil is only about

4 inches deep. The subsoil seems to be of the same earthy materials, but lighter colored, for want of as much vegetable matter and perhaps atmospherical influence gained by exposure.

I have ceased as a useless waste of words to argue with any person on the propriety of using the plough. There seems to be almost a universal prejudice against it in these lower districts, in the minds of the most intelligent & well-informed planters, as well as of the most ignorant. Few persons are more intelligent & none have had more of planting experience than Major P.; & I have met with no one who is more opposed to the operation of ploughing, except so partially in aid of the hoe-ing, & so shallow, as to be of very little importance whether it be done or not, & the hoe be relied on entirely for corn, as it is everywhere hereabout entirely for cotton.

There are two things in regard to this embankment swamp worthy of observation. One is, (& which serves Major P. as a strong ground for his opposition to any but the shallowest possible ploughing,) the long time before the subsoil is clothed with grass, or has any growth, after being put on the embankment, although it seems to the eye to be rich. I saw where the bank had been given a heightened surface summer before last; & to this time there were very few & widely scattered spines of grass or weeds. The other fact is, that the flood let in by the great breach above described covered over some acres of the best soil with gravelly sand, seemingly absolutely barren, & of all depths not exceeding a foot. Yet these parts, including the most deeply covered with sand, have been the best producing land ever since.

The mansion house of Mexico is near the Santee canal, & within two miles of its entrance into the river. There is a beautiful prospect of a part of the canal, & a bridge over it, from the house.

About a mile from the house & half a mile from the commencement of the summit level of the canal, its course crossed (much below the surface) the narrow strip of enormous fossil oyster shells, which extend to & across all Georgia. I had heard of this fact long ago, & had endeavoured to obtain some of these shells, but it was supposed that all had been removed by different persons curious enough to take them. I looked in vain on the frequented side of the canal, finding there but few fragments. But on passing around to the opposite side, & digging into the sloping face, I found where the shells had been laid in the original excavation, & there I soon obtained as many good specimens as I cared to have, & many other inferior, which I left at Major Porcher's. These shells have been called *ostrea*[1] *gigantica*. They are enormously large. . . .

This new embanked part of the Santee swamp was the principal fortress of Gen. Marion[2] & his followers in the revolutionary war. I was shown the ground of his camp. The whole breadth of this great body of swamp, including the low outer parts not embraced in the bank, is from 3 to 5 miles according to the irregular course of the river. Here, he was safe from the pursuit of the enemy, no matter how much superior their force; & ready to sally forth & fall upon them whenever there was an opportunity to strike with effect.

These higher swamp lands having been cleared for indigo, when that culture was abandoned, & nearly all culture because of the increasing freshes, the land became a rich pasture for cattle, & presented a scene of great beauty before it began again to grow up in trees.

In the evening, we rode 4 miles to the plantation of Mr. Layton on the upper springs of Greenland swamp, to look for the calcareous bed. The stone had been burnt there in former times for lime, & the pits remain where it was obtained, but were under water. Another place examined presented calcareous nodules thrown out of the ditches, which as I think indicate the solid bed of marl to be not far below.

1. These were oysters, edible marine bivalve molluscs, living in shallow, muddy shores and estuaries, which were at their peak of abundance, distribution, and diversity in Cretaceous. In his report Ruffin records one Santee canal fossil weighing twenty-one pounds, ten ounces, and another on the Savannah thirteen and one-quarter inches long.

2. Francis Marion (1732–85), the legendary Swamp Fox, conducted guerrilla assaults on the British from swamp hideaways. He was a powerful politician and soldier in the immediate post-Revolution period.

23 MAR. Went to the Black Oak club house (adjoining the church which was the limit of my excursion from Cooper river,) to meet the Agricultural Society. Rode on horseback along the side of the canal, while Mr. Mazyck Porcher's servant drove my buggy along the carriage way. The summit level of the canal is 37 feet above the Cooper, & about 8 feet less above the level of the Santee at lowest water. It is fed partly by springs cut by the canal—but for much the greater part by shallow ponds of great extent, & together covering thousands of acres along the margins. These ponds would be enough to make a healthy region sickly;

but here I presume it is considered that the health of residents cannot be made worse. Indeed there are no proprietors who remain residents during the sickly half of the year.

About 15 or 16 members of the Agricultural Society met. Our conference & my examination in regard to calcareous manures were conducted as heretofore; but I think that much less general interest seemed to be felt in the matter than at Edisto Island. Still, there are particular individuals here whom I deem my most valuable new disciples, & who will do most service by their operations.

In the evening, went home with Mr. Philip Porcher. His invitation had been accepted the previous day, upon the condition made by me that I should find his carts carrying out marl, which he promised & complied with. He had begun to dig it some days before.

The weather has still continued remarkably cold, & this day, though the sun shining brightly, has been severely cold. 19 miles.

24 MAR. This morning, at 15 minutes after 6, the Thermometer was at 22° in the open air. Ice remained in the south piazza at 1 o'clock, & perhaps much later.

After seeing Mr. Porcher's marling operations, & giving him such suggestions as seemed requisite, we went to bore in his swamp for marl, & found it (soft) at 4 feet below the surface. This was what I counted on; & few as have been the observations yet made, I feel confident that the bed of marl is near to the surface of all Biggin swamp & its extensive branches, of which this is one of the highest, & all of which are the head tributaries of Cooper river. . . . I have also learned from Mr. Henry W. Ravenel that he lately saw the softer marl exposed by the blowing up of a tree in the middle of the Wassamasaw swamp, which is the head source of the Ashley river, & which connects with the Biggin swamp. The close proximity of the marl to the surface of the swamps, I think caused the superior quality of the rice of the inland swamps to the tide lands—which is a fact universally admitted, & which applied especially to this in which we bored today. When cultivated in rice formerly, its product was in demand for seed for rice lands on the tide waters. Upon hearing this opinion Mr. Isaac Porcher, the brother of Philip, took the auger to bore in a narrow swamp in his field a mile distant, & at first trial, found it at about four feet deep. This saved him a mile of hauling, as he had prepared to get marl from his brother's pits.

I consider this as one of my days of labor most usefully devoted to public interests, though directed merely to instructing Mr. P. Porcher in the humblest details of his new business. He is just such a man as will succeed the best; & who will profit by every aid my experience has enabled me to afford him. His difficulties are great. His marl yields water, slowly indeed, but from every part, & is quite hard, & is made much worse to dig by the continual access of water. His laborers have to learn every thing of the new operation & the tools are altogether unfit, except the shovels. The implement for digging is what is called a grubbing hoe, but does not deserve the name, & with which a man could not dig in three hours as much as in one with a proper hoe. Every thing else needed improvement, & all will be promptly improved by Mr. P.'s zeal & judicious apprehension of my instructions. I remained a second night, for the purpose of rendering as much service in this way as was needed.

In the forenoon we rode to visit several localities of marl on the neighboring lands. Some of the gentlemen whom I had met yesterday dined with us today; it is arranged that I shall have the company of two of them on my next trip down the Santee, which is to be next Monday morning. It will be to the plantation of Dr. John Palmer & his brother Col. Samuel Palmer, both of whom have already been marling with success, & who came a good day's journey yesterday to meet me with the Society.—12 miles.

21. The Cymbee of Woodboo (25 March)

25 MAR. Still cold though much less so than yesterday. Made an early start to the house of Mr. H. W. Ravenel's, who was ready, by previous appointment to go with me to examine some interesting localities east of the canal. In all this neighborhood there are numerous sinks, subterraneous passages of considerable streams of water, & of the limestone boiling springs, such as were described about Eutaw. The sinks however are more rare & less deep than in the upper neighborhood, indicating as I infer greater depth of the lime-rock below the high land, or a less cavernous interior. The springs of this kind, when deep in the visible opening & throwing up large quantity of water, are known by the name of *fountains*, which term is thus limited, & not of general

application as it is correctly. One of the most remarkable of these fountains for size & beauty is at Woodboo, & was the first object visited. It bursts up from two separate apertures about 8 or 10 yards apart, the smaller of which only could be then seen, as the reservoir or pool, which is several feet deep some 15 yards wide & 40 or 50 long to its narrower outlet. What is the depth of the wide opening which discharges the greater jet, there was no means of ascertaining; for though there was a little canal on the water, it was unfit for use. But the smaller opening, which is close to one side was measured, & found to be about 14 feet deep, of what is a funnel shaped mouth; but besides this depth, which is clearly exposed to the eye through the perfectly transparent water the passage to it from below, through the broken ledges of rock marl seen at the bottom, though much narrower, must be of no small size. I tried to obtain a specimen of the body by the borer, but it being only 11 feet, I could not reach the marl any where, except on the top of a very thin ledge standing perpendicular, & there I could not enter it enough to obtain a specimen. It is however certainly of the softer & lower bed, & not of such hard & stony texture & fracture as frequently shown, & such as the Eutaw spring, & others I have seen force their way between, without being able to wash a large cavity. This difference of form would alone prove the softer texture of the rock of the Woodboo fountain, & all others like it in this respect. This is at the edge of the swamp, through which the water flows in its channel on a level, without visible current. This still & placid appearance, the surrounding tall trees & the dense swamp forest on one side, give to this fountain a gloomy & solemn beauty altogether different from that of the Eutaw. The general bottom of the tiny lake is of black mud; & the water appears so dark as to be almost black at first glance, although its crystalline transparency is evident by the clearness with which the marl is seen in the bottom of the opening whence the water comes up, & near which the trout are seen swimming almost as distinctly as if they were floating in air. The continued upward rush of the water keeps these apertures open & clear no matter how deep.

The Wadboo [sic] plantation belongs to Mr. Mazyck. When it was formerly a family residence, (which it is no longer,) the margins of the fountain were kept in neat order which made the place more beautiful than now. It however is yet a place of great beauty, without the aid of ornament or care.

It was at this fountain that I first heard reference made to a superstitious belief universal among the negroes, which of all their superstitions is the only one having any touch of romance & deep interest. Each of these fountains, of considerable depth & size, is believed to be the habitation of a kind of water sprite, or supernatural being called a Cymbee[1]—(I doubt whether the word was ever written before, & therefore I know not its spelling—). Each fountain has a different cymbee, the size, appearance & habits of each varying some what from others. The negro head man, or *driver* as called in S.C. who accompanied us, had never seen the cymbee of the Woodboo fountain; but reported from the accounts of others that it is web-footed like a goose—a feature certainly but little compatible with the beauty of a water nymph. Another old negro who *had* seen a cymbee at another fountain, when he was a very small boy, told me that she was seated on a plank which was laid across the water, & that the long brown hair of her head hung down so low, & so covered her face & whole body & limbs, that he saw no other feature; nor could he answer to my question whether she was a white or a negro cymbee, except as may be inferred from her long hair. The descriptions were very loose & meagre. The latter admitted that he was so young at the time, & so much frightened that his recollection of what he saw is very indistinct. After seeing her but a few seconds, she glided into the water & disappeared, as may be presumed in the deep cavern from which the waters of the fountain rush to the surface. But however few appearances of the cymbees are heard of, they are nevertheless believed by the negroes to be frequent & numerous. For it is another part of the superstition that it is bad luck to any one who may see a cymbee to tell of the occurrence, or refer to it; & that his death would be the certain penalty, if he told of the meeting for some weeks afterwards. Still, terror, which could not be hidden, falsehood, or other causes, have brought out many reports of appearances of cymbees above the waters of their deep caverns, & all under different circumstances, showing different habits of the supernatural & solitary resident. The Cymbee of the Woodboo fountain is seen only when the sunshine is "right up & down", as the old driver said. At other fountains, they appear only in the night. They are usually seen in a sitting posture, on any low bridge, or plank, crossing the water, or on the margin of a steep side. But at Pooshee, the plantation of Dr. Ravenel, (next to Black Oak & on the canal,) a negro boy who was so unusually bold as to go to the fountain for water late in the night, was frightened almost out of his senses by seeing the cymbee running around & around the fountain. What a fine subject would this superstition furnish to another Scott, with such a residence for the cymbee as the dark & gloomy pool of Woodboo, or the bright crystal stream & subterraneous channel of Eutaw—or still better, when this fountain, as formerly filled its broad & deep basin & overflowed its brim!

As mentioned previously in regard to *matters of fact* of these fountains from the lime-rock, and which facts are in strict accordance with the cavernous foundation which I suppose, these fountains sometimes suddenly disappear entirely, & in other places, new fountains burst out. When the former occurs, the negroes believe that the cymbee has died, or has been offended & abandoned her residence. When Dr. Ravenel enclosed his fountain with masonry & confined & raised its water, an old half breed Indian of the neighborhood, who was half negro in blood, & wholly in habits & superstition, remonstrated with him, upon the ground that the cymbee might be made angry & leave her haunt, & that then the spring would be dried. Unluckily for the story, the fountain continues to flow as previously.

1. In Indian folklore dwarves, water babies, and old women variously inhabit springs and other wet places.

22. By the Santee Canal (25 March–27 March)

The course of my road through Woodboo, & below to Mr. Robert Mazyck's plantation, (4 miles from the south lock,) was nearly parallel with the canal, & with the upper part of my first ride along its margin from Cooper River. So that my observations hereabout connect with those formerly made, in my first excursion. Among sundry specimens of the marls & harder lime-rock brought for my inspection by different persons to the meeting of the Agricultural Society on the 23rd, I was struck with the singular dark greenish gray color of one from Mr. Mazyck's plantation, which he had recently observed & opened, & is carrying out for manure. Upon rough examination, I found that it was weak in calcareous earth, compared to the ordinary marls, but is largely intermixed with *green-sand*. . . . This is a very interesting deposite; & it is much to be regretted that the shallowness of the green & softer layer, & the unusual hardness of the inferior & more calcareous layer, will probably prevent its extensive application.

. . . After dining with Mr. Ravenel, we bored in the edge of Biggin swamp near his house, & reached the marl at 6 feet depth. It is overlaid by scattering small & hard calcareous nodules, which were all that Mr. R. had previously seen. I believe that these nodules are very often to be found in the clay above the solid body of marl.

Having left my auger with Mr. Dwight on Ashley river, (who by its aid found marl in every low place where it was sought,) because it was deemed too cumbersome for land travel, I found so much need of its use, that I had another made lately, of less weight, 11 feet length, & the stem 3/8 of an inch square, with an inch bit. This being laid in a groove in a light wooden stock, I can carry it attached to my buggy; & I have already found it very useful to myself & to others. Several persons have already had augers made, on this plan, for themselves—& I believe that in a very short time there will be as many marl augers in S.C. as I have ever known of in Va.

At night went to visit Mr. Wm. Cain, 4 miles off my course where I had been in my first excursion.—26 miles.

A letter from Mr. Whitemarsh B. Seabrook informs me that he had visited the site of a so-called marl exposure 16 miles from Beaufort which had been reported to me when I was in that place, & which nothing prevented my going to see (as designed) but the subsequent report of the proprietor, Dr. Verdier, that the place was then under water. I requested him to obtain a specimen for me as early as possible, & send it with a description of the deposite. Mr. Seabrook could not see the "marl": but from the description heard of it, & his finding remaining fragments of oyster shells lying about the surface, he had no doubt but that the deposite was *artificial;* & I doubt not that it was like the Port Royal bed, the lime used & thrown away at an old indigo vat. There must be many thousands of such deposites on the former indigo plantations. Yet this "marl", though never dug or used, the proprietor had procured a speciment to be analyzed by a distinguished chemist. . . . And this is the way in which these planters act in regard to their supposed marl beds. The fact of the discovery is perhaps published in an agricultural journal, & at any rate talked of for a long time—perhaps reported to an agricultural Society—an analysis is procured by a chemist of reputation—& yet none of the so-much vaunted marl is used as manure, & therefore it is not found out that the whole bed might have been exhausted in a day, if not in half an hour's work!

26 MAR. Sunday. A beautiful bright day, & the weather becoming warmer, but still much too cold for the season. In the forenoon rode with Mr. Cain to pay a morning visit [to] his neighbor Mr. Frederick Porcher, whom before I left so sick, & who is not yet quite well.—3 miles.

27 MAR. Steady slow rain which began in the night & continued through morning. This is a very inconvenient disappointment of any designed setting off this morning, as the gentlemen who were to accompany me were to join me on the way at Pineville.

The cotton fields are every where bedded up ready for planting, & waiting for warmer weather. After the two hoe operations the tilth & condition of the soil is good, & the appearance neat, when well performed. But all the tilled mould is drawn together in a narrow ridge; & all the subsequent hoeings will only scrape more from the alley to heighten that ridge. This mode of tillage seems to me utterly opposed to reason & to the nature of the plants; & only justifiable, if at all, where such ridges & alleys are necessary for *drainage*. Yet they are universally used, for all crops & all soils.

The lands of this part of the country are of better quality than is usual, & many may be called good. There is much of large oak mixed with the larger quantity of pine. There is very extensive manuring generally practised, & more land manured by the planters than I ever heard of elsewhere, except where they use salt mud & marsh grass as the principal materials, & of these on the sea islands only. Leaves raked from the wood-land furnish the almost entire vegetable material here, except the highly prized cotton seed, which are used alone as manure. The corn stalks are left in the fields, & picked by the cattle during winter. The cotton stalks are also left, (which I approve) & "listed in" if the land is to be cultivated again the succeeding year. The oats furnish very little litter, & no rice is made, except small patches for family use. Therefore the woods' litter forms the almost entire litter for stock & vegetable part of the stable & cattle-pen manures. The quantity made is very great, & the application, if not made to every crop, is needed & would be made if the supply of manure served. Mr. Thomas Porcher of Walworth, is said to be a good planter, or crop-maker, & the most extensive & indefatigable manurer of his lands in the parish. He cultivates annually 400 acres, & manures of it 250, with from 24 to 30 large single mule or horse loads of the prepared manure. His plan (and the usual plan everywhere, when fully executed,) is to rake & haul the litter from the woods through all the year, spreading it continually in the stables & pens for his cattle. This goes on, & in the cowpens the mass is not moved until autumn, when it is perhaps 5 or 6 feet thick. This "compost" as it is called is then carried out, spread in the alleys, & "listed in", which operation is finished by planting time or long before, when the manure-making operations are recommenced for the next years crop. Mr. Th. Porcher keeps 80 head of ordinary cattle, & about 20 mules & pleasure horses together, (rejecting as usual nearly all ploughing,) & their excrements are all that add anything to the enormous amount of poor vegetable matter. The universal practice & opinion prove the value of this kind of manuring, & I deem that result of much importance. Still, all that is done by these continual applications is just to keep up the productiveness of the land to a certain mark, & that mark in no case higher than the original power of the land, & rarely so high; & if the manuring be intermitted but a single crop year, it is seen that the land has fallen back in its production. I think this attention to collecting manuring materials from the woods as highly commendable; but still the material is certainly poor, & the labor of transportation is the chief cost of the application. Therefore I would prefer to cart it from the woods directly to the field during its year of rest, (& would require such rest, though it is not generally given,) spread it, & leave it to manure the growth of weeds & grass, & to be rotted by their shade & moisture, & thus be ready for the next crop. This would save all the second hauling, (from the compost heaps to the field.) The inert & little-putrescent leaves would be reduced to a fit state to act as food for the next crop, & the wasteful & heating process of summer fermentation in heaps would be avoided. I have seen the unprepared leaves & rakings from the woods at once listed in for the ensuing crop. This was an interesting fact, as proving the value attached to this material for manure. But I am confident that it must, when thus used in water, do some harm as well as good to the next crop by its open & unrotted state—& sometimes, if in a dry spring, more harm than good. If laid in winter, or spring on the surface of land not to be cropped that year, (though such intermission is rarely allowed here, as a regular system,) there could be no injury to manure or land, & no waste. For as the leaves slowly decomposed by aid of the shade & moisture of overshadowing weeds, the soluble parts would be carried into the soil by every rain, & be immediately taken up by the roots of the growing weeds & grass, & in their increased bulk, reconverted to manure, with the large addition drawn from the atmosphere. The remainder of rotted leaves & all their product in weeds & grass should next winter be ploughed under, or "listed in" with hoes, as no other putting in is thought of here; & I am sure that, besides the mere benefit of a year of rest to the field, & its own proper product thereby of growth, that the manure applied in the leaves & their product in weeds, would be much greater than if the same had been passed through the stables & cow-pens, & the labor much less. I am not counting in this estimate the excrements of the beasts; for they could be as well saved in other modes, & using more litter. As practised, the penning of the cattle adds but little to the value of so great a bulk of manure—& it & other attendant treatment is so injurious to the animals, that perhaps they lose as much of value as they give to the manure. Still cattle are kept, & in large numbers, almost entirely for their manure; & so badly are they kept, & usually so poor, that even their manure must be of less value than would be supplied by one-third of the number, well fed, & productive of milk, butter & meat, as well as manure.

It rained nearly all day, & much fell. A short intermission before night enabled me to bore & to find marl at the edge of the swamp (a large branch of Biggin,) at 6 feet depth—or about 5 below the level of the swamp.

23. The Abandonment of Pineville (28 March)

28 MAR. Still more rain this morning; but after a clearing-off heavy shower, I was enabled to set out by 9.

After going over the ground to the neighborhood of Deveaux' mill-pond, and travelling a few miles more through pine barren, I reached Pineville. I was prepared by previous report to see a deserted village; but still the reality was more melancholy than the anticipation. The dwelling houses were neat & commodious, & some of them large two-storey houses; & though having the appearance of being long unoccupied, still do not exhibit ruin or much of dilapidation. They stand separately, yet near enough together to be mostly seen at one view, under the pines which cover the site almost as thick as the usual open forest. In all the central portion, & indeed only excepting the first house passed in the out-skirt, I saw not a human being, nor any evidence of any being there, until I saw some pigs & poultry around what seemed to be a negro's house. Here I called, & having brought out a negro woman, learned from her the locality of the "Ball Room", the place designated for the meeting with my companions, Mr. Porcher & Dr. Joseph Palmer. But before reaching it, they espied me, & we soon were proceeding on our journey.

Pineville was the pine-land residence of the neighboring planters, during the sickly season; & for many years after being resorted to for health, it was found to answer the object sought. But in 1834 & for some years after it became so sickly, that it has been almost abandoned since. Ten families only now live there, whose permanent residence it is still made either because their plantations are near, or for other peculiar circumstances. Before the breaking out of the general & very fatal congestive fever of 1834, there were about 60 families who resided here during the summer & autumn. The influence produced by the then state of things at Pinevillan society was very important; & the loss of the benefit is deplored by all who had enjoyed the benefit. However inconvenient & costly is the annual moving of planters' families from their homes, the bringing them together at the summer retreats has many & marked advantages; & these were felt no where more than at Pineville. By such annual meetings for 5 or 6 months together, the selfish feelings & prejudices which secluded life is apt to induce, are removed, & social feelings & social virtues are cultivated. It is *too inconvenient* for such close neighbors to be hostile to each other—& they either become friendly, or appear so, from necessity if not choice. The society was gay, yet the social enjoyments innocent; & refinement of manners & improvement of mind & of habits were necessarily produced. Every planter's business & opinions became known to all. In riding every day to visit their plantations, they frequently went in parties to visit each other; & thus every man had the benefit of the example & the opinions of those better informed. The careless & the indolent were stimulated also by the good-natured jeers of his neighbors; & each one was anxious to avoid such censure, & to command applause. A sensible improvement in agricultural operations was seen. But other things were still more improved. A good miscellaneous library was established by the residents at Pineville, & a good school found abundant support. Nearly all these benefits have been lost there, and found less perfectly supplied at the other retreats to which the former residents have since resorted.

No satisfactory cause has been assigned for the change of Pineville from healthiness to sickliness. For whatever cause is supposed had long excited except perhaps the greater richness of the ground, from being the long continued residence of so many persons & their servants & domestic animals. Latterly, it is asserted to be again healthy; but none of those who abandoned it are willing to try it again. The few remaining & permanent inhabitants reside in the out-skirts.

We pursued the excellent & but seldom travelled road along the course of the Santee & reached the residence of Dr. John S. Palmer between 4 & 5. The house is in the pine land, a mile & a half from his cultivated land on the river, or rather on the river swamp, which lies between. Very little of highland borders this river from Eutaw to its mouth—& the swamps are usually very wide—seldom less than a mile & sometimes 4 miles wide.—37 miles.

24. Santee Marlers (29 March–30 March)

29 MAR. This day viewed Dr. Palmer's land & his marl, which he & two of his neighbors have used so advantageously—& which use, no less than his invitation induced my visit. My remarks will be given upon the whole together. Col. Samuel J. Palmer & Dr. Robert Gourdin joined us at dinner. 7 miles.

30 MAR. Rode over Col. S. J. Palmer's land, which is the adjoining tract, & dined with him.

On this side of the river, the marl, or lime-rock first reappears on the high land at the lower point of Mattassee lake, & adjoining on Canteys land, & next at Coutourier's, all on the Santee. It is also on Wedboo creek, & at Savannah creek. Indeed it is found so generally that it may be inferred to be accessible on every plantation with these limits, to this place, which is next above Lenuds ferry. Below it is at every place to

Echau Church & Warren's, just below Echau creek which is the lowest exposure known. I learned these facts from Dr. Palmer, as I only saw his & his brother's beds. This whole stretch, in a straight line on the map is 22 miles. The general character, I suppose to be much like what I saw there. The two large plantations of the Messrs. Palmers are much of the same general character, though Dr. Palmer's, the upper one is of higher level, & more sandy. Col. Palmer's more clayey & lower—though generally at least 10 or 12 feet above the swamp. Still it has had a large portion of its arable land covered by the highest freshes of the Santee. These lands, or their river parts, were of better than ordinary quality, & mostly of oak & hickory growth. They are underlaid with the lime-rock not very deep below the surface. Sinks are very frequent, & most if not all of them communicate by hollow passages in the lime-rock with the river, & probably the passages extend much further into the interior, serving to carry all the water, & to make it all limestone water. Not a single rivulet from the higher pine land reaches the river. They all sink some where on their passage, & come out again near the river or swamp. Whatever rain-water collects in these basins, soon disappears by sinking—unless when their bottoms are higher than the height of the flood in the river, & then the water in every one of the sinks soon rises as high as the river, & afterwards falls with it. In Col. Palmer's cleared land, there is scarcely as much as 25 acres together in which there is not one or more of these basins thus evidently communicating with the river. On the opposite side of the river the same features exist over the marled lands, as I was informed by Dr. Gourdin. At one place, 3 miles from the river, the water in a sink rises & falls with the freshes in the river. At another, the "Weeping Fountain", in Williamsburgh district, the water from a pine-land stream sinks underground. At other successive basins below in a line, the stream is seen passing; & finally, at the swamp side it again bursts out, large enough to turn a mill.

These facts alone would prove that the lime-rock is full of crevices & open passages. But besides this, whenever attempted to be dug below the level of the surface of the river, such crevices invariably admit water from the river, & prevent the digging deep. This, & the frequent & uncertain risings of the river are great impediments to the otherwise very convenient excavation. . . . Generally in the river bluff, at the old site of

Jamestown the Huguenot settlement (of which village not a trace now remains,) the marl shows about 3 feet high on the face of the bank. . . .

The deep sinks here are also called "fountains", although these in no case furnish streams of water, but serve as receivers of water instead of suppliers. The deepest of them, being lower than the level of the river at low water, are permanent & deep pools; & when these are large, or are connected with a wider space of shallow water, they are generally inhabited by alligators, which are cymbees of a very different character from those heard of above.

The trials of marl made here were the earliest of any extent made in S.C.; & have shown results more valuable so far (merely because being older,) than any others. Still even here, the manure has been allowed but a poor opportunity to act; for the general course of culture is to take a crop every year—& this was done for some years before the marling (or ever since being cleared) as well as every year since, except when the contrary will be mentioned in the following concise statement of the principal & more important circumstances of the trials—all of which trials were induced by the reading the "Essay on Calcareous Manures".[1]

Col. S. J. Palmer made a small experiment with lime spoilt for cement, in 1834, of which it is enough to say that it was as successful as the later experiments, & that it has continued to act without any diminution of its best effects. This encouraged him to set to work more largely for the crop of 1839, of which 24 acres were marled at the rate of 100 bushels, & planted in green-seed cotton. This land was partly light, part stiff—quite poor when cleared—had been cultivated the two first years after clearing, & the second year (which is always the land's best crop, other circumstances being equal,) this land made only 140 lbs. of ginned cotton to the acre. Because of its poverty, it was then left to rest 2 years, 1837 & 8, in 1839 was marled & planted in cotton. The product was from lbs. 270 to 280. The benefit was on both the sand & clay soil, but thought greatest on the latter. In '40, the crop was much shortened by both a fresh & bad season, & by early frost in 1841; but still the superiority to its former state was maintained, allowing for these disasters. The last named crop on this piece made 220 to 225 to the acre. No *equal* part had been left out for comparison; but the *richer* part had, because being subject to freshes, it was not thought worth being marled. This part, even when producing well, has been regularly inferior in product to the other marled & formerly poorer parts. Similar results found on subsequent applications—no effect being seen the first year, & good effect after. 60 acres marled this year for the ensuing crop—& altogether more than 100 acres. No rest yet allowed to any.

Dr. John S. Palmer in 1839 covered 50 acres of his cotton land for 1840. This, like the preceding, placed *on the list,*[2] after this first process, & covered by the after-bedding. (This is not quite so bad as laying it in the alley, before listing, but scarcely more than that made permits any mixing of the marl & soil the first year.) No benefit seen. Next year oats, and the beds reversed, & the good effect of the marling manifest. In 1841, cotton again, & the increase full 50 per cent. The subsequent applications of 1842 & 3, have made the whole near 100 acres.

Dr. Robert Gourdin in 1839 marled at the rate of about 100 bushels a piece of very poor new ground, stiff, naturally wet & sour, but which he had drained pretty well. It had lain out 2 years before being marled & cultivated. Part through the middle was left out for comparison. By neglect, the comparison was not made, nor any difference noted during his absence in 1840. In 1842, the separate products were weighed carefully. The unmarled part produced at the rate of 63 lbs to the acre, & the marled at 128. Dr. Gourdin's marling now extends beyond 100 acres.

I obtained the promises of all three of these gentlemen to write statements in full of their marling operations, & the results, & report them to the approaching meeting of the Black Oak Agrl. Society, which, if published, at this juncture, I trust will attract attention, & do good as example.

In the evening crossed the ferry (Lenud's) to Dr. Gourdin's, where I lodged.

A passage across the Santee is not the quiet & quick affair seen in crossing most narrow rivers. Though the river is not exceeding 180 or 200 yards, it is necessary to ascend at 1¼ miles to reach firm land; & even this is the shortest passage possible for 40 miles above. The ferry boat is a large flat, propelled by the poles of two men, aided by the steering oar of another in the stern. Neither could the poles, long as they are, reach the bottom of the river, and if they could, the current would be too strong now to be stemmed. To avoid this, the flat is pushed along narrow natural passages of the water through creeks & lakes (so called) when they offer, & when not, along narrow channels made by cutting down the trees of the swamp. Over the swamp lands the water stood

4 or 5 feet deep, near the river side. The swamp is thickly covered by large cypress gums,[3] & other trees suitable to such land, all hung with drapery of the long grey moss. The passage of our large & unwieldy boats through this dense & sombre forest was a novel & interesting scene. When necessary to cross the river, instead of continuing to ascend, the flat necessarily drifted down fully 150 yards before the crew could enter the swamp on the other side. We landed a little after sunset, & I soon reached the house of Dr. Gourdin, who had met me on the way.

1. Edmund Ruffin wrote the essay in Prince George County, Virginia, between the early and mid-1820s and published it in Petersburg in 1832. It is a prime text in nineteenth-century agricultural chemistry. See the Belknap Press edition of 1961, edited by J. Carlyle Sitterson.

2. The list was the small ridge formed by throwing two furrows together, producing alternating beds and depressions. (See p. 120, n. 1)

3. Ruffin probably means the bald cypress.

25. Hunting Alligators (30 March)

Yesterday & today I made my *debut* in a sport which I had been extremely anxious to enjoy, viz. shooting alligators. I have no fancy for pursuing small game of any sort—such as hunting foxes, shooting partridges, or catching small fish. Even deer hunting never attracted me enough to make a single trial; and it is 12 years since I remembered to have fired a gun at any thing. But I have always felt desirous to join in such sports as hunting bears or wolves, catching sharks, or shooting alligators. Yesterday the bright & warm sun, after such continuance of cold & often cloudy weather, made it almost certain that the alligators would be basking on logs, & at my request I was guided to the places where they were most likely to be seen. The first place visited was one of the largest "fountain" ponds above described, nearly two miles from the cleared land which we had been examining. As soon as we reached the margin, Dr. Palmer saw an alligator on the land, & pointed it out to me. It was seen over an intervening slight rise of the ground, so that its curved back & upper part of the side only were visible; & to me, who had never seen one of these animals before, it would have been very doubtful whether it was not the upper part of a pine log, with rough bark. The deep fountain lay between, & there was no getting nearer, or obtaining a more full view. So I took aim & fired as low as the earth permitted, to stand the best chance to strike the side—for the armor of the back is impenetrable. To my surprise, the alligator did not move; & as I could not believe it possible that I had killed it outright, to make sure, I fired again, when it still remained motionless. Then we went round the water to secure the prize—& found him killed indeed, but it had been done before I had seen him. His skull was beaten in by a blow given not very long before, as his tail still moved when he was struck by my hatchet over the spinal marrow. Three of my buck-shot only had struck, one of which only had entered, & the other glanced off. The animal was only 6½ feet long. This shooting of a dead alligator would have been a capital joke upon my apprenticeship to the business, if the Dr. had known the true state of the animal & guided me to it. But, practiced as he was, he was as much deceived as I was. I went to several other ponds in the woods without seeing any. But taking the negro who accompanied us for a guide, I went to another, which, like the others was thickly set with tall trees, but most of which had been cut down. It is no easy matter to distinguish an alligator in such places, among so many standing & fallen trees. My guide soon found several; but whether his eyes were too good, or mine too bad, I could not distinguish either; & indeed could not see but one of the things which he said was one. This was on a log, & near enough; but it rose but little above the log, & if indeed an alligator, must have been but a small part of one. To my eyes, it appeared as much like a terrapin, or a large coiled snake, or any protuberance on the log. However, on Andrew's positive assurance it was really an alligator, I fired, & as he also assured me, certainly killed; & indeed the place was so well raked by the shot, that the object must have been struck. If killed, he sank instantly out of sight. However, as soon as I fired, an alligator which was across the water about 60 yards from me came flouncing & splattering through the shallow water, which he had to traverse full 15 yards before he found depth to conceal himself. This appeared to be about 9 feet long. It would have been of no use to discharge my second barrel at him, at even his nearest distance to me, as the head & back are impenetrable; but in truth surprise at his unexpected appearance, though his resting place had been in full view, would probably have prevented my firing quick enough, even if a penetrable part had been long enough exposed. I now sent off my guide, & took my seat to wait for the possible reappearance of the alligator. After more than an hour he rose just enough for me to know his presence by the motion of the water, & to see the bony protuberance above his eyes, & the end of his snout barely above the water. He then remained [?] perfectly still, as if on the watch, so long that I had no hope of his coming out; & expecting every instant that he would sink, I fired at the lumps above his eyes. He sank instantly & silently, & no doubt without being hurt—as nothing but striking the eye, & that through water, could have hurt him. Upon returning homeward, at another pond, still another & larger one was seen by Dr. Palmer; but before my eye could light on him he had slipped from his log into the pond. This one was again seen in same manner the next morning, & again was gone before I had a glimpse of him. At the pond where I had shot at the large one, Dr. Palmer only saw him rise in the water, & took a fair fire at his head, the only part visible. But of course without effect. Going on to Col. Palmer's, in his field we went to a deep fountain pond. The Dr.'s practised eye soon saw an alligator, & he tried for some time in vain to make me see it. At last I caught sight of it & immediately fired. He plunged in the water in such manner, & my aim was so fair, that we were almost certain he was struck severely. I supposed this to be about 5 feet long. In the afternoon I came back again with Dr. Palmer, without whose eyes to aid me it seemed as if mine were almost useless. He soon apprised me of seeing the alligator, & when I got sight of him, I had a fair fire at the middle of his side, & think that I could not have failed putting part of the load in his body. However he again sank, with a tremendous splashing, & was not again seen. He had been lying in the edge of very deep water, where the bank sloped steeply down; which position was so unusual, that Dr. P. thought it indicated that he was before severely wounded by my morning's shot. But these animals are as hard to kill as mud-turtle, even when their coat of mail has been penetrated, & the most terrible & seemingly desperate wounds inflicted.

I hastened then to take the ferry, & had no thought of getting another chance for a shot—and lost the best possible, by not being provided with a gun. Just before sunset, when poling along one of the narrow passages through the trees of the swamp, I saw an alligator ahead, on a small log which lay just on the side of our track, & parallel to the course. His head was directed up the stream, & as the flat came close by him, with all the noise of the poling, & rattling of the loose flooring over which the men walked, the alligator neither moved his head nor made the slightest motion. From my yesterday's experience of dead alligators, I began to think that this must be another—& was just about asking the ferry-men whether it was not dead, when I was restrained by the manifest absurdity of such a supposition, in such a place. When close by the alligator I told the polers to suspend their pushing & the flat slowly glided within less than 30 feet of him. Still there was no sign or appearance of life, & indeed he looked as little like being alive as a complete suit of ancient armor would seem like a living man. His deeply seated eye could not be seen. When about to pass by, one of the men struck at him with his long pole, & might have touched him, but that before the blow descended, he leaped into the water as quickly & actively as can be imagined. This alligator appeared full 10 feet long. His appearance there evidently caused not the least surprise, or exclamation of the ferrymen; & they said that he was seen in the same spot every day in proper season, & had been so for some years. This was doubtless an exaggeration; but these animals are remarkable for keeping in one basking place until driven from it by some continued annoyance.

It is surprising to me that so little is done to hurt & destroy these animals, when it may be done so easily. For although almost proof against buck shot, or even rifle balls, except in their most vulnerable parts, & at short distances, & though able to recover from & scarcely to suffer under wounds that would cause instantaneous death to large quadrupeds, alligators are easily caught by those who will take a little time & trouble. They will swallow a baited hook; and even a short stick sharpened at both ends, & enclosed in the body of a dead squirrel or chicken, will hold them by turning across the stomach, when they may be drawn out by the cord made fast to the middle of the stick. They may also be found in their deep burrows at the edge of ponds in which they hybernate [sic], & caught & pulled out by a large iron hook, attached to a long pole. They are cowardly, & rarely attack or resist a man if they can escape. They commit much destruction on hogs, calves & dogs, when finding them in or at the edge of the water.—11 miles.

26. Santee River to Waccamaw River: Pine Barrens and Poor Whites (31 March)

31 MAR. Examined the marl & hard limestone at the ferry landing, where the former is seen 9 feet high where it has been dug for manure. There are holes or crevices seen, large enough for a fox to crawl into, which are coated by the yellow clay left by the river water. . . .

The calcareous bed has been seen in detached places by Dr. Gourdin in Georgetown Dist, to 7 or 8 miles below Lenud's ferry in a direct line, in the high lands bordering on the river swamp—for there is no bluff—& higher up on Cedar creek which enters the river about 4 to 5 miles below the ferry. It is also known at several places up the river, to Boggy branch which enters the Wittee lake, with Wittee Branch, about 6 miles above the ferry. . . .

The tide rises regularly at Lenud's ferry, in low water of the river. . . . In extreme low water, the swell of the tide is felt much higher than this ferry. The rice planting, which requires a certain "pitch of tide", & freedom from freshes, begins some miles below the fork of the Northern & Southern outlets of the Santee—& ends some miles before reaching the sea, where enough fresh water for flooding cannot be commanded.

The morning was lowering & threatened rain—but I was anxious to get on, & set off for Georgetown before 10. After several light & passing showers, steady rain, & pretty heavy, fell during the last 6 miles of my journey—which including 2 miles of backward movement to the Santee, was 30 miles. Put up at the best tavern, which is very plain & ordinary in its fare.

The whole extent from within a mile of the Santee swamps to Georgetown is miserable pine barren, interrupted only by narrow swamps. The first part was clay, & red clay near the surface; then some 6 or 7 miles of black moorish or peaty land, which but for its cover of stinted & scattered pines & broom grass, seemed to approach nearer to the description of the moors of Scotland than any land I have seen. The soil is full enough of vegetable matter to show black, & is rather sandy & full of moisture, owing to its flatness & a retentive clay sub-soil. There is no under-wood; & the pines are so thinly set & the surface so level, that a cow may be seen at half a mile's distance. In other places, so much thinner is the growth of trees that the land approaches to the appearance of the savannas[1] of N.Ca. Afterwards, as it would seem when passing the impenetrable ridge which divides the waters of the Santee & Sampit creek, the ordinary sandy pine barren begins. The previous part of the road was very wet & bad; but the after part good. Along the whole distance from the Santee lands to the dividing ridge, there was no appearance of cultivation, of residents, nor of civilized inhabitants in the neighborhood, save the road itself & the few traces of wheel carriages upon it. A few mean & poor cattle only were seen grazing the recently burnt wood land. The only industry pursued by many persons is the raising of cattle; & as they never feed them at any time, the possession of land is not considered as a necessary part of the capital required for the business. There are persons owning from 500 to 1000 head of cattle, & in some cases residing at a distance, who own not an acre of the lands which their cattle graze & roam over. These capitalists hire some of the poor population of the neighborhood to pay some attention to their stocks. In the latter part of the summer or best grazing season, all the fat cattle are driven to market, & the poor allowed to become much poorer, & many of them to starve, during the next winter & spring. The milk & butter from the stocks are part of the remuneration of the herdsmen. These belong to a peculiar class. In about a space of 10 miles square of Williamsburg & Georgetown districts, the inhabitants are wretchedly poor, have no slaves, scarcely any other property, & almost no agricultural labor or regular industry of any kind. The men live by hunting, often helped out by petty pillage or trespass on the neighboring properties. These people, & even the more industrious & honest of the poor in lower S. Ca. are a most wretched & worthless population—compared to whose condition that of the slaves is one not only of greatly superior comfort & happiness, but also of respectability & dignity. They suffer every year & greatly by sickness—& are generally too sick and always too lazy to work. I have but rarely seen a few of these miserable people coming as beggars to the houses of their wealthy neighbors. Otherwise I know them but by report; & were it not for such report, it would have seemed to me that all the country residents of S.C. were of rich class, or at least possessing comfortable & competent property. And so far as the residents of the lower districts have come under my observations, they are in better circumstances, are better informed in mind, & have more appearance of health & physical power than any people I have ever seen. All these beneficial results they doubtless owe in part to their universal practice of removing to healthy places of residence during the sickly months. But though seeming to enjoy excellent health, when purchasing it at so high a price, there are few persons who live to old age. There are however some remarkable exceptions of persons attaining great age. The negroes enjoy good health, so far as regards malaria, on other than the rice plantations, & may live to great age. But the domestic servants though born & raised in this country, who have for years accompanied their masters' families to healthy summer residences, are thereby made unfit to remain on the plantations during the sickly months, & when attempted, they have always suffered by severe illness the first season, if not afterwards.

My partial wetting by the rain furnished me with a good excuse for returning to my room immediately after my dinner, & having a fire, which the returning coldness of the weather made necessary, & I spent the remainder of the evening in writing my journal.

1. Such areas of flat grassland were common in the Southeast. Fires, natural and deliberate, fed off inflammable grasses, limiting pine and oak growth and promoting conditions for bird hunting. Wetter stretches often resembled everglades.

27. Georgetown Sandhills and Ocean (1 April–2 April)

1 APR. The mail brought me two letters from home, forwarded from Charleston, but the latest date the 14th of March.

My first object was to cross the Waccamaw, to visit Col. John Ashe Alston, whom I found was at his fathers estate Rose Hill, about 3 miles distant. There being no public ferry, a traveller who is compelled to cross any of the several rivers which unite here is a choice prey to the negroes who act as boatmen, & their charge is graduated by the necessity for their services. Having still to write, & not being in a hurry, I waited for my market price to fall; & when of the sundry offers to convey me (alone—leaving my horse &c. at the tavern) had fallen from $3 to $1.50, for two oarsmen & their boat, I went at the latter price, immediately after taking the early dinner at the tavern, & arrived more than an hour before the later dinner hour at Rose Hill. In the afternoon rode with Mr. J. A. Alston to visit some of the neighboring rice lands which form a broad continuous margin all along the sides of the Waccamaw.—6 miles.

2 APR. Sunday. Went to the neighborhood church, a few miles distant where I met & was introduced to some half dozen of the neighboring planters, who with their families formed the congregation. After the service, continued our ride across the narrow peninsula which separates the Waccamaw river from the nearly parallel shore of the ocean. A cold & pretty strong wind was blowing on shore, which served to raise high billows crested with white foam, & presented a scene altogether different from the placid surface of the ocean upon the only previous occasion of my visiting its shore. I cannot say which most excited my admiration & delight, the wild sublimity of the present ocean view, or the beautiful as well as sublime scene of its calmest repose.

On the sand-hills immediately bordering the ordinary tide marsh, are the summer dwellings of the rice planters of Waccamaw, who thus have the great advantage of having perfectly healthy retreats within a few miles distance of their rice-fields, & which therefore they can visit every day, & be as much upon as they would probably desire if they were on their plantations. They are not collected in a village, as on Edisto & elsewhere, but each planter's sea-shore dwelling is at the spot nearest & most convenient to his plantation, & thus are scattered a mile or two apart along the beach side. Except for the always magnificent & often varying ocean view, these are dreary & uncomfortable sites for a continued & almost solitary residence of five or six months of every year. Not a tree & scarcely a shrub serves to relieve the eye from the dazzling whiteness of the sand which forms the entire surface, & which, above high tide is drifted by the winds into sand-hills, the surface of which is kept moving continually by every high wind, & thus gradually the entire hills are moving land-ward. A sudden change of direction, & severe wind will sometimes change the position of a sand hill in a night, & the entrances to the houses have sometimes to be dug out from the obstructions thus made by the shifting sands. When walking on them, the yielding of the sand to the tread, as well as its color, present a considerable resemblance to deep snow drifts somewhat hardened by time & compression, & recently covered by fine recent & dry snow. But comfortless as is this region of sand, in its privations of all the beauties of the land, it offers abundantly all the riches as well as the beauty of the ocean. A cool & bracing & healthful breeze from the ocean is always felt. The pleasures of fishing, from sharks in the ocean to small fish in the back creeks, are of the highest order; & the finest table fish & wild fowl are the most common & abundant of food. The close neighborhood of the Waccamaw offers all the tributes of fresh tide water, as the sea & sea creeks do of the salt waters.

These beaches are mostly islands at high tide, as different creeks flow behind the greater extent, which are separated only by broad & low sand flats, covered by every flood tide. The nearest firm land on the main, is very low, though dry after being ditched, & rises imperceptibly to the ridge or middle of the peninsula, where the land is level, full of slacks & wide shallow ponds in all but dry seasons. The larger growth is of stunted & scattered pines, & shrubs indicating very poor soil. Next to the sea side only, these lands are cultivated & yield to the rice planters a part of the provisions for their estates, all of them having to buy a large proportion from abroad. Next to the broad alluvial marsh lands on the Waccamaw, all of which are embanked & made highly productive rice fields, the high land rises by a rapid slope to some 40 feet of height, & forming an undulating surface, which together with its formation of loose & almost barren sand, indicate that these eminences formerly bordered the sea shore, & were formed by the winds of the fine & pure sands dashed up by the waves, in the same manner as an outward range of sand-hills have since been formed bordering the present shore of the ocean. The sparse & low natural growth of the sandy highlands of the river consists of dwarfish cedars, oaks, & other shrubs, which do not hide, & very slightly shade the general surface of naked & loose sand. The mansion houses of the proprietors are placed along these eminences, which offer beautiful situations in respect to elevation, & extensive views of the water & the rich & extensive rice-fields. And by planting & care, live oaks & other trees to which sandy soil is suitable, have thriven enough to make grounds pleasing to the eye, so far as can be with the entire absence of grass & of all ornamental shrubbery. The pleasure ground of Rose Hill, where I am now, with every care for its adornment by trees, walks, & laying out, is now otherwise a bare sand. On such land however, sweet potatoes are grown well; but no other field crop.—10 miles.

28. The Pee Dee Delta (3 April–4 April)

3 APR. Recrossed the water to Georgetown, calling on the way at & walking across the lower rice plantations of Waccamaw Island. It had been my intention to proceed this day on my journey up the Peedee. But this being the commencement of the District court session at Georgetown, I met there most of the neighboring planters, & I was urged to delay my departure, & to return to Waccamaw. Accordingly, in the evening, accompanied Mr. J. I. Middleton in his boat to his residence on the highland shore of the Waccamaw, about 8 miles from Georgetown. The weather still continued cold & raw. A little rain had fallen at intervals during the day, & worse weather threatened. Our course was altogether against a strong head wind, & for half the distance also against tide; & it required from 5 to 8.30 O'clock to make the cold & disagreeable passage.—11 miles.

4 APR. During the forenoon, Mr. Middleton took me to see all the different establishments for & operations then going on of his extensive rice-planting. The pounding mill, as is usual on the large plantations is worked by the tide, a portion of the rice field serving as a tide-pond during winter & early spring & which is cultivated in rice afterwards, being prepared and planted after the other lands. Mr. Middleton's . . . cultivated rice fields, (about 600 acres) occupy the whole of a small island, & a part of the large Waccamaw island, which fills all the space between Waccamaw river & the Great Peedee. Walking across the latter island to the Peedee, brought us opposite Mr. Poinsett's house, & rice land, & within half a mile of Black river, flowing back of the Pedee [sic]. The Waccamaw, the Great Peedee & Black River all mingle their waters below & about the great Waccamaw Island, by several navigable "thoroughfares" or straits, as well as at the great junction opposite Georgetown, where those waters usually muddy uniting with the clear but black water of Sampit Creek, passing by Georgetown, together widen in to Winyaw Bay. Nearly all the land on this great island called Waccamaw, which is in fact made by the thoroughfares a number of separate islands, is rice land of the best quality. According to the map this island, from its southern point to Bulls' creek, the highest & largest strait, (& which in truth is the main discharge of the Peedee,) is 25 miles long, with a very irregular breadth, but which may average 2½ miles.

The tide rises commonly 4 feet; & it serves well to drain all this land. But the upper 6 or 7 miles is not cultivated or embanked, because made more hazardous by the freshes of the Peedee. Besides this immense body of tide rice land in the island, the other sides of the Waccamaw river, & the Peedee, & the neighboring Black river, have all along wide borders of the like lands, & which reach some 4 or 6 miles below Waccamaw island, before being too much subject to "salts" to forbid profitable rice culture, to which every acre is subjected, & almost every year. The whole is like the delta of the Nile, formed of alluvium brought down by the rivers from above, as rich as land could be, almost level, &, by aid of the numerous outlets afforded by the rivers, creeks, & thoroughfares, the whole capable of the most perfect system of drainage & irrigation, both of which are essential to proper rice-culture.

As soon as my interesting walk over Mr. Middleton's land was finished, at my request he sent me in his carriage 7 miles to the house of Mr. John H. Tucker. My horse & buggy, & heavier baggage had been again left at Georgetown, to be sent on to Col. Ro. Allston's on Peedee, where I am to join them by crossing the rivers higher up. Mr. Tucker is a zealous agriculturist, & President of both the Agricultural Society of Charleston & that of Peedee. He is now confined to his house by serious indisposition. His friends whom I saw at Georgetown urged on me so much to visit Mr. Tucker, for the pleasure that it would afford him, that to comply was one of my objects in returning. And their expectations seemed to be fully realized, in my reception, and his conversation was not less gratifying to me.—11 miles.

29. Georgetown Shells, Fish Pestilence, and Coastal Floods (5 April)

5 APR. Dr. E. T. Heriot, who had been detained at Georgetown on the Grand Jury, according to his invitation & arrangement there made, joined me before breakfast; & immediately after, we left Mr. Tucker's, on our designed first excursion. The first object was the sea-shore, which is here less than 3 miles directly across the high land, as the peninsula gradually narrows instead of widens, as the distance increases from Winyaw bay. But we obliqued northward to Oakley inlet, a small sea-creek, in which I hoped to find a new marl formation similar

to that found in the creeks of Edisto & Lady's Islands. But in vain, both here & afterwards in Murray's inlet. But I could only see the creeks near the ocean; & higher up would be the more probable places for this peculiar deposite. A little north of the mouth of Oakley Inlet (now a narrow & shallow creek,) Dr. Heriot showed me near the sea-beach some places where the sand was a little more elevated above ordinary high tide, & which had been covered formerly by sand hills 15 or more feet high. These sand-hills had since been levelled by the winds, & had served to fill up the former miry tide marsh of the creek, immediately in the rear, & connected it to what it now is a wide extent of level & firm sand, but moist from its low level, & which is quite naked, at all times. The foundation of the former sand hill presents some remarkably large shells, to which my attention was asked as being enormous clam shells; but in which, as well as in several other kinds there found, I recognized *fossil* shells, not belonging to the present coast. When stating my impression to Dr. H. & pointing out to him the differences, he was satisfied that the shells are different from all recent shells now furnished on this coast. I believe it to be a bed of the more recent of the ancient formations, in the bottom of the present sea, but not belonging to it; & of which the shells have been dashed up & mixed with sand, as are the recent shells. There were fragments of calcareous rocks, or stony concretions of shelly matters, which were pierced by recent pholacles[1] or boring shell-fish.

This beach & also far both north & south there occurred about 1818, a singular & unaccountable pestilence & mortality of the fish on the coast. The circumstances I learned from Mr. Tucker & Dr. Heriot both of whom witnessed the facts. The fish, of all kinds known on the sea-coast & the creeks, & of some species which had never been seen before, seemed to be affected by some fatal & unaccountable disease, which impelled them to seek the shore, on which the waves of the sea threw them up to die in enormous quantities. If, when seen faintly struggling in the shallow water, they were taken up & thrown farther into the sea, as was done for experiment, they would be seen to return immediately, & either be left on the beach by the retreating tide, aided by their own exertions, or thrown up by the waves. Many of the fish were of large size (especially bass.) & many young whales, some of which were 20 feet long, besides the smallest sizes of all common kinds. The neighboring residents feared that by the putrefaction of such enormous quantities of fish that pestilence might extend to themselves; & under the apparent necessity, they buried them in the sands, to avoid the consequences. No cause of the mortality was ever suspected. If the people had known how to profit by this singular supply of materials, they might have as well avoided the stench of putrefication, & secured an immense amount of richest manure, by composting the fish in alternate layers with the shell beds which lie close by in great quantity, & which neither in that or any other way have ever been thought of for manure.

There occurred another memorable event on all the coast in the autumn of 1822. Then, much more than now, the summer residences of the neighboring planters were made along the margin of the sea shore. A terrible gale of wind raised the tide 5 feet above its highest ordinary flood, & drove the waves against & over most of the dwellings with a force that none of them could resist. This was at the time for low tide; & this flood both rose and fell again in 40 minutes, causing all its destruction in that time. Nearly every house was either destroyed, & floated off, or moved from its position & reduced to a wreck. In the rear of the beach, there is generally either tide marsh, or creeks, or sands lower than the sites of the houses, & which for half a mile or more was dashed over by the utmost fury of the breaking billows; & before the residents knew of the necessity for their retreating to the main-land, all possibility of retreat was cut off, except as directed by the force of the wind & waves. Many lives were lost; & many more were barely saved by the occurrence of some remarkable accident, or rather of unlooked-for & mercy of God. The house of Mr. John Allston near the mouth of Santee was swept away from its foundations, & nothing saved it from being carried off by the ocean, & all his family being drowned but for the house being stopped by a tree. Another residence & all its out-houses, including the carriage house, carriage & horses, were all swept off, & nothing left except a few bricks of the chimney. All the family & every living being on the place were drowned. This terrible visitation caused the sea-shore to be abandoned by most persons. And besides the possible recurrence of such a storm, the ocean is generally though slowly encroaching on its shore, so as to make the ultimate loss of buildings near the shore, or on the sand hills, a more than probable event. On this account, Dr. Heriot, who has for 20 years used a sea shore summer residence, has just built on the out-side of his mainland on the water & marsh of Murray's Inlet, a mile back from the shore to which place we next proceeded. Here, as on many other places along the coast, there are very extensive banks of clam-shells, left by the Indians. Those at the creek shore of Dr. Heriot, are much broken & disintegrated by age, & the very little of any other earth in the heaps is black vegetable mould, & a little sand carried by the winds. Altogether, these heaps offer an extremely rich & valuable resource for calcareous manuring; of which not the least use has yet been made, or even thought of being tried; but which, I trust, my representations will prevent being left much longer useless. On Dr. Heriot's shore alone, there must be enough of this material, perfectly accessible & convenient, to manure several hundred acres.

Returning across & up the peninsula, we called to see the high bluff at Laurel Hill on the Waccamaw, where I had heard that shells were to be seen, & where therefore I hoped to find the great calcareous bed visible, notwithstanding that the proximity to the ocean opposed the truth of any such visible exposure. In that I was disappointed. . . .

From viewing this bluff, we returned down the course of the river two miles, & crossed it to the residence of Dr. Heriot. His house is beautifully situated on an eminence close to the river. The site is a small high-land island, not more than 10 acres in size, separated by a narrow piece of tide (which now like all the other alluial [sic] swamp,) is cultivated in rice. Across this, two causeys join this small island to the large & high Sandy Island which stretches far along the rear, & is the only considerable extent of originally high & firm land in all the great Waccamaw Island. 15 miles.

1. *Pholadidae* is a family of rock-boring bivalve molluscs with thin, elongate shells, commonly called piddocks or angel wings; they flourished from the upper Carboniferous to the present.

30. Sandy Island (6 April–8 April)

6 APR. The planting of rice is now proceeding rapidly, & the consequent rapid & extensive changes of dry land & water are very remarkable to the observation of a stranger to the culture. Nearly every acre [of] the great region in sight is of rice land. As soon as a field is planted, it is overflowed & kept flooded, so that nothing but its narrow

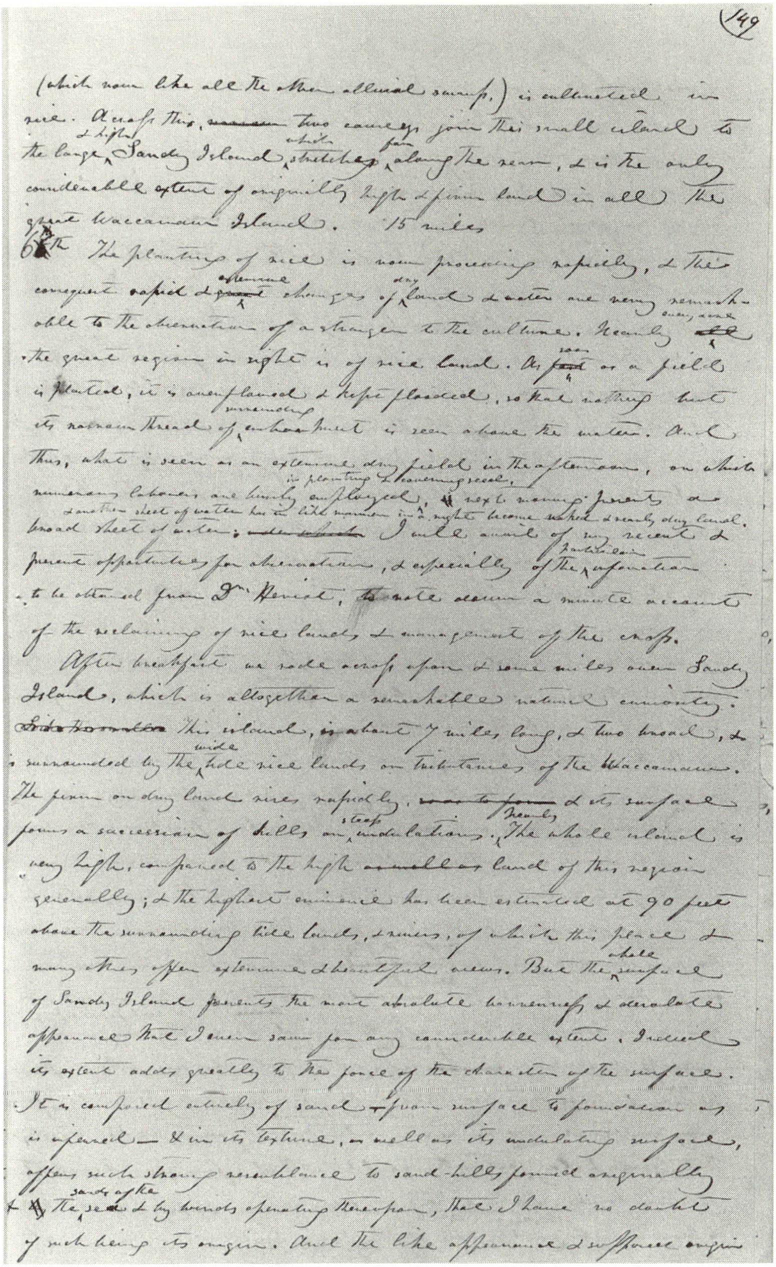
(149)

(which now like all the other alluvial swamps) is cultivated in rice. Across this, two causeways join this small island to the large & higher Sandy Island, which stretches far along the river, & is the only considerable extent of originally high & firm land in all the great Waccamaw Island. 15 miles

6th The planting of rice is now proceeding rapidly, & the consequent rapid & extensive changes of dry land & water are very remarkable to the observation of a stranger to the culture. Nearly every acre of the great region in sight is of rice land. As soon as a field is planted, it is overflowed & kept flooded, so that nothing but its narrow thread of surrounding embankment is seen above the water. And thus, what is seen as an extensive dry field in the afternoon, on which numerous labourers are busily employed in planting & covering seed, next morning presents a broad sheet of water; & another sheet of water has in like manner in a night become naked & nearly dry land. I will avail of my recent & present opportunities for observation, & especially of the particular information to be obtained from Dr. Heriot, to note down a minute account of the reclaiming of rice lands, & management of the crop.

After breakfast we rode across upon & some miles over Sandy Island, which is altogether a remarkable natural curiosity. This island, is about 7 miles long, & two broad, & is surrounded by the wide tide rice lands on tributaries of the Waccamaw. The firm or dry land rises rapidly & its surface forms a succession of hills or steep undulations. Nearly the whole island is very high, compared to the high land of this region generally; & the highest eminence has been estimated at 90 feet above the surrounding tide lands, & rivers, of which this place & many others offer extensive & beautiful views. But the whole surface of Sandy Island presents the most absolute barrenness & desolate appearance that I ever saw for any considerable extent. Indeed its extent adds greatly to the force of the character of the surface. It is composed entirely of sand from surface to foundation as is inferred — & in its texture, as well as its undulating surface, offers such strong resemblance to sand-hills formed originally by the sands of the sea & by winds operating thereupon, that I have no doubt of such being its origin. And the like appearance & supposed origin

Manuscript page from Ruffin's diary that includes his impressions of rice planting along the Waccamaw River (courtesy of the Virginia Historical Society)

thread of surrounding embankment is seen above the water. And thus, what is seen as an extensive dry field in the afternoon, on which numerous laborers are busily employed in planting & covering seed, next morning presents a broad sheet of water: & another sheet of water has in like manner in a night become naked & nearly dry land. I will avail of my recent & present opportunities for observation, & especially of the particular information to be obtained from Dr. Heriot, to note down a minute account of the reclaiming of rice lands & management of the crop.

After breakfast we rode across upon & some miles over Sandy Island, which is altogether a remarkable natural curiosity. This island, is about 7 miles long, & two broad, & surrounded by the wide tide rice lands on tributaries of the Waccamaw. The firm or dry land rises rapidly, & its surface forms a succession of hills or steep undulations. Nearly the whole island is very high, compared to the high land of this region generally; & the highest eminence has been estimated at 90 feet above the surrounding tide lands, & rivers, of which this place & many others offer extensive & beautiful views. But the whole surface of Sandy Island presents the most absolute barrenness & desolate appearance that I ever saw for any considerable extent. Indeed its extent adds greatly to the force of the character of the surface. It is composed entirely of sand—from surface to foundation as is inferred—& in its texture, as well as its undulating surface, offers such strong resemblance to sand-hills formed originally of the sands of the sea & by winds operating thereupon, that I have no doubt of such being its origin. And the like appearance & supposed origin had struck me all along the river hills below. But what is much more remarkable here, is the great height of the island, surpassing by far every part of lower S. Ca, & for many miles higher up the rivers—and also the entire separation of this island from any other high land by the intervening river & alluvial tide lands of perhaps a mile of least breadth. The soil of the island contains barely enough of any other earthy matters than pure sand, to present a little difficulty to scratching a hole of any depth with the fingers. There were formerly on it some long-leaf pine trees of considerable size, & furnishing excellent timber. But these have been removed for use. The only present cover is of a few small & stinted pines, & numerous small & ragged scrub oaks,[1] rarely above 10 feet. These though standing thickly do not prevent the sight extending far over the barren & otherwise almost naked surface.

No fire had been over the land for several years to thin its scant cover. Yet all the remaining leaves, collected by the winds in different spots, did not prevent a general appearance of naked white sand, glowing & dazzling to my eyes. The only small growth seen were a few lichens, or mosses such as are found only on the most barren lands, or on rocks.

On this island of barrenness surrounded & in absolute contact with rice lands as rich as land can be, there reside a few only of the proprietors; & of these only two, Dr. Heriot & Col. Belin have families. For the former resides on another small island, yet from its close proximity, & identity of character & present artificial junction with Sandy Island it may be deemed a part of the same. These two residences have everything desirable in elevation, prospect, comfort & elegance within, & decoration without, except that which no other place of residence wants, but which no care or expense has heretofore been able to command here, viz: soil for the poorest growth of grass & common crops of grain or vegetables on the high lands, even where putrescent manures have been put in the greatest quantity & waste, for many successive years. Lime with such manure would probably serve as the remedy, but lime & clay together certainly would. And for grounds about mansion houses, & where there is every year to apply (& heretofore to throw away or otherwise as much to waste,) the straw from hundreds of acres of rich rice crops, as well as the dung of all the animals kept for pleasure or for use, it surely would be worth while to *manufacture* some 20 acres of good soil, as the process would be to make it here. Before Dr. Heriot began lately to find that his rice lands were benefited by manuring, & which has shown a use for his straw, it had been for many years before mostly spread out to rot, on the ground back of his barn. This place never was thereby made able to produce any perfect crop, & is now as always heretofore a barren & almost naked sand. The utmost result has been to sow early rye to furnish a scanty supply of grazing for sheep before the destruction of the growth by the burning heat of the summer's sun. . . .

On this island there are many low basins which form ponds in wet weather. The soil of their bottoms is peaty & worthless for cultivation or grass, even after draining. Liming is needed for them.

The well at Dr. Heriot's residence (on the small island,) affords remarkable proofs of the openness of texture of the sand in which it is dug. It is about 18 feet deep & 32 yards from the nearest rice land on one side of the island, & 72 yards to the high-water mark of the river, on the other side, between which & the well the high land rises to about 30 feet high. The water of the well is clean and good when there is no flowing of the rice grounds; but soon after the water is put on, the water of the well begins to show a faint tinge of coloring extract; & if in summer, & after much flowing of the rice grounds, it acquires a disagreeable flavor. The tint I saw, & therefore inquired & learned the cause. And whenever a fresh raises the river for some time, say a week, the water in the well rises also, & lowers after the fresh subsides.

On our return from viewing Sandy Island called by to see Col. Belin, who came & dined with Dr. Heriot.—6 miles.

1. *Quercus margaretta* form barrens with other oaks and pines on sandy land in the coastal plain.

7 APR. Went again to the sea-shore with digging implements to examine whether the fossil shells found there indicated a bed just below. But on digging below, nothing was found but the usual pure sea sand, without a shell of any kind, & water soon rose & stopped the useless labor. It seems as if the many shells there which I believe to be fossil, were broken up from a bed some distance out at sea, & driven by storms to the shore where they lie, above common tides. There are many kinds of concrete shells, or calcareous stone, some of 10 or more pounds in weight. Gathered specimens of all the shells & stones to send home for examination.

Called to visit Mr. Tucker, whom I found better. Returned with Dr. Heriot, to dine with Col. Belin. Found him busy applying a few bushels of lime, which he had on hand, according to my advice, on ground before his dwelling.—14 miles.

8 APR. At last a warm day. Except for a few short intervals, rarely if at any time making a full day each, the weather has continued remarkably cold. Very few kinds of forest trees are yet opened, except of early kinds.

I had appointed to cross the Peedee to Col. Robert Alston's on the 6th, (where my horse & carriage have been sent to wait for me,) but learning of the relapse & dangerous illness of one of his children prevented & still prevents my going. I have used the time to write out an account of the embankment & culture of rice land, from the verbal information obtained from Dr. Heriot principally. . . .

Dr. Heriot cultivates no high-land, though owning sea-side lands; & buys all his provisions. His rice straw from more than 200 acres of land annually under that crop, has been until lately lost as manure, & for all other purposes, except the little eaten by his horses & few other animals. But he has latterly found great advantage from applying his straw to the rice-land, but not cultivating it in rice that year, but in peas, & by dry culture of course. He has found that the land so treated gives a double amount of rice the next year. But much of this may be owing to the rest, or alternation of crops. Rest is seldom given to any rice lands, & manure to them as seldom. As to rotation or alternation of crops, I have never heard of the least approach to it, except this recent & partial trial of Dr. Heriot's. I infer that in the general neither rest or putrescent manure is requisite—& that alternation or rotation of crops with wet rice culture is not only not necessary, but scarcely possible to practise.

Further opportunities for observation which I have had have presented more strongly the remarkable features of this long & narrow strip of high & dry land that separates the rice swamps on the Waccamaw from the sea shore. I have before spoken of its elevated & undulating surface, near to the river, the still increasing sterility of the midland & the low & poor, though better lands near the sea, where there is almost the only cultivated high land. These are the general characters, though with variations Mr. Tucker's highland along the river rice land though very light is superior to the quality of the range in general; & by very high & wasteful manuring, he has made on that field 50 bushels of corn to the acre. But it is strange & unaccountable that the high lands bordering on the sea formerly furnished all the cultivation of the peninsula; & that many planters there lived in comfort & some grew rich by raising indigo & corn on these lands, which are now so poor, & appear as if they had always been poor. With calcareous manures, there is no doubt that these lands may be profitably improved, & afford subjects for very profitable application of the quantities of rice straw which the proprietors have to waste, & of the richer mixed manures which they might make. But if used alone, I doubt whether the latter would be worth applying on these lands, or that they would be worth cultivating without liming.

31. Black River and Lower Pee Dee (9 April–13 April)

9 APR. Sunday. In the evening left Sandy Island & crossing the Waccamaw rice lands through the Wando-passo Thoroughfare, & then the Peedee river, landed at the residence of Mr. John H. Allston, one of the best rice planters on the river.—6 miles.

10 APR. After breakfast set out to visit Black river, & search at its bluffs for lime-rock. The road to the lower ferry (Pringle's) which I crossed, passes through very poor land, or pine barren, except on leaving the Peedee & arriving on the rice lands of Black river. A mile or two past the ferry, & up the river brought me to Mr. Stephen Ford's plantation, to whom & to his two nephews, Stephen C. & Reese Ford, Mr Allston had given me a letter of introduction. These gentlemen, as well as some others on Black River cultivate rice on the "all water plan." . . . Their rice lands on Black river are generally much more of vegetable composition than those of the Peedee, which are more earthy & stiff. The former kind of soils require more water, & are less in danger from its excessive use. Mr. John Allston from his reminiscence of similar soil on the Pan-pon thinks that the Black river planters are right in using the more extended watering cultures; but on Peedee he prefers & practices the more usual method.

I examined the several bluffs (20 or more feet high) on Black river near & above Pringle's ferry, & found nothing but sand; except a sandy blueish clay below high tide. Also examined by the auger the bottom of a tide creek, 11 feet below the surface. The solid bottom was of the same clay—though some loose stony masses lay above, which I could not see to examine.—15 miles.

11 APR. Continued up the west side of Black river to Avants Upper ferry, which is 10 miles from Mr. Ford's, & crossed over to examine the rock reported to be there. It lies in a solid horizontal ledge, not more than 2 feet thick, & a few feet above high tide. It is so hard that it was difficult to break off specimens with my hatchet. It contains marks of shells, & is calcareous throughout, but apparently not rich enough for manuring. . . . The poor appearance here, & not hearing of any thing more promising above, left no inducement to proceed higher, & I returned by crossing the other Ferry a mile below, & by a different

road, over Snow's Mill, returned to Mr. John H. Allston's house on Peedee.

Roads travelled today when off the river margins were through land of the same poor quality, & level surface as seen before. Altogether pine barren, generally with sandy soil on clay subsoil. Where wettest, from the level of the soil & impervious texture of the subsoil, it approached to peaty land, & a wide extent had scarcely any living trees remaining, though showing the remains of many that fires had destroyed. Such lands approach to the character & appearance of the *savannas* of N.C; & nothing is wanting to complete the resemblance but more retention of water, & continued annual action of fire on the brown sedge[1] & other grasses.

Though there has been much more appearance of spring latterly than before, the winter weather still lingers in part. There was white frost both yesterday & this morning.

In the afternoon crossed the river & walked around the banks of Mr. Allston's extensive rice field. The hands were just finishing planting one division, & the bank having been opened, the water was rapidly covering the surface. There is an increase of fresh in the Peedee, & the very high tide was within 6 inches of the top of the embankment in many places, & expected to go over in the next night flood tide. But this caused no uneasiness, as the firmer texture of the embankments prevent their being washed materially even when the water passes over.

I saw here a means of preserving banks from washing which appeared so effectual, that I am surprised it has not been extended beyond the sides of a single crop bank. It was made by setting out roots of the tall sedge grass which grows commonly on the fresh tide marshes here, & universally on low fresh marshes in Va. It had spread nearly all along, & had prevented all the usual washing of the bank by the action of the winds on the water.—27 miles.

1. These sedges were various *Carex*, coarse, rushlike grasses growing in wet areas, of no value as fodder.

12 APR. The sky over-cast generally, & some times a few drops of rain, as well as at others transient gleams of sunshine. This much opposed the pleasure of an excursion in which I was conducted by young John Alston the son of my host. After walking on the banks across

the extensive intervening rice-lands, we embarked, in a small but stiff canoe which had been sent ahead, on what is called Indian Lake, but which is one of the sundry straits or "thorough-fares" which intersect Waccamaw Island. This is about 5 miles long, & passes through the uncleared swamps alongside of Sandy Island, & connects the Wando-passo Thorough-fare with Little Bull creek, near where the latter discharges part of the Peedee waters into the Waccamaw. Turning up the bolder course of Little Bull Creek, we through it entered the main & then rapid stream of the Peedee, swollen & muddy with the fresh. All along Indian Lake & L. Bull Creek, & also for several miles down the Peedee with the exception of one bluff, all was tide swamp, & in its natural state of a thick forest of tupelo, gum, cypress, maple &c.[1] of large & sometimes of enormous size. The liability of these lands to be inundated, prevent their being safe enough for rice-culture; & no one seems to think that land is worth embanking for any other crop. Thus the immense extents of swamp lands above regular & full tides, or where exposed to freshes, on the Peedee as of all other rivers of S.C. will probably remain a nuisance for a century to come, & held at scarcely any value except for timber. Yet land a few miles lower down, no better as to constitution & fertility, & even worse, by long culture, is valued at $150 to $200 the acre. Last winter, a body of 360 acres of embanked rice land on the lower part of Waccamaw Island, without any building, was bought at $200 the acre. The same purchaser had previously bought another piece of 40 acres which lay convenient to his other possessions at $250 the acre. Yet these lower lands are in such course of being washed away by the waves of storm tides, that they can scarcely be deemed "*real estate*." Their margins are all gone, & in many cases the first outer embankment has been so washed away as to be abandoned, & a new line made within. Still the enemy comes on & both by sapping & breaching, must finally prostrate all such defences as are now used. Every addition thus made to the breadth of the water gives more power to the winds & waves, & increases the probability of their final & complete success.

Notwithstanding the very unfavourable weather this day for alligators, which now leave the water merely to bask in the sun, we saw four of them out. My companion, who is an excellent marksman, struck the head of a swimming alligator with a rifle ball; but though he was evidently struck, as appeared by his floundering under water, the ball glanced & probably caused no injury beyond a violent shock, & pain for a short time. I shot at & *missed* another. This Indian Lake is a great resort for these animals, & if the sun had been bright & warm, there would have been a fine show of them.

The rapidity of the swollen current of the Peedee soon brought our canoe home, which we reached in the evening just in time to escape a storm of wind & rain.

1. Tupelo gum, *Nyssa aquatica*, is found in coastal swamps of the eastern and southern United States, commonly alongside cypress. It tolerates shade, acidity, and permanent wetness and provided a white wood for utensils, furniture, and wagon wheels. The cypress was probably another *Nyssa* (e.g. *sylvatica*). The maple was probably *Acer rubrum*, or red maple, in which galic acid in the bark combined with copperas to give a black dye: the wood was used for furniture and gun stocks.

13 APR. I had waited until this day, & even returned from above when so far on my intended route, for the purpose of meeting the Agricultural Society of Peedee today, to which I had been invited & urgently pressed by several prominent members, including the President. Owing to some interference of the time & place of meeting with those of the neighborhood club, though both bodies were composed of almost precisely the same members to my surprise this morning I learned by accident that no meeting of the Society was expected to be held. This would have made my appearance there to me embarrassing & disagreeable—& inasmuch as I had not been invited & was under no obligation to meet the club, or any other association than the Agricultural Society, I deemed that self-respect required that I should not attend. I therefore proceeded on my journey, & regretted much that I had not done so several days earlier.

32. In Marion: Geological Investigations on the Pee Dee (13 April–18 April)

Last night a heavy rain fell; & as I was told that it was impossible to cross one of the creeks emptying into the Peedee, (except by swimming) I was obliged to give up my design to pursue the bank of that river, & touching in at each of the few bluffs to examine for marl. The only course left was the same by which I returned from Black River, & thence across to Britton's ferry on Peedee. Throughout this weary journey of 35 miles, except where I left the Peedee, & where I again reached it at the close, there was scarcely any sign of cultivation, or indeed of human habitation. Perhaps in all there were some 5 or 6 spots, of a few acres each, under crop, or in preparation for one; & scarcely any other land is cleared. All pine-barren, generally sandy but firm, some clayey & some moorish, varied only by narrow swamps or slashes.

At Stony River, a few miles below the Black River ferries, the same kind of calcareous stone there seen is exposed in the road & on each side of the stream. A little farther on, the like appears, apparently silicified, so as to show no effervescence with acid. The shells also *not* calcareous. . . .

When arriving at Britton's ferry, (which is at the junction of Georgetown & Williamsburgh districts,) the first news I heard was that I could not cross, owing to the fresh being too deep over the wide swamp on the other side through which the road passes for two miles. This alone I cared little for, as I had designed to proceed up this side of the river. But no marl or stone could be seen, the ledge being covered 5 feet deep under water. From the description of the ferry keeper, who however appeared to possess very little talent for observation, the rock was such a thin horizontal ledge as I saw at Black River, but of blue color. Nothing whatever was known of its composition or quality.

Worse news than all this was that I could not proceed up the river, without returning directly backward, & retracing the road just travelled for 11 miles, and which was so wearisome even on the first passage, & when it was taking me towards, instead of directly from my journey's end. As there was no other stopping place near, staid at the ferry keeper's house, where of course, travellers are entertained for their money.

14 APR. After breakfast set out. It was not until having gone back on the road 11 miles, (which road is so far the dividing line between Georgetown & Williamsburgh districts,) that a practicable carriage track opened to the stage road, which then was only a mile distant. Turning then nearly back again, this road brought me to the ferry over Lynch's Creek. There I again heard of marl, but which was covered by the fresh. Thence, after crossing the ferry, & 4 miles, I reached Davis'

tavern & stopped to feed my horse, & get my dinner. I soon learned from Mr. Davis that he had marl in abundance, & had already sent samples for me to Charleston, with a letter inviting me to call on him. I walked with him to the principal exposure which is a high bluff on the Peedee; & to my great surprise, found that the upper part was miocene marl. . . . The whole visible bed about 24 to 27 feet, or 29 to 32 to low-water. I could have gladly spent a day at this very interesting exposure; but designing to return at a time when less hurried, I hastily gathered but a few specimens & proceded [sic] on my journey. My anxiety to proceed was caused by having promised Mr. Jolly (in answer to his letter of invitation) to be at his house before this, & to which I had ordered my letters to be forwarded, & which I hoped awaited me there. I reached his house some time after dark. Found no letters there or at the Post Office.—44 miles of road travel, & 12 miles more of walking to visit the marl.

The road this evening lay not far from the course of the Peedee, & throughout showed a change of surface, improvement of quality & of culture, compared to the high lands of the lower districts. The land is high & the surface undulating, at least along the roads owing to the many streams & valleys passing to the river. The settlements are thick, & the plantations generally small. Cotton is again the chief crop, of which none had been seen in Georgetown or Williamsburgh. The soil, though still far from rich, is much better, & easily improvable by marl.

15 APR. Proceeded with Mr. Jolly & Dr. Harlee & Dr. Gregg, who came early to join us, to examine the neighboring marl beds. We returned 4 miles down my last night's road, & a mile farther to the Peedee, examining sundry of the exposures therein. . . . On Myers' land, in a ravine or narrow valley leading to the river, the section shows 5 separate horizontal layers of stony hardness each from 6 to 12 inches thick separated by thicker layers of soft marl. . . . But what was equally interesting & surprising, the first fossils seen were grypheas (exogya?)[1] & belemnites, showing that this calcareous bed is of the secondary formation. I had before heard from Dr. [E.] Ravenel that such was the case . . . high up Lynch's creek, 20 miles from Camden. But I had never heard of this being the character or age of the Peedee marl. . . .

At Burch's ferry on the Peedee, the like marl shows continuously along the bluff for 150 to 200 yards, forming a perpendicular bluff of 8 to

10 feet high above the present height of the water. . . . Several of us got into a small canoe, & by holding to the limbs of bushes & to the prominences of the rock, as well as by the aid of the paddle, we were enabled notwithstanding the strong current, to examine the whole extent, & to secure many fine specimens of belemnites & gryphea. . . .—11 miles.

1. Grypheas are a relation of oyster, now extinct, mainly from the pre-Cretaceous era. Exogyra are a Cretaceous descendant of *Gryphaea* and relation of oyster.

16 APR. Sunday. Dined with Dr. Harlee, & rode over his farm.—7 miles.

17 APR. Rode to Marr's Bluff on the Peedee, to examine its formation. The land is there very high to the river. . . .

No appearance of marl or any calcareous earth. Yet this place has been stated, in an agricultural paper of this state to have marl in abundance, & of which I had entertained no doubt before my reaching this neighborhood. But before I saw Marr's Bluff, I was convinced that the marl stratum had risen so high as to run out, below that place. It was no doubt the black clay shale which was called marl. It contains no carbonate of lime—but I found dry on its surface, where kept from rain, alum[1] certainly, & I believe copperas.[2] There was also a tasteless white powder, which probably is sulphate of lime.[3]

Returning, I saw the marl on the Clay-wall plantation of Mr. Jolly, about a mile above the mouth of Jeffreys' creek & about midway between that creek & the Peedee. And this I believe to be the highest exposure of this secondary marl formation. We also examined the marl on Gordon's plantation up Willow creek, about 5 miles direct from the river.—26 miles.

This & the two preceding days have been the first of warm weather, & this day has been oppressively hot. On the 13th. I first saw the buds of the oaks generally swelling & opening; & now they & all other high-land deciduous trees are enough in leaf to give them a general green appearance. Much of the corn is up; & cotton planting is going on generally. This is a *ploughing* district, & in that respect the culture presents a marked contrast with the lower districts. It is common here for planters to keep one mule or horse for ploughing, for every two laborers employed.

1. Alum is double sulfate of aluminium and potassium: a whitish, transparent mineral salt, used to tan leather, size paper, and purify water.

2. Copperas is a green sulfate of copper (and sometimes of iron or zinc).

3. Ruffin refers to calcium sulfate or gypsum.

18 APR. Received this morning two long letters from my son & his wife, (forwarded from Charleston) & most welcome they were, reporting that all were well. The last previous date from home was of March 16th.

Left Mr. Jolly's after breakfast, on my return. Called at Myers' to see again and examine now particularly the junction of the secondary marl with the underlying soapstone.[1] The two intermingle a little for an inch or two at the junction & the shale[2] for about a foot lower only is slightly calcareous.

Called to examine Bigham's marl, of which I saw his pit alongside the road as I came up just before sunset, 7 miles before Jolly's. It differs from all other places seen here, in having shale at top, which is not the least calcareous for the first foot, & below that becomes more & more calcareous, until it gradually passes into the marl. . . . At the river side, Cooper's bluff, of the same plantation, the blue marl usual here is seen at 3 feet above the present height of the river. At Jewett's bluff, 3 miles lower, the same is also visible—but I sought it there in vain, having no guide with me.

Got to Mr. Davis's house to dinner, & went with him to examine more fully his marl, of which I scarcely had time to glance at on first reaching his house. He has the finest exposures I ever saw. The lands next the river are very hilly, with many ravines & steep slopes; & almost every hill-side has rich marl from the bottom nearly & sometimes quite to the top, & so near to the surface that the ploughs often cut into it, & every tree blown up brings marl with its roots. On the river, the bluff is a perpendicular or over-hanging cliff. . . . We embarked in a small canoe for better observation, & went upwards of two miles down the river, which had so rapid a course that it required no effort except to guide the canoe. Nearly the whole course the land is as richly supplied with marl as above stated, though sometimes not visible from the shore at all, owing to intervening swamp, & at other times but partially owing to a more sloping & covered surface of the hill being presented. The river at present is here about 6 feet above low water—& Mr. Davis' marl cliffs seem to be 25 to 28 feet above the water. At Godfrey's ferry, on Gasquets plantation next below, the visible marl is all miocene; & the public road cut through it shows its depth to be full 20 feet. At low-water, the blue (or secondary) marl is said to be visible. Next below is Gibson's land, which for a mile is similarly supplied. . . . In Gibson's land there are many lime sinks, & a narrow & low cavern which opens to the river & which has never been explored to its end. Mr. Davis has been in it about 90 feet as he supposes. I went in it as far as I could on my feet—& as the bottom had been recently covered by the river fresh, & was very muddy & wet, I could not go farther. . . .

There has been very little use yet made of marl or lime on or near Peedee as manure, & of that little, most of the applications have been as injudiciously made as generally. Altogether the trials made would not amount to 600 bushels of both lime & marl. I will mention all I have heard of. Dr. Robert Harlee made the largest application. This was 150 to 175 bushels of the caustic lime united with his compost or farm-yard manure, & spread over 15 to 20 acres of land. Of course the mixture was injudicious, & very far too light as to the lime, which was only about 10 bush. to the acre. Still it was believed to be obviously beneficial. Mr. Gordon on Willow creek tried some little marl—how I know not—& believed it to have done no good. Mr. Bigham applied his roadside marl (under the shale) to 1⁄12 of an acre at the rate of 400 bush. per acre. The land was put under its second crop from new-ground state, & full of vegetable matter. The result was so remarkable on the first crop that it was measured, with an equal space adjoining of the land not marled. The product of the marled was equal to 30 bushels the acre, & the other 15. A quarter acre more was marled the next year (1842) & produced equal benefit, but was not measured. Mr. Henry Davis & Capt. Jolly are making small experiments for this year's crops.

Tradition tells that lime was burnt of the Peedee limestone, to use for making indigo, in the revolutionary war. But though this was done here, & in other parts of S.C. under the necessity produced by the want of supply of lime from abroad, as soon as intercourse was restored, the purchasing of imported lime was resumed, & with very few & partial exceptions has continued to this day. S. Ca. has paid an incalculable amount for foreign lime, when it can be burnt here more cheaply than at the northern lime kilns.

Capt. Jolly burnt lime 19 years ago, & used it for building. The blue stone makes excellent lime for cement, & very nearly of pure white. I saw a bulk still remaining which he had burnt a year or two ago. I expect that from this time forward there will be a very different state of things from the almost total disuse of marl, & indeed the very general entire ignorance of its existence in this neighborhood.

1. The soapstone was talc or steatite, foliated, shining magnesium silicate, of greenish color and soapy touch.

2. This was finely laminated clay.

33. Marion-Williamsburgh Borderlands: Trials of a Surveyor (19 April)

19 APR. Rode with Mr. Davis to make the necessary further examination of the marl exposures of this neighborhood. First returned upward of two miles, to see Witherspoons bluff. For this purpose we called for a canoe at the next landing below, to ascend. But we soon found that though with two paddles, it would be hard labor to make very slow progress even for a quarter of a mile; & we had to land & continue the search on land. By good fortune the marl is exposed immediately where the path descended to Witherspoon's bluff; or if hidden there & open to view from the water 20 yards above or below, we might have gained nothing by the effort. No one before trying can appreciate the difficulties of making a thorough examination of such river sides as this, & most others above tide. It is impossible for one person, as little aided as I am by information & by physical assistance & labor. The river sides are generally of wide & impassable swamp, & which now is overflowed. The high-lands back of the swamps slope down gently, & generally have not been under cultivation, & even if the marl be near to the surface, it is not known. It is only at the river bluffs where the water has washed away the high land & formed steep cliffs, that the marl can be found, if the explorer can obtain information of where to seek, & can then descend to the place. For these cliffs are partly of loose yielding earth, too steep & loose to bear the tread, & bare at such places of bushes to which one might cling. If the marl or calcareous rock is exposed, it is either perpendicular or overhanging, so that no glimpse of it can be had from the top of the hill. No place for walking is offered at the waters edge, & which, in times of fresh, is but

a part of the steep precipice, from which a slip would send the explorer into deep water & a rapid current just beneath his failing footing. If water conveyance be attempted, which I could do but rarely & for small distances, there is indeed afforded a good general & distant view of the steep & naked sides of the cliffs. But they cannot be ascended, for close examination; & every little border of swamp cuts off the view, & forbids continued observation. For these reasons, my examinations have been & must necessarily be imperfect. It would require the combined operations of both land & water exploring parties provided with laborers, implements & especially with good ladders, to do justice to this part of the Peedee banks.

The marl at Witherspoons as seen above the height of water, is about 8 feet high, & all yellow miocene—firm & rich—& full of casts of shells but as elsewhere hereabout no shells remaining except a pecten & an ostrea (sometimes with balani) which are very firm & perfect.[1] The numerous casts are but of few species, & principally chance congregation. Here I found two perfect casts of large panopœa[2] [ER], both showing the natural portion of the living fish as burrowing in the earth. Two other casts of bivalves found here not known, & certainly not common in miocene marl.

Just below this land, our ineffectual attempt at navigation against the current had served to show the miocene marl fully exposed along Belin's bank below & each side of his landing place. Yet he did not know of its being there, nor even what marl was.

Returning, we rode again to visit Gibson's bluff from the land side, which I did very imperfectly, & at great risk first of losing my yielding foot-hold, & plumping into the deep & rapid water, & next of not being able to climb up again to the top of the cliff. . . . As I saw no fossils . . . it is only from general similarity of position & appearance that I infer that the upper is miocene & the lower secondary.

A few hundred yards above this place is the cavern, where the marl is all miocene. . . . I came provided with defences against the mud, to permit me to explore this passage, if practicable. I proceeded, pushing a lighted candle before me & crawling on my hands & knees, & soon had to be almost prostrate. The passage was not only so low as to require this, but soon became too narrow for me to turn about any where, & also very crooked &, as it required more space to go backward than forward in such a constrained position, I was afraid that I might get jammed in returning, & have to wait to be dug out. Therefore I did not go as far as I might have squeezed through, but not much short of it. I guess that I went about 60 feet; & if any one ever goes farther, it must be effected by digging out & deepening the bottom, which has evidently been filled higher than formerly by the water washing earth in from above. The sources of this water & earth were seen in two deep sink holes about 150 feet back, & which are part of a longer chain of sinks, which no doubt, before breaking in, were parts of the roof over the interior of an extension of this cavern.

It was not merely my old fondness for exploring caverns that carried me into this. I wished also to see how these passages are formed through marl. The lower part of the marl, through which the passage has been formed, is a beautiful fine grained marl, moderately soft, & without a shell remaining. The stratum above, which forms the roof of the cavern, is a stony concrete of shells (pectens & ostreas only) cemented together with hard calcareous earth, & on which the water had made but slight impression.

The next place below Gibsons seen was Waters' Bluff a little above Port's ferry. The upper hard stone showed from 2 to 3 feet only above the water, & was secondary as appeared by shells found. Mr. Davis said that the appearance was the same at Port's ferry—& all below that must have been under the present water, & therefore unnecessary to be visited. I accordingly returned to dine & feed my horse, & at 4. P.M. proceeded over Lynche's Creek ferry, intending to reach Col. D. Wilson's house on Black Mingo creek. Mr. Davis had heard from this gentleman that limestone had been found thereabout, & therefore I asked for a letter of introduction to him.

Along this part of the Peedee, where rich marl is so abundant, land is as low priced as any I ever have seen. Mr. Davis' tract of 1200 acres he bought for $2000. It is much better quality than lands generally; & I heard one of his neighbors say in reference to Mr. Davis' marling, that "Such land was good enough without marling". Gibson's land tract below, of which the extensive marl exposure has been described, contains 3500 acres, of which not more than 300 is Peedee swamp. It has been offered for $5000. This land is not near as good as Davis', but is not very poor, & is highly improvable by marl. It has been deserted for about 30 years, & all its formerly cleared land has grown up in pines of large size. The tract of which Witherspoons bluff is part contains

2,200 acres. It was bought a few years ago, at private sale, for $1100. The purchaser has made no use of it, & has held it until the interest of his purchase money &c. has raised the cost to him to $1400. He has offered to sell it for $1500, but cannot get that price. This land is not so good as the others named; but is equally susceptible of improvement by its abundant & rich marl.

To vary my route from the mail road before travelled, I took the road towards Indian Town. A little after sunset I had travelled 12 miles, & had about 5 more to go, by a private cross road, which I was told I could not possibly miss, & which, as in almost all such cases, I *did* miss. By dark, I was passing over a dull track, obviously but seldom used, through the regular pine barrens, numerous trees prostrate from burning, & with the usual & only varieties of slashes & swamps crossing the road. Two trees had already been across the road. Luckily they were not too large for the horse to draw over the empty buggy; more luckily, none were found afterwards across the road. The night was very dark, being cloudy. Not a star was visible, & the moon had not risen & would not for some hours. It threatened to rain, which threat only prevented my striking a fire & stopping until morning. Soon I could not see the road, except sometimes immediately under my seat, when made plainer than usual by deeper ruts or by water. I had to let the horse walk, & to leave it to him entirely to choose his way. And well did he discharge this trust; for he seemed to be aware of the danger, & kept the middle of the narrow track as well as he could have been guided in day-light. Two narrow bridges were passed over; & one of them, as I afterwards learned was in very bad condition, & quite unsafe, & was over a deep cut through the high bank of an old rice reserve, along which the deviation of a foot from the track would have thrown the carriage off. But I did not see the holes in the bridge, nor even know that I was elevated on a bank; & therefore went coolly enough through the danger, because I did not know of its existence. At last the track reached a public road into which I turned, & drove on briskly. But still no light was seen for a long time. When one at last was seen, it proceded [sic] from a long house immediately on the road, where I asked for & obtained quarters. This house was 10 miles from my destination, & but 3½ from Lynch's Ferry, from which I had rode 18 miles. I had been riding 4½ hours, & had travelled 22 miles, of which 10 were after dark, & 7½ before reaching the public & visible road. I would have rejoiced to stop anywhere that enough food could be had for my horse & a roof & fire for myself. But poor as my host was, he found me clean & decent bedding, & tolerably comfortable; & though it was placed in a house that was built for a corn-crib, & as open & airy as cribs usually are, & the night quite cool, I slept well. One precaution only I used against the too free access of air. This was to tie a handkerchief around my head, which was the first night-cap I had worn since I was a boy.—34 miles.

1. A pecten is the fan-shaped bivalve *Pecten*, or scallop. Balani, *Balanidae*, were acorn shells.

2. *Panopaea elongata* had recently been discovered by T. A. Conrad in the Great Carolinian Bed.

34. Cattle Country of Williamsburgh (20 April–22 April)

20 APR. After an early breakfast I turned directly backward on the course by which I arrived, (& which was the same stage road on which I before first reached Lynch's creek ferry,) & travelled 10 miles to Col. D. Wilson's; the last 4 miles being across the pine woods to the edge of Black Mingo creek. Here I learned that limestone was in & just above the bed of Indian Town creek . . . above its junction with Black Mingo. Of this stone lime had been burnt & used for building many years back, & which Col. W. knew to have made excellent cement, from having pulled down the brick work of the building. Limestone had also recently been taken from the bed & sides of Black Mingo, just where it crosses the dividing line of Georgetown & Williamsburgh. Part had been burned to lime, & proved satisfactory in the burning, & part remained where brought to be burned at the place to be burned. This stone we rode to see, between 3 & 4 miles, back on my route, (leaving my own horse) as it could be examined there better than in its bed, now but little above the water. The stone had been of the upper part, & in loose masses, & was of different appearances. Some lumps were full of impressions of shells, & other compact & without shells. All were calcareous, but I doubt whether very rich. Took specimens, but, unluckily, that of the compact & seemingly the best stone was left behind by mistake.

After dinner rode on up the country, & stopped early, because of the fatigueing [sic] service of last night, at the comfortable house of Wm. Flagler. Again I am in the region of houses of "private entertainment", where almost every man when required is accustomed to entertain travellers for their money.—26 miles of driving, & 7 more on horseback to & from the limestone.

21 APR. Hearing that stone supposed to be limestone was in the bed & on the sides of Black river, I left the main road, & under guidance of young Flagler set out in pursuit of it. On reaching the house of a poor man who knew everything about the river, by his advice, & in his company, we took another direction through the woods, & rode on horseback & walked 4 miles farther to the river. Here no rock was visible, owing to the height of the water. But crossing the river in a little leaking & very "ticklish" canoe, two separate & large flat masses were found some distance higher, & these only rising about 4 feet above water. It seemed an impure limestone, showing many ostreas, & by a single small exogyna, was known to be of the secondary, & the same I believe as that before found lower down on Black river, & at Stony river.

Returning to my buggy, went on & dined near the lower bridge over Black River, & crossing which, continued my journey through the pine barrens to the Santee road for Lenud's ferry. Reached Dr. Buford's house just after dark.—32 of driving, & 8 extra.

The great business of Williamsburgh is the rearing & keeping of cattle, to sell for beef in summer & autumn at Charleston; & the business is considered very profitable, (netting 15 per cent. or more) although conducted on the abominable system of leaving all the stock to starve in winter & spring which cannot find enough food in the pine woods & bays[1] or swamps to keep them alive. The burnt wood-land already is covered with young grass, making it a perfect & continuous green, which is so beautiful as to greatly moderate my previous strong objections to this plan of burning the woods. On this young pasture the miserably poor cattle which have escaped death from starvation during the protracted & severe cold weather, are already getting their fill of grass; & they will continue to improve & become generally fat enough for beef by July or August. Then all the steers old enough & fat enough are driven to Charleston & sold; & that constitutes nearly the whole year's income from the stock. For they furnish scarcely any milk or butter (compared to their great number,) as little labor, & no manure—

their dung being dropped on the pine barrens, or in swamps, & if dried, burnt up with the leaves or dead grass the following spring. Under this management, I have heard the average loss from poverty & starvation estimated by different persons at from one-sixth to one-third of the whole number. But I presume that either must be exaggerated, unless meant to apply merely to years of mortality; for in mild & short winters but few cattle die. But there is no question that the losses by death are enormous. And in addition, there is the greater loss of all the fat & progress toward higher improvement of flesh, gained by each of the greater numbers of animals during the preceding summer. When hearing two of the smaller planters & graziers speak of their recent heavy losses of cattle, I ventured to suggest the propriety of feeding the stock during winter. My suggestion was heard with looks of astonishment, & they declared the utter impossibility of their feeding their stocks; & even if feeding as many as they could find food for, they had no idea that it could do otherwise than absorb all the profit to be made by the ordinary non-feeding plan. I did not push the matter further. For I suppose that if I had adduced the facts as proof, they would have deemed the furnishing the entire food of cattle for 3 or 4 months as done in Va. & longer in N.Y. as either fabulous, or showing an absurd & losing practice.

The Williamsburgh graziers can in this manner keep cattle to any amount, even without owning any land; or if furnishing land as part of the investment in the business, may have as much pine-land as they choose at 50 cents the acre, or swamp land for still less. And I should think, that, with such good use of marl & lime as would make the application of putrescent manure profitable, they might feed with profit during winter, if only for the manure of the cattle added to the unlimited supplies of pine leaves which may be had where the woods are not burnt over. And even by keeping salting troughs[2] on the planter's premises always accessible, either with or without food, the cattle would be attracted & furnish much manure at little cost. Though a man might not be able to find food through winter for his 500 head of cattle, yet he might, & well, for 100; & by not only preserving all the lives, but much of the previous summer's flesh of his 100 head, they would produce more of meat, milk, & butter through the summer, besides their manure. For the latter, valuable as it is, I would pen cattle only in winter, when they are better off fed & sheltered in a dry & littered pen than out of it. In summer, penning does more harm to the cattle than their manure is worth. And yet much of their manure may be as well saved in summer, if the cattle are merely induced to leave it scattered ever so widely on lands marled so as to retain the manure, & which lands will turn the manure to profit, either in products of grain or grass.

In opposition to the introduction of salting, I heard a respectable & intelligent planter say that he had tried his cattle & they would not eat salt. It may be that the taste is an acquired one; but it is very easy & certainly acquired by permitting cattle to have free access to salt.

The swamps of the Lower Santee in this neighborhood also furnish fine grazing; but they are covered so deeply & often unexpectedly by freshes, that many cattle are drowned, & therefore the swamps are considered as of little value. Here again, the salting plan would operate to bring every cow out of the swamp to the salting troughs once in every 2 or 3 days.

The price of land in Williamsburgh is as low as might be inferred from the grazing system. In my search for limestone on Black River, I was accompanied by an old but hale man, who as an overseer in different places, had had good opportunities to observe & compare, & was manifestly a strong-minded man. Having viewed the pine lands extensively, such as furnish the grazing, I asked him what such lands would sell for. It was his own large tracts we were then passing over. To my question his answer was, "Nothing. If I were to put the tract up now, I could not sell it for anything. I could not *give* it away. People will tell you that such land is worth a dollar an acre. Half a dollar is full price. There is a piece lying next to mine which has no owner & which I might take up if I chose. But I have enough such, & do not choose it, & neither does any of my neighbors." I afterwards heard of land of incalculably superior value on the Santee having abundance of the best & most accessible marl which might be bought for from $1 to 2 the acre for the dry land, the river swamp being thrown in as of no value. If I were 30 years younger, & had my present views, I would choose this or some similar part of this state in preference to any other for improvement of its product & its price by proper means of fertilization & culture.

Where I crossed Black River, pieces of rock are drawn out in hauling the seine which would be taken for calcareous but which are not so in the least. They are full of impressions of shells—but do not effervesce at all with acid. Some miles higher up the river, at Nelson's quarry the stone makes excellent mill-stones, which rarely need picking. They are full of small hollows, which keep sharp edges always. I found these also, of course, free from calcareous matter.

1. By bays, Ruffin means elliptical depressions of uncertain origin, apparently named after bay trees commonly growing therein, usually on sandy terraces; watery, and with peaty soils.

2. These troughs provided livestock with salt, generally deficient in their food.

22 APR. Rode to visit the Weeping Fountain, which was formerly described on report. It is near the road which I had travelled, & a mile back from Dr. Buford's road. It is a large & deep sink, about 30 feet deep, to the water, & about 60 feet across the top. Formerly it was nearly a circle at top. & the sides regularly steep. But one side has since slided in, so as to destroy the regular shape & much of the former beauty. The water drips in over a thick & naked & perpendicular bed of hard dark clay or shale, in horizontal thin layers, like slate, & having a good deal of sand between the layers. The water forms a permanent & deep pool in the bottom of the sink, & which is discharged by subterraneous & unseen passage. In high freshes of the Santee, this pool rises & falls with the river, but never falling below a certain height. The nearest part of the river is from 2 to 3 miles distant.

Though the Weeping Fountain & other sink-holes furnish such certain indications of marl being below, there is none visible higher (in this neighborhood) than the next plantation below Dr. Buford's, Glass, where I saw it exposed in abundance, & of excellent quality. In the afternoon passed Lenud's ferry to Dr. Palmer's.—15 miles.

I had borrowed a gun to take across the ferry, hoping to see the alligator who had so little minded the boat when I passed before. But I heard he had been killed on the same log the day after I saw him, by my companions whom I had left at Dr. Palmer's. However, I saw & shot at & struck a smaller one, but did not kill it. I heard that I had succeeded better in my previous efforts than I had before known. The first one I had shot at was found dead some days after. And the one at which I shot twice in the fountain pond at Col. Palmer's was supposed to have been killed, as he had not been seen since. He probably died in his hole, & therefore could not float. . . .

35. To Charleston and Home to Virginia (23 April–25 April)

23 APR. Set out for Dr. Huger's plantation on the Eastern Branch of Cooper river. As usual, I got lost, & went 5 or 6 miles out of my way south of the river. From misinformation & mistake, I had supposed marl there to be exposed abundantly. This was not so; but one of the ditches at the edge of the rice ground had reached & brought up lumps which I found to be of marl—mostly very hard, but some moderately soft. Reached Mr. Roper's house by dark, where my meeting with him & his lady was I believe very gratifying to all of us.—30 m.

24 APR. Before breakfast walked with Mr. Roper to see his lately discovered marl, & of the labor of digging which he complained heavily, & seemed to think difficulty enormous. The marl had been reached by a ditch, & lay only 3 or 4 feet under the surface of the level high land of the farm. Very little water oozed out, & there could not be desired a better chance for the whole operation than is here afforded, except for the hardness of the marl. But though it is quite hard, it is not stony—& what is dug up in large lumps can be beaten to smaller pieces without much difficulty. When I before visited this place, Mr. R. had no idea that his land had marl. But I was sure it must be there & accessible enough, & advised his digging for it. This he ordered a few days after; & when describing to his driver the kind of substance he wished to find, the man told him that he thought it had been reached already at the bottom of a ditch which had been dug some time before. And upon examination, the bottom of the ditch was of this rich marl, which, hard as it is, is so convenient to use, that I would desire nothing better to cover the nearest 500 acres. However, so much are the difficulties of all new operations magnified by their newness, that Mr. Roper thinks the job excessively laborious & costly.

Travelled to Charleston—30 miles. The greater part of this distance was on the state turnpike, & is decidedly the worst road, for summer, or dry weather, that I have tried in S.C. It is of deep sand, & I had to walk my horse nearly the whole way, & indeed walked myself for 4 or 5 miles to lessen the fatigue of the horse, whose load has become considerably increased by my last specimens obtained on the Peedee.

I have now completed such examinations as circumstances permitted me to make in all the sea-coast districts except Horry, & also Williamsburgh & the marl region of Marion. I would have gone to Horry, which is called the "dark corner" of the state, but for having no expectation of finding any one acquainted with or feeling interested in the objects of my explorations; & without the guidance & information of some such person, it would be hopeless for me to attempt any wild country, where marl had never been tried, & not even heard of, except by some few whom I might not happen to meet. Under these circumstances I deemed it best to postpone visiting Horry to a more favourable opportunity. And before commencing another route through the next higher lying districts, it is most convenient to make a visit to my home & family,[1] which is the object of my course since leaving the Santee yesterday. Accordingly I went to work forthwith to examine & dispose of all the specimens of marls &c. which had been sent to Charleston during my absence; & to put up to carry home such shells, marls &c., as required further examination to determine their kind or character. Besides several boxes & parcels sent by other persons, my own collections had arrived made in Edisto Island, & in Williamsburgh, while unluckily my several boxes collected & put up near Beaufort, Eutaw, Santee canal, & Waccamaw have not yet arrived. By hard work I completed all these necessary arrangements, & also answered several letters of invitation from Agricultural Societies to attend their next meetings, &c. early enough the next day (25th) to embark at 4. P.M in the steamer for Wilmington, on my journey homeward.

1. His two oldest children had married and moved out, but six daughters and two sons remained, aged from around eleven to twenty-two.

II

Early Summer

MIDDLE COUNTRY

36. Back in Charleston and on Cooper River (10 May–12 May)

10 MAY. Returned to Charleston. Arrived before breakfast. Engaged laboriously all day in analyzing specimens of marl & limestone, & packing those designed to be sent to Columbia for the state cabinet.

11 MAY. Continued my labors through the day, & wrote letters until late at night. Was assisted in analyzing today by Dr. J. L. Smith, a young man of talent & much chemical knowledge, & who has invented a very convenient apparatus for analyzing marls, with which we operated. . . .

12 MAY. Rose early to complete the packing of my baggage, apparatus for analyzing &c, & at 8 set out again, in my buggy, for the Santee, to complete imperfect observations, & to explore the upper part of the marl region not yet visited. The day very hot.

The state road, which was so bad with deep sand when I last travelled on it, was now excellent, owing to a very heavy rain which fell on the 9th. Reached Dr. Sandford [sic] Barker's on Cooper river (30 miles) to dinner. After dinner we rode together to Steep bluff, 7 miles distant, which I was anxious to see again, as when before visited, it was hard frozen, & very unfit for observation. We measured the height of the cliff of marl, & found it 25 feet above high water mark, or 31 to level of lowest tide. Except one horizontal layer of stony hardness, about 18 inches thick, the balance is of compact, homogeneous marl, not so hard as to be very difficult to dig. There were seen no fossils in the whole exposure, except some few of very small & thin & fragile pectens, smooth & white, in the stony layer. The marl is exposed to view in part, &

nearly perpendicular for some 80 yards along the river. . . . I was disappointed in not finding fossils at this extensive exposure. . . . Very few shells or casts are to be seen in any of the marls washed & exposed by Cooper river. Our return to Dr. Barker's made 44 miles.

At the next plantation, Mulberry, I passed by where large excavations of marl were making, & at which laborers had been regularly employed since my first visit to the place. But not a bushel had been used for manure. It was used to heighten the banks of the rice land of Mr. Milliken, the proprietor. Yet he was one of those who promised me to try marl this season. The quantity he has since then put upon his banks, if applied as manure, would probably have manured 100 acres. I learn that only five persons of those whom I engaged to do so, have made trials of marl. These are Dr. S. Barker, Mr. [blank] Simons, (Lewisfield,) Mr. Ford, Mr. Roper, & Col. Ferguson—& the largest application is but to 4 acres.

The residents hereabout are beginning to move to their summer retreats, & in a few days most of them will have moved. Very few will not have moved by the end of next week. This is the most forcible instruction to me that it would be unsafe for me to remain in this region longer than the proprietors dare & yet I would gladly continue examining the marl region more fully, & especially as the water may now be expected to be moderately low, for the first time this year.

[A letter of this date to the *Charleston Mercury* is pasted into another part of the diary.]

St. John's Berkley, May 12, 1843

Messrs. Editors—Having finished analyzing nearly all the specimens of marls (and limestone) which I had selected from the various localities recently visited and examined, the results, as stated below, are offered for publication, for the information of all concerned. It is proper to state, that the operations of my last day of these labors were much facilitated and forwarded by the aid of Dr. J. Lawrence Smith, of Charleston, with the use of the very convenient and cheap apparatus for analyzing marls which he has invented, and of which I trust he will publish a description.

Edmund Ruffin

Agricultural Surveyor of South Carolina

Marls &c. from the borders of Ashley River

[*15 specimens cited*]

Marls from borders and neighborhood of Cooper River

[25 *specimens cited*]

Marl (and Limestone) from Santee River and Neighborhood

[13 *specimens cited*]

Marls and Limestones from Peedee River, (Marion)

[20 *specimens cited*]

The above Peedee Marls belong to the secondary formation, and have alternating layers of limestone.

Marl of latest formation, post-pliocene,[1] found near the Sea Coast

[12 *specimens cited*]

All the specimens of Marl from and near the Ashley, Cooper and Santee Rivers, without question are from the same connected and continuous great bed; and which has a horizontal extension within South Carolina, far greater than I have yet traced. The enormous thickness of this bed may be inferred from the interesting fact that the deep boring for water in the City of Charleston in 1824, to 335 feet, was for 189 feet, certainly, and most probably for 242 feet, through this same body of Marl. This may be clearly established by examining the numerous specimens of the earth drawn up by the auger, which have been preserved in the Medical College.

1. It may be Quaternary, which is divided into Pleistocene, 1,000,000–10,000 B.C., and Holocene, 10,000–present.

37. Renewed Eutaw, Santee, and Cooper Investigations (13 May–15 May)

13 MAY. Dr. Barker has agreed to accompany me on part of my journey; & after early breakfast we set out for the neighborhood of the Santee. Passed along routes formerly travelled, by the plantation of Mr. Cain, (whom I called to see for a few minutes,) & by Philip Porcher's, whose labor I would gladly have stopped to see, but had not time. I however met his brother Isaac, who told me that he had marled 7 or 8 acres, since I lately left him, & Philip had marled 13 acres. This is doing well, for such late beginnings, & under unusual difficulties of excavation. Reached Mr. James Gaillard's house for dinner.

Rode in the afternoon to re-examine the marl at Rocks Creek mill pond, & to make sure of the fossils. . . .

In returning by part of the land marled by Mr. Gaillard, in 1837, which is now in corn, & which he thought still had shown no benefit, I had the satisfaction of tracing the outlines of a marled acre distinctly, & of having Mr. G. as well as Dr Barker also satisfied that the improvement was obvious & great. Yet this application was only of 50 to 60 bushels to the acre. The land has been cultivated every year since, (& long before,) & manured nearly every year, with woods litter passed through the cow pens.—36 miles.

14 MAY. Sunday. Attended church, below Rocks creek. In the evening examined the bank of Dawshee cove, which the height of water had covered when I visited it before. The river is now moderately low, though not near its lowest state—not by 5 or 6 feet. The ordinary freshes rise 6½ feet above the present height, & the extraordinary 4 feet higher still. When such as the latter occur, everything below is covered, except Porcher's embankment, & that is barely safe. The rice lands & their embankments low down the river are all inundated by the highest freshes.

By getting in a canoe, I was enabled to see the bank well. The marl on limestone is continuous, for about 200 yards. Compact limestone & pretty hard marl are in alternate strata, & not always regular. . . .—13 miles.

15 MAY. After breakfast to Mr. Sinkler's, Eutaw. On the way, stopped & collected many specimens of fossils from the limestone which rises to the surface in Mr. Gaillard's field, & of which much had been quarried & remained in heaps. By obtaining many specimens of the limestone & of each variety & elevation of marl, it is my object to clear up the obscurity & confusion existing among the geologists & conchologists in regard to this great bed. Morton Conrad & others call it the upper cretaceous, belonging to the secondary, & Lyell deems it certainly the eocene, & tertiary.[1] I did not before take care, as I now do, to keep distinct the fossils of each locality & elevation, where differences of kind exist. . . .

Proceeded immediately to the river, at Mr. Sinkler's landing, & was paddled in a canoe up about 2 miles. Midway of this distance is Nelson's ferry, so famous as a locality of shells, as stated by conchologists—& where I found none, except a few imperfect o. sel., & am very sure that none were found by others. Indeed the marl is visible there for but a very short distance along the bank.

Along the river bank of Mr. Sinkler's land, & above so far as seen, the general character is the same. The limestone is in separate masses, of all sizes from an inch or two in diameter to 3 or 4 feet—sometimes in contact with each other, or with marl below, & sometimes separate. . . . The marl all along has greensand very apparent—& is unusually hard, though its texture is very different, & easily distinguishable from the limestone. Indeed the clear ringing of the hatchet, when struck on the limestone, & the duller sound on the marl can alone serve to distinguish them to a person who did not see them. Obtained not a specimen of fossils. None seen but o. sellæformis.

Afternoon visited the Blue Spring, a very deep limestone fountain in Eutaw creek, & Eutaw spring. In examining on both sides of the high ridge under which the stream of the latter passes, I found coarse granular limestone at bottom & top (full of o. sellæformis at top, & moderately hard marl between).—12 miles. . . .

Nearly all the planters in this neighborhood, are suffering greatly under the effects of drought, notwithstanding their partial supply of the late rain, which was so abundant in Charleston. The cotton was planted late, owing to the cold spring, & probably most of it when the ground was too wet; & so little has come up yet, that entire failure of plantings are feared. It is too late for the black seed cotton to do well after this, even if the seeds yet come up. The fine soil of & adjacent to Eutaw, formerly described, present exceptions. On this land, the cotton has come up well.

This fine & peculiar soil, formerly was all in one possession, & could not be sold at $1 the acre. The great objection to it then was the limestone water, which (besides its other great defects for use,) is unfit to prepare indigo, which was then the universal crop for market. It was long before the land gradually rose, & was properly appreciated. Though its peculiar & great value is certainly derived from some natural admixture of lime, there is not enough, & I saw sorrel growing within a few feet of a protruding mass of hard limestone near the Eutaw spring. By the way—the sorrel of this state is certainly a different species from that of Va—though I presume is alike a correct indication of acid land. This has less leaf & more stem than ours—grows commonly to 18 & sometimes 24 inches high—& the tops, now in seed, are of different tints from pale red, to yellowish & almost white, instead of the general dingy brownish red of our low growing sorrel.

Not having time when I was in the neighborhood before, I had requested of Mr. H. W. Ravenel to make more full observations on two points, which he has done, & reported as in the annexed extract of his letter. I had found marl near the surface of Biggin swamp & its branches wherever I bored, or saw exposures, & thence inferred that such was the case generally with the head swamps of Cooper & Ashley. I had heard an intelligent & scientific gentleman assert that the soil of Biggin swamp would effervesce with acid. This I disbelieved; & my opinions on both these points are sustained by Mr. Ravenel's observations, as well as my own of more limited extent.

[Ravenel writes:]

> "With respect to the general position of the calcareous stratum under our swamps, I have observed in five of the swamps of the neighbourhood, all branches of Biggin, that the calcareous stratum was invariably found at from 2 to 6 ft under the surface.
>
> I have bored in a great many places in order to find a favourable locality for opening marl pits, & I have never failed to find it at the distances above mentioned, from the surface. In two or three instances I found it too hard for the auger to penetrate, but generally the auger would penetrate several inches showing it to be a soft marl. I have almost invariably found a greenish, soapy clay overlying the marl.
>
> I found the marl bed in the Wassamasaw quite near the surface, in one place above the water level of the stream & only a few inches below the surface of the swamp.
>
> With respect to the presence of Carb. of Lime in any of the soils of Biggin swamp, I have examined 17 specimens selected by myself, from 1 to 12 inches under the surface taken from several of the branches of the Biggin & some from the Wassamasaw. In one only have I been able to detect the slightest effervescence when treated with Muriatic acid, & that was taken about one inch below the surface, from the swamp near the road leading from Mr. Cain to Black Oak. It is possible that a small portion of marl may have become incorporated with it accidentally. I will procure specimens from the adjoining parts of the swamp, & test them more fully."

1. The question is of the Great Carolinian Bed's age, as determined by fossil evidence; see Introduction, also 8, above.

38. "Magnificent & Beautiful Exposures of Marl" at Vance's Ferry (16 May)

16 MAY. Left Mr. Sinkler's & proceeded to Vance's ferry. Mr. Wm. Avinger, who resides adjacent, was so obliging as to paddle me in a canoe 4 miles down the river, & back, which took me to the lowest exposure of marl above the point I reached yesterday. After returning & dining at the tavern, I proceeded to Mr. Keating Simons', two miles above Vance's ferry, & we examined from the highest exposure I can hear of on the river hereabout, to Vance's ferry, whence I returned to sleep at Mr. Simons' house.

The water excursions of today have enabled me to see the most extensive, magnificent & beautiful exposures of marl that I have ever seen or heard of before. For four miles below & about 1½ above Vance's Ferry there is a continuous hillside (except where depressions where a few streams enter,) of about 40 feet high, very steep always, often perpendicular, & sometimes the top even overhanging the general face of the cliff. The river washes against this high land except for about a mile of the distance, when a narrow margin of swamp lies between the river & the cliff, which however was often near enough to be seen, & the marl visible, from our canoe. This swamp is no more than about 60 yards wide. The marl is exposed from 20 to 30 feet high all along the face of the cliff, wherever it presents a face of that height. . . . It is said to be visible down to the lowest level of the river, which is 6 or 7 feet lower than now. Though on the face of the hillside the marl is everywhere near to the surface, it is mostly veiled from sight. . . .

The upper part of the marl . . . is of compact but moderately soft texture, homogeneous in appearance, & containing very few fossils. The only shells in preservation are ostrea sellæformis, always of small size, & a few casts of other shells. Below this is a limestone, 5 or 6 feet thick full of large o. sellæformis. Below this the marl to the river again is nearly as soft & similar to the upper stratum, except containing these large shells. . . .

The texture of the marl . . . is different from the universal appearance of the marls of the lower Santee. These are much like the marls of Cooper river. And the limestone is also different. Hard as this is, it is much more coarse-grained than that seen yesterday & all below.

The marl & limestone shows at the surface in parts of all the plantations which this bluff borders. On Mr. Simons' land (which is the most hilly I have seen in S.C.) the marl shows in sundry places near the tops of hillsides of considerable elevation; & the hard limestone (coarse grained) is sometimes in separate masses, exposed at the surface at the top of the marl, as well as elsewhere seen lower in the bed.

In the cliff just below the ferry there is a low cavern open to the river, through which is discharged the stream which is seen to pass under the marl in the rear. . . . Into this the canoe entered its length. Another cavern, still larger, opens a mile or so lower down. Both were bottomed with the deep slimy mud left by the river, so as to forbid my attempting to penetrate them farther.—28 miles, by land & water, though only 11 on my journey.

39. Along the Santee in Orangeburgh (17 May–20 May)

17 MAY. Accompanied by Mr. Simons for a few miles on my way up along the river lands. Every plantation is well supplied with marl. . . . Left the public road, & by private roads, & through plantations in part without roads, visited Pinckneys mill, Felder's marl beds & the sites of Hales' Mill. Along Pinckneys mill-pond, for about a mile on both sides, I was told the marl showed in abundance. Near the dam, the softer upper stratum, as on the lands next below, was naked about 16 feet high above the pond; & when that is dry, the marl shows to its bottom, about 10 feet lower. . . .

At the mill, the section of marl is exposed, & shows the surface to be very irregular. . . .

At the deserted plantation of Adam Felder, about a mile higher, the marl stretches along the steep hill side next the river swamp. Here is another change of quality or appearance. The marl is nearly white when dry, of texture approaching to hard chalk. . . . The proprietor burnt lime here for sale for several years, & disposed of it readily at 25 cents the bushel, at the kiln, & at $2 & sometimes $3 the cask (old flour barrels, holding 3¼ bushels) in Columbia. This continued until some

7 years ago, when he thought it best to move to Alabama, since when the lime-burning has been abandoned & also his plantation. The lime kiln remains, nearly as good as ever. It is dug in the marl in the hill side & bricked. The outlet is dug through the marl, & stands as well as when first made. It is said that the plan answered well. The kiln might have been burnt perpetually—but was generally cleared at each filling. . . .

At Hales' old mill, the next exposure seen of marl above. The upper part is like Felders, & without fossils for 11 feet, then 1 foot limestone with O. Sel. & below, to & under water line. . . .

Continuing up the course of the river by the public road to McCord's ferry, the next point for observation reached was at Cavehall, belonging to Dr. Cheves, but not resided on or cultivated. I had heard of a remarkable cavern here, & with great difficulty was enabled to obtain a guide to it. For since leaving Pinckney's Mill, & my companions at Felder's lime kiln, I had not met any person on the road, & had seen only 3 houses, all of which I went to, & found only women & children, who could give no information. The almost absolute solitude on this & some other public roads is surprising—& a great cause of difficulty to my exploring. Yet this seems to be good land, & there must be much cultivation somewhere about. . . .

The cavern I found about 1½ miles off the road. It is in the side of a high & steep hill-side, of which the lower 20 feet . . . is marl to the surface. In this, at bottom is the entrance to the cavern, about 18 feet wide & 10 feet high, forming an irregular arch. Within, the length from the entrance to the visible extremity . . . appeared . . . 75 or 80 feet. The roof is nearly horizontal, though rugged in surface. The floor is covered & concealed by coarse & loose reddish sand, which has been brought in from a sink-hole at some considerable distance, through which the waters of heavy rains enter, & are discharged through the cavern, bringing along & leaving in it a great accumulation of sand. . . . There is also a small opening downward, as I was told afterwards, which no one has explored, though it probably reaches water within a few feet. This I did not find. My guide was a very ignorant negro boy, from whom I could learn nothing. . . .

When coming out of the cavern & climbing the hill to my waggon, I found a party of huntsmen who had just arrived, among them Mr. Christopher Hampton, with whom I had become acquainted in Charleston, & who resides in this neighborhood. Upon his invitation, I postponed to the next day my intended going to Col. McCord's house, & went with him to the residence of Mrs. Richardson, his mother-in-law, a few miles on my road from Cavehall.

After dining, rode to the river swamp to examine the high hill side exposed by the road. It shows nothing of marl.

Half-way swamp, about 2 miles above Cavehall, has marl all along its hill sides, below, or southward, & none on the other side, though the hills are as high, & present no difference of configuration. Hence has arisen the prevailing opinion that this stream is the limit of the marl & limestone formation. But this is not so, as I learned from Dr. J. R. Cheves.—26 miles.

18 MAY. A heavy rain falling early in the morning, which is most acceptable to all this region. Before 9, the rain had ceased, & I proceeded, with Dr. Cheves & Mr. Hampton to make examinations. As the weather was still threatening we went in a carriage, & were provided with other safeguards which proved necessary, as the day continued cloudy & showery. . . .

On the banks of a small stream running to Stout's Creek from the upper side, & about 3½ miles above Half-way Swamp, where nearest, we saw the exposure of marl of which Dr. Cheves informed me yesterday, & which is especially interesting as being the highest known deposite up the course of the Santee. It is exposed 4 feet high above the stream. The upper part is a rich & soft granular marl, containing a large proportion of green sand. . . .

Crossed Half-way Swamp to Heatley Hall, another of the plantations of Col. Edw. Richardson's estate. The marl rises high on the sloping hillsides. It is similar to that at Cavehall & Felder's, & has very few fossils. . . .

Near Stout's creek, where the public road crosses, there is a bed of hard sandy clay, which is used in lieu of stone for hearths & sometimes for other parts of building. It is moderately soft when dug, but hardens by exposure. . . .

It continued overcast & with slight showers or drizzle all day. After dinner, proceeded to Col. D. McCord's house, 2 miles from the river.

19 MAY. A cloudy & cold day, with slow rain at intervals which prevented the making a designed excursion across the country. It had

been planned that this & the following day should be so occupied, & that then returning here, & leaving my horse, I should proceed to Columbia by the rail road on the 21st. The state convention assembles there on the next day, & as many planters will be there from all parts of the state, it will be a favorable opportunity for me to have personal conference with such individuals among them as may desire it.

The weather permitted me only to see, & under great disadvantage a very remarkable deposite of fossils on this place, & of which Col. McCord had shown me some perfect specimens of his previous collecting. Several enormous ravines, 30 to 40 feet in depth have been made in depressions by the washing of rain floods. . . . At some 20 feet below the surface the fossils referred to are found in a thick stratum of indurated[1] clay. . . . The hardness of the matrix makes it difficult & very tedious to cut out perfect specimens—& my being alone, & in the rain, with the adhesiveness of the wet clay, made my collection very poor. But among those previously collected by Col. McCord, & of which he has given me a part, there are beautiful & perfect specimens, of several shells, & mostly, of a pholas,[2] which appears to be precisely that of the present time & neighboring coast. This always very delicate shell is preserved by its being filled by the indurated clay; & each was also surrounded by a covering, making a perfect cylinder, which was the cast of the hole bored by the living shell-fish. These crystallized & transparent shells, though completely changed in their substance from what they were, preserve every part of their original delicate & beautiful configuration. Even the small prickles which stand out from the outer parts of the pholas, are sharp as in the recent & perfect shells of the living animals.

1. Hardened.

2. See p. 190, n. 1.

20 MAY. Still cloudy until evening, & cold throughout. Rode with Col. McCord through True Blue, Mr. Singleton's plantation, to Totness, a summer settlement, & returned to dine with Dr. Thomas Stark on the Santee, & thence back to Col. McCord's.

I had heard formerly from Dr. Ravenel, & from others since my arrival there, of shells found in rocks at Mr. Singleton's place, which I went there to examine. They are in fact found in many places in this neighborhood, above the limits of the marl. A rich supply was found at Mr. Singleton's, but in rock recently hauled there for building from the adjacent plantation (Gates'). It is a coarse sandstone, containing numerous shells, mostly small, but perfectly preserved. They are like the rock, entirely non-calcareous. The shells, like those found yesterday, are silicified & mostly also crystallized. . . .

These specimens are beautiful & abundant, though difficult to extract entire; & I regretted that our arrangements for the day did not permit my remaining more than two hours to break stones & gather specimens.

In passing to Dr. Stark's (Belville) on the Congaree, we passed by the site of Fort Motte, so noted in the revolutionary war. The entrenchments are levelled, & barely to be traced; but the mound thrown up whence to fire on the fort, & from which the arrow was shot to fire the house, still is left.[1]

The banks of the Santee at Fort Motte, & some distance below, are very high, & as precipitous as any can be not formed of rock wholly or in part. The views from the tops of these banks are magnificent, & would appear to be of mountain scenery. The crooked & narrow Congaree is seen in the distance as well as almost immediately under the observer. The bordering swamp, covered with a dense forest of tall trees, lies spread out below for many a mile, offering no obstruction to the view from so high a point, & showing an expanse of the richest verdure. The High Hills of Santee are seen high above & far distant, appearing like mountains compared to the low-lying swamp forest.—21 miles.

1. Fort Motte, a loyalist depot, and the house of Rebecca Motte, were attacked by a force under Francis Marion and Charles Lee in May 1781, using burning arrows. Motte supplied the bow and arrows herself and afterward entertained the opposing officers at dinner.

40. Excursion to Columbia (21 May–24 May)

21 MAY. At 2 o'clock P.M. entered the train on the railway, which passes within 2 miles of Col. McCord's, & soon reached Columbia 30 miles distant. The Santee loses its name in those of the Congaree & Wateree, which rivers unite a few miles below Fort Motte. The railway from near Fort Motte crosses the Congaree & its swamp of three miles width. . . .—32 miles

I remained in Columbia until the morning of Wednesday the 24th, when I returned to Col. McCord's, & immediately resumed my journey. During my stay in Columbia, I met with some twenty intelligent or zealous planters from different parts of the state, who sought acquaintance & conversation with me—& by which I trust that I have received profitable information, as well as made useful impressions that will yield good returns. All these persons were very solicitous to have me in their neighborhoods as soon as possible, & as their guests for the time. I also met a greater number of acquaintances made previously in the lower districts; & heard most gratifying accounts of the interest excited even where I had thought there was the least—& of the extension of discoveries of marl far beyond such as I could make, & merely begin in person, & indicate the course to prosecute them. The post-pliocene, which until I discovered it in Edisto Island & Distant Island, has remained totally unnoticed, has been found in many places near the coast, & in great abundance; & especially for an extent of 5 miles under all Doctor's Swamp in John's Island. I also learned that applications of marl, of more or less size (but none large, as the late time of commencing forbade,) have been made by many persons in every place I had labored to introduce the practice. Indeed, the most "wise in their own conceit" are already crying out against & ridiculing what they call the "mania" that has been produced; & this I am glad to hear, taking the reproach as one of the best evidences of success.

41. Between the Santee and the Edisto in Orangeburgh (24 May–27 May)

MAY 24. Rode to Gates' Quarry, accompanied by Col. McCord & Dr. Darby. The bed is of shell-rock entirely, though some parts containing more shells than others—nearly all hard stone, but a little, at surface, earthy, as if disintegrated by action of the air &c. The shells, as before described very perfect. The mass entirely without calcareous matter. . . .

Next to the deep cut for the passage of the railway through True Blue. The upper part, for 10 feet or more the ordinary red subsoil of this neighborhood, changing to sand as descending. At bottom, a perfectly white clay—which was one of the many kinds of earth which have been erroneously called marl. Across a deep ravine, & in the next rise of high land, the excavation for the road-way exposed a silicified tree, nearly 3 feet in diameter. . . . Returned to spend the night with Dr. Darby.—44 miles.

25 MAY. Proceeded for the central & western side of Orangeburgh. Mr. Starke goes on horse back, & his present indisposition will make me exchange my easier place in the buggy for his horse nearly all of each day's journey.

I had designed to pass down the course of the Four Holes swamp & then ascend that of the North Edisto, provided I could on the route find any precise information of localities of marl, or obtain guides through successive parts. On nearing the swamp, I met a man who luckily proved to be one of the persons I had been directed to seek, Mr. Th. Zimmerman. He returned with us as far as his plantation, where marl in small quantity had been dug, for burning for cement only. . . . This spot is near the edge of the Four Holes Swamp on its eastern side, & 3 or four below its head. It seemed useless to pass farther down its course, as Mr. Z. had never heard of limestone or marl lower down; & when I formerly crossed it below at the bridges, near neighbors knew nothing of it. However, it is not the less certain that it is along the swamp generally, judging from the general feature of holes, or sinks, as well as from other report.

Returning to the road for Orangeburgh C.H. crossed the numerous small streams which form the head waters of the Four Holes Swamp, & reached the long established & noted lime kiln of Dr. Jamison formerly, & now the property of John A. Tyler. It is by the side of the road, & west side of the upper part of the swamp. This business of burning lime was commenced by Dr. Jamison many years ago, to supply lime to the indigo planters, & was carried on to large extent; but latterly there is little demand for indigo making, & the lime burning has almost ceased entirely. The lime was not usually used for cement (though proved by enough practice to be excellent) & never for manure. A large quantity of the material left at the kiln, furnished full choice of specimens. It is generally & mostly a compact marl, such as usual on & near the upper Santee, but soft enough to be dug & used for manure without burning. A small portion much harder. The marl lies beneath the surface of the earth. There are old pits, now full of water, showing continuous & extensive digging down the swamp, which I walked along for a considerable distance, & heard afterwards extended a mile. The only remaining fossil seen, is ostrea sellæformis, generally very large, but always broken as if by compression. Numerous casts of sundry other shells, but very imperfect.

From this place, (after eating our road-side dinner, brought along) we struck through the pineland, & by private paths, for the Cawcaw swamp, a branch of N. Edisto, on which I had heard of limestone. By good luck we first reached Manuel Pooser's plantation; & with the guidance of an ignorant little negro boy, we found an old & long abandoned digging near the swamp. This marl also lay too deep to be seen, though specimens were found around the pit. In all my inquiries made in the neighborhood afterwards, I met with no one who had ever heard of lime on this place: which I therefore found only by chance & the inquiries made of negroes.

Proceeding 2 miles higher up the swamp, we reached it again at the old mill of J. Rumph. A single large mass of stone lay on the road side, containing some marks of shells, but which is not calcareous. I afterwards learned that this was brought, (for building a culvert under the rail-road,) from a quarry a mile above & that it was understood to be lime-stone. I also found whitish stones brought up & exposed by the breaking of the mill dam, & the washing away of the earth below, which showed some impressions of shells, but which were also non-calcareous. Deeming these to be indications that I was beyond the limit, or upper line of the accessible marl, turned back. On the public road from Orangeburgh to Columbia, & 5 or 6 miles from the former, in ascending a red hill of considerable length & height, I saw extending across the road a broad belt of what appeared to the first glance the ordinary hard marl, but which I was sure could not be, as its underlying stratum was visible, & was the red subsoil of the neighborhood. Upon closer examination, the resemblance was even stronger. The color & texture was not only much like marl, but a part was found full of remarkably perfect casts of shells. Still not a particle of calcareous matter was remaining. It struck me that this earth must be reported to be limestone—& so I found it was. And I guess that this is the locality which has been referred to so often, of "6 miles above Orangeburgh", & even in books as of the most noted presence of limestone thereabout.

Reached the small & pretty village of Orangeburgh, about 6, & took quarters at the tavern. Found letters from home.

Gen. D. Jamison soon called on me, & arranged an excursion for the next morning. I soon was sure that the expectation of seeing extensive exposures of marl in this neighborhood had been founded on false or ignorant report.—31 miles.

The land of Orangeburgh district, as seen from Vance's Ferry to some miles above the railroad near the Santee, & thence to the village of Orangeburgh, is generally much better than most of the high land of the lower country. The marl, (with one exception before stated,) is found no further than Halfway Swamp. Soon after begins a large body of singular red land, very much in color like the red lands of Goochland & Albemarle in Va, & much better in quality than the ordinary gray lands. It is partly red at first; but on approaching nearly as high as the junction of the Congaree & Wateree, all the land is red. Still this is but narrow near the rivers—but widens backward to the head sources of the Four holes Swamp & to the railroad. All this red soil is high, & with its extent of table land, has much of long & gentle hill-sides, & of undulating surface. The soil is often gray & the subsoil red. The red land, though called clay, is not so in fact, though more clayey than the gray lands. It is open & loose enough to let surface water sink rapidly, & to dry soon after heavy rains. Oak & hickory formed a large portion of the original growth. It is in this red land that the curious & perfectly silicified shell rock was always found; & it seems that the existence of that formation & this singular land are in some manner connected.

26 MAY. Rode to Gen. Jamison's plantation on Cawcaw, & to some other places in the neighborhood. The only spot of marl to be seen was at Wannamaker's, at the edge of Cawcaw swamp, where it had formerly been dug to burn lime for sale, & which was carried on to but small extent, & like Pooser's Kiln had been long abandoned. . . .

I learned from Gen. Jamison, & from his brother more of their father's limestone & its use, that [sic] I could otherwise have found out. The stratum was reached no where nearer than several feet below the surface of the earth. Excavations had been made from the kiln, to a mile below along the side of the swamp. . . . All this lime burning was to supply the indigo makers of this district, as most planters of that crop required & used 20 or 30 bushels of lime in the preparation of the article each year. And it is surprising, that all this lime flung away every year, often being used to prepare indigo, did not even by chance indicate to some of the planters its value as manure. For the latter purpose, not a particle has ever been used in all this part of the district, except two very small applications, very improperly made, (with quick-lime put in the hills with the seed,) & from which no effect was seen at first, & probably never looked for afterwards.

Indigo, once the almost sole sale crop of S.C. has long been abandoned every where except in Orangeburgh district; & until this year it was still made by many of the middling planters, on the poorer lands. The crop will grow on poor land. When the prepared indigo sold for $1 the pound, which formerly was the ordinary price, there could be made sometimes (I suppose under best circumstances of product,) $400 to the hand. Since the begining [sic] of the revolutionary war, & the production of better indigo in India, the price has been so low that it was abandoned universally in S.C., except as stated in part of this district. But even here it will soon disappear, as there is scarcely any sale for the article & some planters now have their two last crops on hand.

The indigo prepared, & latterly cultivated, is not the exotic & annual plant formerly introduced, & in general use, but a native wild indigo, which is perennial, & needs new sowing of seed only once in 6 or 7 years.

I have been so often led astray by false accounts of marl exposures, & have so often found the most entire & generally prevailing ignorance & carelessness in regard to them, that I have ceased to be surprised at either. But this particular neighborhood & this day's excursion have presented stronger instances than have been lately noticed. Every supposed & undoubted body of marl, or supposed indication of its existence (with the few exceptions mentioned,) had but to be seen, to be dissipated. In every case, the application of a few drops of acid, or the digging for half an hour, would have removed all question, & proved without my aid, that the belief existing was groundless. But in no case had either of these means, or any other trouble, been taken to test the facts.

After dining with General Jamison, which made the departure later than desirable, set out at 3 P.M. for Mr. Jacob Stroman's, on South Edisto, where there was reported to be some appearance of shells, & which at any rate would enable me to strike above the upper limit of the calcareous formation. After crossing the North Edisto near to Orangeburgh, where it is a narrow but bold stream, furnishing good navigation for rafts of lumber, our road passed through pine forest for the whole ride, with the exception of a few scattered & small spaces under cultivation. Though the land is much better than the pineland in general of the lower districts, indeed pretty good, the only value of this, as estimated & used, is for its timber, of which the supply is almost inexhaustible, & the amount of the annual products very great.

Night was approaching while yet a considerable distance of our journey remained, & we should certainly have been lost, but for the kindness of a resident of the neighborhood, who rode along with us, & went out of his way to guide us. We reached Mr. Stroman's house, not much short of 8 o'clock, & found a most kind & hearty welcome. From Orangeburgh 22 miles, & with the mornings exploration.—33 miles.

27 MAY. Examined the shelly deposite at Mr. Stroman's saw-mill on Rocky Creek. . . . The application of acid showed however that not a particle of calcareous matter remained. . . . The silicious character . . . was only what I expected . . . in a belt of several miles width lying partly over & always beyond the upper limit of the calcareous formation. But it was a particular subject of regret to me that this should not be marl, because of the great interest & anxiety felt by the proprietor. And thus it generally happens. Those persons who are out of reach of marl are most earnest to use it, & willing to incur great expense for the object; while the greater number of those who have it the most abundant & cheap to apply, are careless about it, & think the smallest expense too great for its use.

Every stream of any size in this region is dammed to turn a saw-mill, & immense quantities of lumber are sent from them in rafts down both the North & South Edisto. The rafts are made of boards (or other timber) clamped together so as to form a close mass 20 feet long, 10 wide, & about 18 inches deep. When a sufficient quantity of timber has been prepared at one mill to send to market, a number of these single rafts are constructed, & put into & confined in the stream. When all are ready, they are started together, & the floating aided by letting loose water from the mill pond. The rafts float on with only one man to attend to each 8—& often half of his charge are out of sight of the conductor. He uses one of the rafts as his conveyance, & shifts from one to another as his aid is required, keeping always on the hindmost of his squadron. The current of the narrow river carries the rafts down quite fast enough. But they often are stopped by getting aground on sand bars, or on snags.[1] The first raft that thus grounds is but slightly attached, & would be easily removed. But often it is struck & forced farther upon the obstacle by another, and that again by another, until a number of them are fixed, & require the united force of several hands to heave them off with hand-spikes. A raft usually floats with its surface a few inches above the water. The one bearing the conductor, being kept clear of all obstructions, easily overtakes those ahead, & he guides it with his pole to any other one that requires aid, & shifts his position to the latter. At a point down the river, lower or higher on the river according to the height or low state of the water, it is necessary to collect these loose rafts, & put them under more strict guidance. A hand goes ahead in a boat, & draws a large rope across the river, tightly at or near the surface. This stops all the rafts, & each man binds his eight together in one, which he thus conducts by ebb tides to Charleston. This operation is sometimes delayed too long, so that some of the loose single rafts are lost, & swept out to sea. It must be an exciting scene to witness & accompany for a time, a large number of these rafts, before they are connected.

Taking the road down the river, we reached Johnson's Bridge over the South Edisto, as far as which we were guided & accompanied by Mr. Stroman. He had known that an ineffectual attempt had formerly been made to burn lime of a rock thereabout; but it was not known yet whether the failure was because the earth was *not* calcareous, or for other reasons. At the bridge the stratum sought was found to be calcareous. . . . It seems to the eye to be much inferior in quality to the marls on the more eastern rivers. Mr. Stroman knows that this same stratum extends full half a mile higher up the river. It is not easily accessible except by a boat; & I have requested him hereafter to mark & inform me of the highest exposure.

1. Snags are sharp pieces of broken branch or trunk.

42. Barnwell District: Springs, Marl, and a Rattlesnake (27 May–4 June)

Crossing this bridge into Barnwell District, we proceeded down the river along the public road. The plantations are small & very thickly placed, on both sides of the road to Midway, on the railroad. Here Mr. Starke left his horse, which had become lame & could be barely got along in a slow walk for the last 7 miles. This mishap will compel Mr. S. to go tomorrow to his father's residence near Hamburg, by railway, taking the lame horse along, to provide himself with a proper conveyance & meet me at Silver Bluff, Gov. Hammond's residence, on June 7th.

Continued downward to Mr. N. Roach's, upon insufficient information of calcareous rock, & to no purpose; whence I resumed & went to the house of Dr. B. H. Sweat's, to whom, among sundry others, I had a letter of introduction. This gentleman's kind & efficient attentions, & the zealous interest which he took in my pursuits, made it a matter of congratulation that I had directed my course to his house.—28 miles.

The last few days have been very hot & oppressive, & I have been warned by sundry residents' words, as well as by the general removals already made of families to their summer retreats, that it was full time for me to leave this low country. But I do not entertain much fear—& am anxious to see as much as possible of the marl before I leave the region. . . .

28 MAY. Sunday. Mr. Starke set out homeward. Accompanied Dr. Sweat to the Baptist Meeting House. There were not more than 25 white persons in the congregation, & 50 to 60 negroes. The latter was a pleasing circumstance not often witnessed. Returned to Major A. Patterson's. After dinner, rode over his low grounds on the river, in cotton & corn, & examined, in vain, for marl on the river side. Indeed it could not be otherwise, all the land bordering on the river here being alluvial.

This evening felt somewhat indisposed, which admonished me not to go much lower down the country. There has prevailed an unusual drought, & the earth is extremely dry, & the crops suffering for rain. Much of the cotton hereabout is not yet up; & very few fields of corn manage 10 inches high. The smaller swamps & slashes in the high woods are dry, & smell very offensively in my passing by them.—10 miles.

29 MAY. With Dr. Sweat & Major Patterson went down the river side as far as Stokes' (or Walker's) bridge, where I had crossed in

March. The rock, then covered by the fresh, is now 2 feet above water, & was found by the auger 8 feet below it. It is calcareous; but obviously poor. . . . The same stratum, & nearly similar, was seen 2 miles higher, adjoining the river swamp, on the land of George Tucker. There are signs there of its having been, in former time, dug & burnt for lime. . . .

Learned from Capt. May, who was present, that this same exhibition of rock is at his plantation, & at Applebee's, 6½ miles by the road down the river. . . .

Returned to Major Patterson's. The road today as that travelled on Saturday from Johnsons Bridge, mostly of deep sand, & making very heavy draught to the horse. In the afternoon, a very heavy rain, which cleared up early enough for me to return home with Dr. Sweat. 21 miles.

30 MAY. Set out for Barnwell village, & arrived there at 2 & put up at Bracon's Tavern. The roads would have been heavy with deep sand yesterday, but are made excellent by the heavy rain.

Found no indications of calcareous earth by the way, except the existence of some large ponds; & more directly, on the Little Salkehatchie, a few miles below its head, a marl was found which appears very poor & is also probably inaccessible. . . .

One pond which we passed today, is some 40 acres in size, & very shallow. It was formerly dry land, except in very wet spells, & was thence called a "savanna", & was used as a battallion [sic] parade ground. But for the last 8 or 10 years, the crevices through which the water sunk have been stopped, & the water is now always on the ground. Other ponds are deep, & so clear as to be supposed of limestone water.

At Barnwell, Major J. Brown, to whom I had a letter of introduction, & several other residents, gave me all the information they had, which was very little as to *reports* of marl. . . .

31 MAY. To the Boiling Springs, 10 miles from Barnwell, & near the Lower Three Runs. This place is the permanent residence of 7 or 8 families, & the summer residence, for health, of one or two more. It derives its name from a strong boiling spring[1] in the valley, of weakly impregnated limestone water. Not long since, in digging for clay, marl was first exposed within two feet of the surface, near the foot of the

sloping hill by the spring. It appears of excellent quality, & similar in texture &c. to the marls of the upper Santee. A few hundred yards lower down the stream another exposure of marl has still more lately been made, in opening a spring at the foot of the hill. . . .

Two miles below the Boiling Springs, & on the main stream of the Lower Three Runs, the narrow range of the bed of *ostrea gigantica* crosses & forms the bed of the stream. This is the nearest exposure seen to the Santee canal, under which it passes. . . .

Two miles still lower is the Roaring Spring, formed of several other strong boiling springs of limestone water, & of which the principal one is very strong. Upon one of our company thrusting his arm down one of the smaller apertures whence the water boiled up, he drew out with the coarse sand fragments of ostrea gigantica, showing that this peculiar belt passed here also, which is about due south from the upper exposure, & that the water comes up through these shells. . . .

In the afternoon, went to Gen. Erwin's mill, 4 miles from the Boiling Springs village, to see the largest exposure of marl known thereabout—& indeed the only one known of to any one, until the recent exposures made at the Boiling Springs. The mill is on a branch of the Lower Three Runs. . . . This marl is granular & very soft—& in texture like that at the Grove low down on Cooper river. . . . Returned to the Boiling Springs & passed the night at the house of Col. Hay, to which I had first come & had been kindly entertained, & the host & his sons had been my guides & companions in the day's excursions.

1. The deeper the source of water in the limestone, the higher its temperature; each thousand feet might add twenty degrees Fahrenheit.

1 JUNE. To Gen. Erwin's house . . . below the Lower 3 Runs. . . . Not finding the proprietor at home, nor any other source of information or guide, was not able to make any examination thereabout for marl. Proceeded up the course of the river to the plantation of the Rev. Elliott Estes, to whom also I had a letter of introduction from Gov. Hammond, & in whom I found a countryman, who gave a most cordial & kind welcome. From all heard in answer to my inquiries as to the nature & exposition of the land, I had very faint hope of finding marl, & yet no where had I desired it more earnestly, for the sake of

my worthy & friendly host. Mrs. Estes, incidentally mentioned a place where a stream disappeared under the earth for a short distance, & it struck me that such a place, if any, might be of calcareous formation. Accordingly I requested to be guided to this spot, & found the subterraneous passage to be through marl, which was visible under the stream for some distance below. This stream passed along the foot of a hill-side, & through a piece of narrow low-ground emptying into the Savannah about a mile below. I inferred that, though visible only in & beneath the water of the stream, the marl lay horizontally under all the low-ground; & upon boring near the opposite side, it was reached at 4 feet below the surface. . . .

On a part of this highest low-ground, there are some Indian works, for defence most probably. Next to the swamp, where it is now so low that it appears formerly to have been the bed of the river, a ditch passes around a piece of land nearly rectangular, of about 120 yards long by 60 deep. The ditch is filled with water from the river during every high water—& probably is filled several feet deep with the deposite of mud. It seems to have been about 10 feet wide at first. Next the side towards the river are two circular mounds. The larger is about 35 feet across & 12 to 15 high. The other is about one third less in size.

While looking for marl this evening, we came very near to a rattlesnake, which was lying extended across our way, & remained motionless until after receiving the first blow. Though I know by report that these most dangerous & deadly animals were to be met with in all lower S.C. near the swamplands, I had not before thought of the danger as affecting myself, & had groped through bushes, & along steep hill-sides, in search of marl, as carelessly & with as little thought of danger, as if at home. But the meeting with the rattle-snake brought the risk fully to my view, & it will cause me to be much more cautious hereafter. This snake, though but small in comparison to others of the same kind, often met with in this state, appeared large to my eyes. It was about 4 feet long & as large as my wrist at the middle of the body. It had only 7 rattles & the button, or closing part. Upon opening its mouth, the four keen fangs in the upper jaw were easily examined. They are crooked, slender, & as sharp as needles, and more than half an inch bared. The forward two are fixed; the others, immediately behind them, are moveable, & lie flat except when in the act of striking the deadly blow, when these fangs are erected in the position to be most effective. They have channels (. . . it is said) through which poison is conveyed & ejected. The two sacs of poison lie at the roots of these fangs. On opening one with the point of a knife, the fluid flowed out in considerable quantity. It is yellow, & appeared like oil.

The rattle-snake though one of the most dangerous & deadly enemies that can be made, is as magnanimous & forbearing as brave; & unless attacked, or approached so nearly that it supposes attack is meant, it never seeks to do injury. This one, though approached before seeing it at first near enough to become fully aware of it, did not move in the least, either to threaten or to flee, while a weapon was sought for to strike it safely. It was only when attacked that it exhibited its rage.—19 miles.

3 JUNE. Hoping from the unexpected success here to be able to trace the calcareous bed in other streams, emptying into the Savannah, I continued the route up the river by the road running nearest, & parallel. But though it passed several mills, the sites of which offer the best means for observation, no further trace of marl could be seen or heard of. At Four Mile Creek, the limit of my day's journey, a former breach in Dr. S. J. Bailey's mill-dam had scooped out a hole 15 feet below the sheets, & clay & sand only had been seen. I therefore deemed the search hopeless. Yet this spot is within 7 miles of Shell Bluff in Ga, the celebrated & enormously elevated body of marl. The bluffs on the Ga side of the Savannah are very high, while the land on the Ca. side opposite is comparatively low, & what are called bluffs are only of the higher alluvial formation, which of course can contain no marl. It was useless to proceed further, especially as I was within a short distance of Silver Bluff, to which I shall go by appointment on the 7th or 8th, when Gov. Hammond will have arrived.

The days ride was heavy, owing to the deep & already thoroughly dried sand of the road. Reached Dr. Baileys house at 3, & learning that he was at his mill, 2 miles farther on, proceeded there immediately, supposing that there was the best ground for examination. A heavy rain began just as I reached the place, which kept us under the shelter afforded by the mill-house, until 8, when, having procured our great coats & umbrellas, we returned to the house, though still through slight rain. Much hail had also fallen, & during our previous confinement, we were in a very damp as well as uncomfortable cool atmosphere. When we sat down to supper, it had been 13 hours since I had eaten anything, & yet felt no hunger.—30 miles.

Dr. Bailey is also a Virginian, from Charlotte county. During my ride yesterday, inquiring as usual for exposures of marl or limestone, & as usual also, finding that most persons knew nothing of either, in their neighborhood or of the substances in general, I heard of a reported large & new exposure of the gigantic oystershells, on Harley's land, & near his mill, near to my point of departure the day before, the Boiling Springs. This induced me to take that route back, but by another road from Steel Creek. Upon arriving at Harley's bridge & Mill, on Lower Three Runs, I found no sign of marl, but in the bed of the stream a quantity of soft sand-stone in large masses which had formerly made a continuous stratum. No person could be seen to give any information. Passing on, I gathered enough to infer that the person sought was Holly, a small planter on land just below the exposure of gigantic oyster shells before visited. On going to his house, found that I had arrived only two hours too late to see the exposure, as the flood had just broken his embankment, & overflowed the place, which is in the low ground of the Three Runs. As he described it, it is the ordinary marl, such as I showed him specimens of, which is exposed by the plough in different parts of a piece of 3 acres. He had never heard of marl, nor thought of this substratum thus exposed as of any account, until a few days ago: when a preacher from a distance called to visit him, & walking to see his fine corn, observed the appearance, & suggested that it was marl. This, together with the preacher's taking specimens, & also some of the large oyster shells from the stream above, & exhibiting both elsewhere, caused the erroneous report which had brought me back. However, I did not regret the trouble to ascertain the truth; & if I now know it, this is a very important extension of the exposed marl & I trust the poor & laborious proprietor is enough impressed with what I told him of its value, to make good use of his treasure. He promised to obtain & send me specimens, when the subsiding of the water shall permit. Returned to Col. Hay's. After dinner visited Rocky Point 4 miles below on the Three Runs. The quantity of stone there in masses is very hard sand-stone. No appearance of marl. But fragments of the gigantic oyster have been thrown out by former floods, higher than the present, which circumstance indicates that the bed must extend in the stream much nearer to this place than the lowest point (the Roaring Spring,) where I saw it exposed.—29 miles.

My route along the course of the Savannah was generally within from 1 to 3 miles of the termination of the high land, or edge of the river swamp. All the soil light & sandy. Much of it of good quality—the poor being the spurs or extensions of the higher land. This kind even, & the general table land of the district, has less of barren, & is generally of much better quality than the bulk of the upper & interior lands of the other & lower districts. The surface is mostly undulating, & there are numerous streams of water, running briskly, instead of the still stagnant or sluggish waters of swamps most common below. If having the benefit of calcareous manure, there is but little of Barnwell district that would not be good & profitable land. In this district there is a great change of kinds of wild plants from those of Va. Many new, & some curious & beautiful plants are here common, which I do not know, & some seem not to have attracted notice of but few inhabitants. Among the latter is a large white lily,[1] which floats on the surface of deep & still ponds, on a very long & flexible elastic stem. This is one of the most magnificent flowers I have ever seen, & its structure admirably contrived to suit its peculiar location. The sensitive brier plant[2] is common in the most sandy lands, along the road sides, & in the fields not under tillage. It is very long since I have seen this curious plant; & was much struck by its beauty, as well as by its remarkable quality of sensation. This plant must have the sense of feeling.

1. Probably *Nymphaea odorata,* water lily, an anchored aquatic with floating leaves that is often found in lime sinks.

2. *Shrankia microphylla,* a leguminous plant, grows in places throughout the state, especially on barrens and sandhills.

4 JUNE. Sunday. Went to the Presbyterian Church with Col. Hay's family, at the Boiling Springs—one of his sons being the minister. Returned to Barnwell village.

43. Barnwell District: Sales-Day Marlers and Malarial Mill Ponds (5 June–6 June)

5 JUNE. This is the monthly "sales' day" of the district, when all the property under execution for debt is put up at auction by the sheriff,

which always brings together a large assemblage. It was for this that I was persuaded to wait & as on some former occasions have found the result not worth the waiting for. From all that I learned from report, the entire new subject of marl, & especially the recent discoveries in this neighborhood & the district, formed the general subject of conversation of the day, but more, as I believe, as matters of gossip & ignorant *wonderment* than of proper interest & curiosity. I had been told of the great anxiety of sundry persons to see & converse with me on marl. Well! Some 6 or 7 persons, who appeared to seek it, were introduced to me in the course of the day, & who made inquiries on this subject. Perhaps a dozen more were introduced, but apparently without their seeking it, or in reference to my pursuit. It is true that I was mostly in the private sitting room of the tavern where I lodged, conversing successively with the few who sought me. My presence & my objects of course were known to all on the ground, & I presumed that all who desired to confer with me would come to find me. But lest some might omit it from shyness, or other cause, I went out & strolled through the crowd enough to afford all who desired it an opportunity to meet me. If this had been my *earliest* experience of such things, I would have inferred that my labors in this district were nearly thrown away. But, on the contrary, notwithstanding these appearances, I believe that a very unusual sensation & interest have been excited, & that hundreds who have not shown to me any interest in the matter, will search for marl, & if finding, will not neglect to supply it. Received Holly's specimens, (from the flooded low ground,) & also one sent me from Edisto river 1½ miles higher up its course than Johnson's Bridge.

Desiring next to examine the upper streams of the Upper Three Runs, (where I had heard of indications of marl,) Mr. Charles Hay, one of the representatives of the district made me acquainted with Mr. Jesse Cherry, a plain old planter of that neighborhood, who invited me to his house, & to accept of his aid in my examination. Reached his house, across Tinker's Creek, before dark—18 miles.

The route crossed the upper streams of the Big Salkehatchie, & the Lower Three Runs. The latter were in two bold & rapid streams, which were as clear as to indicate that they must be of limestone water. The land generally of the wide ridges between different waters, & poorer land than usual elsewhere. The surface undulating. Craig's pond, by which the road passes, is a large savanna, said to be 3 miles in circumference, which always has more or less water, according to the season.

There is but little water now, & the surface is that of a beautiful green meadow, grazed closely by the numerous cattle on it.

The streams in this district are navigable very high. Not only the Salkehatchie, to where my road crossed it, above Barnwell village, but Tinker's Creek to 5 miles above, & the Lower 3 runs to within 3 miles of the road, carry rafts of lumber to the Savannah. The streams appear much too small for this; but when a number of rafts are started together from a saw-mill, the flood-gates are hoisted, & a transient overflow produced, which serves to convey the rafts to sufficiently deep water. The great demand for lumber some years ago caused saw-mills to be constructed on every one of the numerous sites afforded by the streams. And as the fall is generally small, say not often exceeding 8 feet, & the ponds are partly dry in summer, the sickliness of the inhabitants has been obviously & greatly increased. On Four mile creek, quite a short stream, there are 5 mills, fed by as many successive pondings of the same stream. No wonder that its neighborhood is deemed among the most unhealthy in the district.

6 JUNE. Rode to examine the reported indications of marl on Tinker's creek. Below Mr. Cherry's house, found only silicified shell clay, not calcareous. . . . Of course no marl can be supposed lying lower. . . .

I had designed to go on this evening; but a slight prospect of rain, & the hope of rendering some service by advice to my worthy & kind host, induced me to accept his urgent invitation to remain longer. Walked to see his swamp land on Tinker's Creek, of which he has ditched 40 acres, & as yet to no purpose, because it is too little elevated above the water in the main stream, & also subject to be overflowed by it whenever the mill-owners above let loose water to float down their rafts. The stream has enough fall & a rapid course. It has been made navigable by cutting & removing the upper logs which had obstructed its course. Still many remain lying deeper, which, if removed, would quickly deepen the bottom, as it is of loose sand. But this, though required for navigation alone, & a cheap improvement even for that end alone, will not be done by the combined action of those interested. I advised Mr. Cherry to clear out the logs well just *below* his swamp, so as to deepen the channel enough there to lower the stream & make very shallow & bad navigation *above*. This will compel the lumber-men to deepen all above.—10 miles.

44. Silver Bluff, Shell Bluff, and Savannah Swamps (7 June–8 June)

7 JUNE. Crossed the Upper Three Runs for the Savannah at Silver Bluff. The public road from Savannah to Augusta after the first 6 miles, the most sandy & deep yet tried. Reached Silverton P.O. where I had hoped to meet letters from home, but was disappointed. Gov. Hammond, whose country residence is close by, had not arrived. Took quarters at the P.O. where "private entertainment" is offered. The house & accommodations are very neat, & mostly comfortable. Yet there is not a glass window, (or one designed for glass,) in the house. Still the general appearance of things indicate competency, & regard to neatness & even taste as well as to comfort & abundance.

After dinner rode on horseback, 5 miles, to Silver Bluff, (Gov. Hammond's plantation,) to confer with Mr. Barns [sic], his overseer about my getting to Shell Bluff, down the river. It was with much interest that I first saw the marled fields of which I had published the report.[1] The entire field of cotton is marled, except 2 separate acres left out for experiment. The difference between these & the surrounding land, marled last year, speak plainly enough for the fact of beneficial effect. And so I would infer of the whole crop, compared with the crops of the country in general. Still, the difference is not near as great as I have been accustomed to see in Va; or as it would be here if the land had had more rest. The operations here in marling present a remarkable & admirable contrast to the general neglect of marl in this state. As late as the summer of 1841, Gen. Hammond remained not only totally ignorant of the value of this improvement, but a scoffer & ridiculer of what was said of it by those who believed & talked but did nothing with it. He then first met with & read my "Essay on Calcareous Manures", & became fully convinced. By November of that year he had made his expensive arrangements & commenced to bring marl from Shell Bluff; & already he has 700 acres marled, & enough marl heaped on the bluff at his landing to cover several hundred acres more. (Between 40 & 50,000 bushels, at this time.) A large flat, requiring 11 able men to pole it up when loaded with 1100 bushels, makes two trips a week, the boatmen also digging the marl, & loading & unloading. As the landing place is too little elevated to be safe from the high freshes of the Savannah, two single mule carts & two men are also continually employed in conveying the marl to the top of the hill or bluff.

In viewing the crop & the marl, the evening was thoughtlessly permitted to pass away, & it was an hour after dark before I got back to my lodgings. This exposure to night air, at this season, is generally deemed imprudent, & I wish to avoid it. Still it is the almost universal custom to sit in the piazza until bed-time, & I have no choice but to do as others do.—22 miles.

1. The report is in the *Farmers' Register,* 1842; see Introduction.

8 JUNE. Before sunrise set out on horse back, accompanied by Mr. Myer, my host, & taking early breakfast with Mr. Barns, we all proceeded to take passage in the marl flat at Silver Bluff, which immediately was cast off to drift down the river. Two very clumsy oars only were used, & the rudder, all the other hands being disengaged during our voyage of more than 3 hours, & amusing themselves with cooking, eating, laughing & talking, or sleeping. But on the return trip, all are required, & good service from all.

The lands on both sides the river for 12 miles are low. The higher (which is at Silver Bluff,) 20 to more than 30 feet above the present low level of the river, & that not always & entirely free from the injury of highest freshes. More generally, the cultivated lands are 10 to 15 feet above, & more or less endangered. On the lower cultivated lands, is estimated that only one crop in three is made & saved. A still larger proportion is low swamp, which though now 6 or 8 feet above water, & appearing firm, is left undisturbed under its dense & tall forest growth.

At Shell Bluff, the land all at once rises to a great elevation, much greater indeed that [sic] the eye from the river can realize. The perpendicular & sometimes over-hanging cliff of solid marl against which the river washes, appeared to be 50 or 60 feet high, &, doubtless is the same marl to the bottom of the river & perhaps 200 feet lower. Apart from all considerations of utility & value, the scene is most interesting for its beauty & sublimity.

Designing to return again & give more full examination to this place I remained at this time only about two hours. It had rained during the passage, from which we were barely protected by the roof of the flat. And there was strong indications of a renewal of the rain, & my exposure to it, as there were no accommodations for staying on the Ga. side. Our horses had been brought by a negro who knew the way through the swamp to the opposite side, where we passed over in a canoe &

mounted. Except a slight shower as we were about to return, there was fortunately no more rain. But even without that addition, which would have been dangerous as well as uncomfortable, I had the most fatiguing ride that I ever experienced. Of the 12 miles to Silverton, 8 or more were through the uncleared river-swamps, thick with trees & bushes, & the flies so numerous as to be distressing to the horses, & requiring continual labor of brushing to relieve them partially. In some parts there was a road (leading to some field) in others only a foot or cow-path, & in others pathless & trackless, but for the morning's passage of our horses. The first mile was through the low swamp, of the latter description, & sometimes boggy withall. The greater part of the remaining 9 miles was a very crooked passage through Big Back Swamp. But this name would convey an erroneous idea of this wide body of wood land. Though subject to high freshes from the river, & truly a swamp of water or mud at some other times, the ground was now firm & dry. And if a road had been made it would have been the best in this region. Different from the usual very sandy soils, this body is very stiff, close, & also poor, so that very little is cleared or deemed worth cultivation. It seems in no quality to be like alluvial soil.

The course of the Savannah for the 12 miles is extremely crooked, & the width varying from 120 to 60 yards.

45. At Home with James Hammond (9 June–15 June)

9 JUNE. For the first time since my service was begun, I had nothing to engage my labor & attention. A letter received from Gov. Hammond yesterday informed me of his expected arrival this evening & I had nothing to see to fill up the short interval of time, or to do, except a little writing, & to read the lightest of all possible matter in old magazines which I found in the house. The weather continues threatening to rain, & a thunder shower fell. In the evening went to the Governor's house, which is nearly 5 miles from the river at Silver Bluff. Besides the family, found Mr. Starke there, who had arrived from his father's residence near Hamburg.

10 JUNE. Rode over the extensive cotton & corn fields of the plantation. The land is in successive broad & level terraces. The soil sandy & light. Altogether land better in quality than the ordinary, &

which will be all that can be desired when having derived the full benefit from marling. The only doubt of this would arise from the absence of sorrel, of which I have not seen a plant on the ground now at rest. This was also entirely absent, or nearly so, on the red land of Orangeburgh. It showed again in Barnwell—but seems generally to be most fond of the sandiest land, & not showing in all such, as I would suppose to be acid.

First saw here today the crop of *pindars,*[1] or ground peas, planted on a large scale. . . . They are raised almost entirely for hogs, for which purpose it is a productive & valuable crop.

The now abandoned brick mansion house at Silver Bluff, close to the river, is small & rude in structure. Before the revolution it was a noted trading post, to which the Indians used to resort in great numbers. During the revolution it was enclosed in Fort Dolphin, which was one of the many small fortified out-posts held by the British forces. This was captured by a body of Americans commanded by Col. Lee.[2] The passage of a cannon ball, from a field piece, is still seen through the opposite outer walls, & the partition between.

But there are much older traditions of this place, & far more interesting to the present generation of the neighborhood. It is said that the Spaniards from St. Augustine penetrated to this place, & found silver ore in the bed of the river along the part since called Silver Bluff & some distance below; & they remained undisturbed long enough in possession to work the ore, & obtained much silver bullion. While thus engaged, they began to dig what is still known as the "Spanish Cut", a narrow & short passage across the low ground peninsula on the Ga side opposite Silver Bluff & the supposed silver deposite, for the purpose of turning the current of the river, & permitting better the working of its previous bed. This cut now conveys a rapid current at high water & might, by being enlarged, be easily made to change the river's course, or shorten it several miles. At last, (so goes the story,) the French (supposed the colonists sent by Admiral Coligny) came up the river, & attacked the Spaniards, & made them flee across the country to Florida.[3] When the necessity for their retreat became manifest, they buried part of their silver, & sunk the balance in an iron chest just off the landing at Silver Bluff. There has been much digging since for the supposed buried treasure, & dragging & diving for the iron chest in the bottom of the river. It is said that the latter was once found, &

raised nearly to the surface of the water, when the iron loop by which it had been hooked broke, & all was lost. What makes the search now hopeless, is that the bed of the river at this place is now filled up 8 or 10 feet more than formerly. As to the digging on land, there will probably be found fools to resume it from time to time for another century.

After dinner, commenced the job of analyzing all the specimens in hand of marls collected since the last operation in Charleston. Besides the larger specimens of all carried to Columbia, & left at different places on my route to be sent there, for the cabinet, I have kept small portions of the same, for testing. Gen. Hammond, since beginning to marl, has also had much practice in analyzing marls; & by giving me his assistance, enabled me to get on much faster with my work.

1. *Arachis hypogaea,* a peanut, was used for nuts and forage.

2. Charles Lee (1731–82), English-born soldier of fortune, commanded revolutionary South Carolina troops in 1776 and later supervised the defense of South Carolina and Georgia. See also p. 230, n. 1.

3. The French, Spanish, and British disputed the area up to the eighteenth century. The conflict referred to here was that between Spaniards, in Georgia from Florida, initially with Fernando De Soto in 1540, and French Huguenot settlers arriving from Europe under Admiral de Coligny in the early 1560s.

11 JUNE. Sunday. As the two preceding days, intervals of threatening clouds, & some slight showers.

12 JUNE. Had designed a sunrise start for Shell Bluff; but light rain detained us later, when we set off. Other showers fell afterwards, from which we were protected by shelter in the marl flat & in our carriage. This trip, under better guidance, showed me that I had before very erroneously judged of the way. For we travelled in the carriage & mostly by a good road, except for its being narrow & bushy, for the whole distance, except the last mile through the low swamp, which we rode on horse-back. Between the wide & miry river swamp, & the still wider Back Swamp, there is a narrow strip of higher land, partly in small plantations, & through the length of which runs the road we travelled today. It was almost 9 oclock when we got back, after having spent a laborious, but most agreeable day.

With the assistance of Gen. Hammond, I carefully examined the different strata & varieties of the marl of Shell Bluff. . . . The actual exposed & perpendicular height of the marl at one place, above the water at its present height, (which is low, but not the lowest state,) was 72½ feet. But this is far short of the thickness of the section. For besides that at this spot some of the upper strata were absent, the depth below the river is said to be more than 30 feet certain, & is reported by Mr. Cotting to be 5 [?] feet. There can be but little question that the entire thickness, if it could be known, would not fall short of 300 feet. But it is enough that 72 feet are here visible & available, & may be used for the lands to 12 miles up the river, & 24 down, as easily & as profitably as is done in the only but extensive use yet made of it, for the Silver Bluff plantation. . . .

After reaching the top of the marl, the land covering it rises with a steep slope to more than 100 feet higher. The marl cliff is seen along the river & up the adjoining Boggy Gut Creek, (a narrow but navigable creek) perhaps half a mile. But this is uncertain; for it is impossible to walk alongside of the foot of the cliff for one quarter of the distance together—& so crooked is the outline, & so covered with trees, that but a small extent can be embraced by the eye at any one view.

The stratification is more obvious here than seen elsewhere, & especially of the lower part, say 20 to 30 feet above the river. . . . The remarkable bed of ostrea gigantica 4 to 6 feet thick, overlies the whole, for about 100 yards separated from the marl by a bed of loose sand & then clay & sand in thin layers, of from 6 to 10 feet thick. . . .

This remarkable bluff was doubtless made by a breaking & upheaving of a part of the great bed of marl, lifting the Ga side some 60 feet above the portion in S.C. The line of breakage is probably the bed of the Savannah river. If this opinion be correct, the character of the great deposite of S.C. may be studied best at Shell Bluff, though out of SC. & the strata examined for 40 feet lower than the lowest visible above the rivers any where in that state.

But enormous as is the exposure here, it is but as a portion of the great marl region of Ga, which as is reported extends from Shell Bluff, & a few miles above & below through the whole of Ga & also into & through Florida. The western limit is a line running parallel to & not many miles below the falls of the rivers; & nearly the lower half of Ga, is probably underlaid with marl. Yet no one thinks of using this as manure; & I hear that a notion is even prevalent that all these lands are already too much charged with lime. A most erroneous & absurd notion truly!—25 miles.

13 JUNE. Continued analyzing marls, throughout the day, assisted by Mr. Starke. Tried Dr. Smith's apparatus, with 12 specimens similar to as many analyzed by Davy's[1] apparatus (which I use) & found the former to show results so different that I cannot trust to it, or to my awkward use of it. Indeed, the obvious escape of ammonia & consequent alteration of the degree of strength, seems enough objection to the test, after its having been kept prepared long. Still, I have no doubt of the plan being good & correct, under different circumstances.

1. Sir Humphry Davy (1778–1829), distinguished British chemist, was responsible not only for such devices but for first alerting Ruffin (in print) to the issue of soil acidity, in the 1810s.

14 JUNE. After more or less of moderate or light rains every day since I arrived in this neighborhood, last night & today it has rained heavily, & repeatedly. After dinner rode to Mr. Wm. Bush's residence, on some report of marl on his plantation on Upper Three Runs, but found no specimens ready, & no certain report. That which I found on Tinker's creek, & which seemed to the eye to be 25 to 30 percent. I found to be only 7 percent. & of course not worth using. It is no doubt in part silicified.—8 miles.

15 JUNE. Finished analyzing the specimens of marl, (for which I had to wait to obtain a new supply of acid,) & wrote a continued statement of them & sent it to the newspapers. Rode again to Silver Bluff, & over the fields of corn & cotton. I would infer, by comparison with the two unmarled acres, & also with all other cotton crops seen, that this crop is throughout & greatly improved by the marling. And, which is more to the purpose, Gov. Hammond & his overseer are both of that opinion, & perfectly satisfied with the supposed degree of effect. This evening Mr. Bradford called to see me; & before he returned engaged 5000 bushels of marl of Gov. H. at 2½ cents landed at Silver Bluff wharf. This is gratifying proof of the extending of favorable opinions of marling.—Rain again, but only very light sprinkling.—11 miles.

It was well that I examined Shell Bluff on Monday, the first available day; for if that had not been made use of, there has been no chance since. Today the river is 6 feet above its previous level, & was still rising rapidly.

Mr. J. J. Barns, now Gov. Hammond's overseer, formerly served in the same capacity 3 years for Dr. Jamison, & also superintended his lime-burning during that time. From him I learn some additional & interesting particulars. The *whole* depth of digging was about 12 feet, (& not that depth in the marl, as I had inferred,) of which only 4 or 5 was in the marl, there being 7 or 8 feet of earth to remove to uncover the marl. After this they reached a stratum too soft & crumbly to use for liming; & from the description of its color, texture etc. I have but little doubt that it is a layer rich in *green-sand,* though the describer had never heard of this earth. [E R Footnote: The lime was sold *slaked,* or if before slaking, 1 bushel was counted for 2 slaked. Thus the price of 25 & 31 ¼ cents the bushel was in fact twice that amount for the unslaked.]

It was not until after my last visit to Silver Bluff, that I learned from Gov. H. that Bartram[1] had also described the place, & that he had copied the part, from which the following is taken. Since 1773, the date of Bartram's visit, it is supposed that 100 yards width of the bluff has been washed away. But in what remains there is no sign of marl; & I am confident that even if he found marl there, that there were no "bellemnites" there. Speaking of Silver Bluff, he says—"It is a considerable height upon the Ca. shore of the S. river, perhaps 30 feet higher than the low lands on the opposite shore, which are subject to be overflowed in the spring & fall. This steep band rises perpendicularly out of the river discovering various strata of earth; the surface for a considerable depth is a loose sandy loam, with a mixture of sea shells especially of ostrea. The next stratum is sand, next marl, then clay again of various colors & qualities, which last insensibly mix or unite with a deep stratum of blackish or dark slate colored saline or sulphureous earth, which seems to be of an aluminous or vitriolic quality, & lies in nearly horizontal laminae or strata of various thickness. We discovered bellemnites, pyrites, marcasites, & sulphureous nodules, shining like brass . . . as also sticks, limbs, & trunks of trees, leaves, acorns & their cups, all transmuted or changed black, hard & shining as charcoal. We also see animal substances, as if petrified, or what are called sharkes teeth."

1. William Bartram (1739–1823), Pennsylvania botanist and ornithologist, was the son of John Bartram.

46. Wasted Time in Augusta and Edgefield (16 June–19 June)

16 JUNE. Left the hospitable & agreeable place where I have been for a week, & proceeded to Augusta, Ga.—15 miles.

Though the clouds were threatening, there was no rain—the first day without for 9.

A few miles after crossing the Edgefield line, the land becomes undulating & rises higher. The flat country is left behind. Crossed the Savannah (by ferry) 3 miles below Augusta. Thence, the road was through high alluvial & very rich land. Here saw the only good oats seen this year. This crop in S.C. is wretched, & always poor—so that it is strange it should be so generally cultivated on even a small scale. Gov. Hammond turns his hogs in his oats as soon as it is barely ripe, & reaps none except for seed. Yet even where thus grazed & dunged by hogs, he has seen a manifest inferiority to the lime, on the oat land, in the succeeding crop of cotton.

17 JUNE. As I could not make any other more convenient stopping place, on the S.C. side, I made use of the opportunity to see Augusta, and at the same time to purchase some few articles needed. I am sorry for having gone there. Like all other small towns where one has no acquaintance, & where there is nothing else to create interest or excite & gratify curiosity, such a town as Augusta, or even Petersburg is worse for a stranger than a country tavern. Left it this morning, through Hamburgh, for Edgefield C.H. I had been detained by vexatious delays of matters ordered until 10, & it was near 4 before I stopped. From Hamburgh (immediately across the Savannah from Augusta,) the road ascends, by long hills separated by flat terraces of sandy land, for more than three miles, & to a great elevation. The land is still sandy. The granite crosses the river 3 miles above Augusta, but is not seen on this road until after 10 or 11 miles, after which granite, mostly disintegrating, is seen at every hill. All along over the visible granite the land is red,

& very high & hilly. More corn is cultivated than lower, in proportion to the cotton—& both are in something like graduated rows running alongside the slopes of the hills. It is badly done however, & further imperfect in the effect, because that the rows are made small beds instead of being kept in flat culture, as would best guard against washing. The pines are here fewer than oaks.

I had come to Edgefield C.H. in consequence of one of the several invitations which have been sent me to meet Agricultural Societies, on a fixed day, or if that could not be whenever I could. The fixed day had long passed (indeed had almost arrived before I rec'd the invitation;); &, having promised to do so, I had written to Mr. F. W. Pickens, (one of the Committee, & who had also added his urgent invitation to his house,) a *private* letter stating that I should now pass through Edgefield, & could delay two or three days at most, for any purpose desired of me. But adding that if a meeting was not convenient, or now desired, I would be very glad to have it dispensed with. Upon arriving here, to my great satisfaction I found that no such meeting could be held, as not only Mr. Pickens but the President was absent, & the time otherwise impropitious. Mr. J. Terry, another of the Committee came from his residence to meet me here, & wished me to go home with him; but I excused myself, & remained at the tavern to write.

From Mr. Terry I learned more of the mode of tillage. There is a rotation of crops getting into usage, viz. 1. cotton, 2. corn, 3. small grain (oats, wheat, or rye,) broadcast & this land grazed the balance of the year after reaping; & then the same thing over again, without rest. Indeed, it is very rare to rest land at all. In this way, the uplands are worn out in about 10 years of tillage.

A heavy shower of rain at dark, & a change to very cool weather.—23 miles.

18 JUNE. Sunday. Finished writing letters. Dined by invitation with Mr. Terry, at his house a mile out of the village.

19 JUNE. Immediately after breakfast, set out for Columbia. Stopped to dine at Mrs. Lee's house, 27 miles, in the edge of Lexington district & determined to remain through the night. A plain but neat house & fare—indeed the best public house I have seen, either in town or country.

I have often had occasion to remark & wonder, that, with all the hospitality of the S. Carolinians *in* their own houses, & in dispensing to me as a guest, their entertainment, carried even to too lavish & troublesome excess, that so little of ordinary attention & civility should be shown to me, in respects more important to me as a stranger, & necessary to enable me to perform the objects of my office. But in this village of Edgefield C.H. the failure of such attention was so marked, that it became ludicrous by its extremity, so that I was more amused by it than vexed. I came to the district, & very unwillingly, merely because of having been invited to meet their Agrcl. Society—& besides the letter of invitation, in form, signed by all the Committee of five, I had two individual & pressing letters of invitation from two of the most prominent members, resident here. I wrote the notice of my coming so as to get off from the proposed public meeting, & was very glad that my wish was readily responded to. And, judging by the results, I infer that the thus letting me off was fully as agreeable to the inviters as to myself. So far all was well—& I should not even have grudged my useless journey. But when I arrived, I found my two individual inviters, & also the President of the Society & Chairman of the Committee, had all gone on journeys after receiving the notice of my arrival within two days thereafter. No doubt these gentlemen were called away by pressing business, which I should have been sorry that either should have neglected to await my arrival. But not a line or message of apology or regret was left by either. Mr. J. Terry, one of the committee, to whom had been left the duty of making known to me the state of things, was as polite & as hospitable as he could be. He had a servant waiting for my arrival, with a letter of invitation to his house—which he came to renew personally in the evening, & kindly pressed upon me much more offers of attention & hospitality than I could accept. But excepting himself, there was not another individual, resident of the village or member of the society, who addressed to me a single word, or offered to [sic] desired to have the slightest voluntary intercourse with me. The gentleman only, who happened to be present, was casually introduced by Mr. Terry, & a few common-place words passed in the few minutes we were together. Yet there were sometimes a dozen residents in the piazza of the tavern, while I was sitting there—& who as at all other occasions stared at me with all their eyes, though none sought to speak to me.

Yet full notice had been given (contrary to my wish & request) of my expected arrival, & of course every individual must have known of it, as afterwards of my stay through 40 hours. I had written to Mr. Pickens as *private,* to prevent any obligation on the *Committee* to act, if action were not necessary—& requested that the purport should be made known only in the case that a meeting would not be dispensed with. Still, my notice had been made known to the committee & they & probably all others knew of my intended coming.

I saw in the village a lot of about half an acre of excellent red clover[1]—proving clearly that *climate* is no bar to its good yield. The land had had 7 or 8 horse cart loads of *ashes;*[2] so that it was well-limed ground, & which is the secret of the success.

The road from the village is excellent. Hard & smooth, & not too hilly. For the first 14 miles, the land undulating, & red soil, or at least red subsoil showing continually. The crops are mean—no corn seen over knee high on an average—indeed few stalks higher—& the greater part much lower. Cotton not more than 6 inches, & rarely so high. After 14 miles, reached the level "*ridge*", which divides the southern tributaries of the Saluda from the head waters of the Edisto. For some half dozen miles along this ridge road, the land & cultivation are both much better—presenting indeed the best I have seen. It is called the Watson neighborhood, all the plantations being owned by members of one family & of that name. Some corn was 5 to upwards of 6 feet high, as far as I could see the field along the road. The soil is a brownish red, & the surface nearly level. After crossing the first stream, the land becomes more sandy & loses its reddish tint; but continued better than the ordinary.

Corn is more extensively made than cotton. Those two crops cover nearly all the open land, for scarcely any is rested. There is but little grown in potatoes. The wheat is more plenty, some fields of 10 to 20 acres being seen today. Nearly all has been reaped, & is in stooks on the fields. The reaping has been very badly done. They have taken unnecessary trouble to cap the stooks.

1. *Trifolium pratense,* broad-leaved clover, indigenous to Europe, is a valuable nitrogen-fixer and fodder crop.

2. Burned plant and animal remains have a notable lime component and are used as fertilizer.

47. Columbia: South Carolina College and Millwood (20 June–23 June)

20 JUNE. To Columbia, 30 miles, over what is correctly termed "a heavy road." But that character, which elsewhere would be inferred to be owing to wet clay or mud, here means dry sand. Such was nearly all the road, & nearly the whole way my horse did not go out of a walk. The first few miles hilly, & the land pretty good, & wheat more largely made. Afterwards still undulating, & mostly very poor sand barren, of pine & small scrub oaks. I was told that as approaching the heads of the Edistoes, the land becomes still poorer, & the pines shorter & of but little value; & that there is not a habitation, or any exercise of labor of any kind for a space of 10 or 12 miles across.

After passing Lexington C.H. the streams show granite in very large masses. The Congaree at Columbia is crossed by a very long covered bridge, over the falls. The river there is much wider than the Santee seen any where lower down.

Put up at Maybin's Hotel. The evening's mail brought the publication of my last analyses. . . .

Visited the College Library. Was sorry to find that Prof. Ellett had gone to the north, as I was anxious to see him & to converse more at leisure than we had opportunity for during my last hurried visit to this place. I found Prof. Henry, the president, attentive & agreeable. I brought away a few books to which I needed to refer—& if my stay shall permit, will again visit the library.

21 JUNE. Having some time to spare, before going to Fairfield, & yet not enough to undertake any distant & new exploration, I determined to use it for commencing my report. With this view, & wishing to be in privacy & undisturbed, I came to this place, & shall remain incog. as much as possible, until my writing is well advanced, & which I commenced this morning. I have found it impossible to direct my mind properly to this or any subject requiring study or close attention, while on the road, or at any private house or country tavern. At the latter, any stranger is an object of curious inquiry & examination to all who may hear of his arrival. And if I were to stay at such a place for a week, with no ostensible employment out of my own apartment, I should excite

a perfect fever of curiosity & conjecture throughout the neighboring country.

After making a commencement, which is to me always a difficult & disagreeable task, I have no difficulty in furnishing matter in abundance for a report. But I fear it will be too long at any rate; & my great trouble & difficulty will be to afterwards retrench & abridge.

23 JUNE. With Dr. J. B. Davis (of Fairfield) went out to breakfast at Millwood, the residence of Col. Wade Hampton, 5 miles from Columbia, & to see the large & splendid plantation. I had promised the visit to Col. H. before; & having accidentally met with Dr. Davis last evening, we agreed to go together.

The neatness & perfection of all the processes of Col. Hampton's management are noted throughout this country. No care or cost is spared to have everything he has, or does, the best of its kind. The chief value & beauty of his wide domain is the embanked low grounds of the Congaree. The quantity of this land is 5000 acres in one body, all more or less subject to freshes, & to guard against which the whole is surrounded & protected by an excellent embankment, which is about 8 miles in length. As most of the land is high, & barely exposed to the highest freshes known, its embankment is comparatively low. But everywhere the embankment is kept 4 feet higher than the highest fresh known. Where the land is lowest, the embankment, to be up to this general level, is 22 feet high & of 90 feet base. This however is but a short stretch.

Col. Hampton has extensive & large stocks of sheep & hogs, the former of the Bakewell & Southdown breeds, & the latter Berkshire.[1] His most costly & valued stock, that of high-blooded & race horses, I care nothing about, & have not taste for; & viewing them as a nuisance & great evil, I am not qualified to appreciate them or do justice to any particular animals. In the afternoon, returned to Columbia.—16 miles.

1. Bakewell was improved by English breeder Robert Bakewell for early maturity and fatty mutton. Southdown was an improved old southern English breed with dense, short wool and high-quality mutton. Berkshire, a British pig, yielded high-quality meat and was of largely black coloring, useful in a hot climate.

48. Richland, Statesburgh, and the "Richardson Settlement" (24 June–25 June)

24 JUNE. Leaving my heaviest baggage, travelled from Columbia to Statesburgh, in Sumter, 35 miles, on my way to search for marl on the Santee. I should have done this from Columbia, when there in May; but for seeing several of the wealthy & intelligent proprietors from the very place, & not one of whom knew or had heard anything of the existence of marl in Santee. I then urged them to make preliminary examinations—& hope (though faintly) that something may have been done to prevent my beginning entirely *de novo*.

The lands through Richland, of middling quality, or rather below. Surface moderately undulating, & sandy mostly. Oak growth more in proportion to the pine than lower. No very good culture or land seen by me, except on the Congaree low ground of Col. Hampton. No good mansion houses, except near Columbia. There are however of both, & in considerable proportion bordering the river, out of my course.

The ferry across the Wateree is not more than 100 yards long, & the river not 60 wide. Yet the ferriage of my light one-horse buggy was $1. And for riding over a shamefully bad road through the swamp in Sumter, I had to pay 50 cents more of toll. Ferriages & tolls are to me and my conveyance a heavy expense. But this is the dearest charge yet found. My vehicle having 4 wheels, I am charged every where the same ferriage as the heaviest loaded 6 horse wagon.

Besides the narrow swamp on the Richland side of the ferry, that in Sumter is 4 miles wide. On the river, it has been imperfectly embanked, & is under corn. Mostly not the best quality of swampland.

Statesburgh, a small village, is two miles from the swamp. The site very elevated, & healthy. Put up at the very paltry tavern. Afterwards invited to Dr. Anderson's house, where I spent the night, very agreeably. Dr. A.'s house is mostly of *pisé*[1] construction, put up by his own negroes under his own direction, & on the French plan published in the American Farmer. It has stood & answered perfectly well; & being very well rough-cast, the walls are as beautiful as any I have ever seen, short of hewn-stone.

1. See Biographical Supplement, entry for Anderson, Dr. William Wallace.

25 JUNE. Sunday. Rode 15 miles, to the Church in the Sand hill village.

The land at & about Statesburgh is red like that on the other side of the river in Orangeburgh, before described. As going down the river road, the quality improves & the surface becomes more level, & with Mr. Macrae's & Mr. R. Singleton's plantations begin the fine reddish brown soil & beautiful level surface of the lands next the river swamp. Next, after passing the dilapidated remnant of the small village of Manchester, the road enters the "sand hills", which region is better marked here than elsewhere, but which may be traced generally in a line across the State. I saw & described the like appearance, but for a much smaller width, near the Peedee, a little below Marr's Bluff. The sand-hills here extend for about 10 miles up & down the course of the river, & 6 miles across, from the river swamp. The surface is rolling, & the hills sometimes even steep, but never long. The soil is of deep sand & very poor. The growth pine intermixed with small scrub & other oaks. The general appearance is like that of Sandy Island, except not so barren & naked, & the oaks much larger. Indeed, at the residences, & where the pines have been cut out, the oaks are coaxed up to a respectable size. For 5 or 6 miles after entering the sand-hills, the country seemed as desolate as possible. Not a creature was seen, nor any mark of man's neighborhood, save the deep sandy track in which I was riding. But afterwards the summer residences of the planters are reached, forming a widely scattered village of large & handsome, & some very costly mansion houses. They are surrounded, & partly concealed by the mature growth of trees, & the scene is very pleasing to the eye, though still on the most barren & sandy part of the sand-hills.

The church service had just begun before my entrance. The church though in miniature, is a very neat & tasteful structure—& I was surprised to hear an organ in this wild & apparently solitary place, & the chants of the episcopal service well performed. After the service, I delivered a letter of introduction to the Hon. John Richardson, (late Governor) & accompanied him to the house of his brother Col. R. to dinner, & to his sisters, Mrs. Manning, at night where he was staying. The residents of the village are the proprietors of the plantations below, for some 10 or 12 miles; & nearly all are members of the Richardson family, & this part of Sumter is known as the "Richardson settlement".—A considerable rain in the afternoon.

49. "Indigo Marl Deposites" in Sumter (26 June–27 June)

26 JUNE. Rode down the river to seek for marl, on & near Jack's Creek, where I felt assured I should find it, from reports that seemed positive & general, & have long been current. After reaching Gov. Richardson's plantation, our party of anxious & interested seekers (besides myself) made six in number; & having obtained saddle-horses, & being provided with two negro men, & digging implements, we proceeded to examine the marl deposites. It is needless to particularize—as the grand results were all the same. The marl in the low swamp, on Dungan's lake side, & the still more noted deposite on the firm bank of Jack's creek, was of artificial formation, & very small extent—probably single boat loads of imperfectly burnt lime, or of marl attempted to be burnt to lime on these spots. The character & small extent were made manifiest [sic] to all by a few minutes digging, or boring instead. To come to this conclusion, we travelled, from & back to Gen. Richardson's plantation house, 10 miles, several of which were through the lowest parts of the great Santee swamp. We got back to dinner, barely in time to escape a very heavy rain. In the evening, made examination in a back swamp, where indications of calcareous matter had been suspected—& again to no purpose. Accompanied Dr. Felder to his pine land residence, 12 miles lower down the country, which we did not reach until an hour after dark. All the proprietors of river estates have quitted them as residences some time ago, & are living in the sand hills, or otherwise in the poor pine lands, some 4 or 5 miles back from the river swamp. Dr. Felder's summer house is somewhere about opposite to Eutaw Springs by a line crossing at right-angles the general course of the river between.—34 miles.

27 JUNE. Returning several miles upon our last night's route, visited Wright's Bluff, (the only bluff, or high land on the Santee, in Sumter district,) & Scott's lake adjoining, where Dr. Felder's plantation lies. What is called Scott's Lake was formerly the only bed of the river, which has since cut through a narrow neck, & taken a shorter route, leaving the lake dead water. It forms nearly a circle of three miles, one end of which is still open to the river, & the other is closed across by a sand bar. The surrounded swamp is Bull's Island. The whole is rapidly filling up by sediment left by the freshes. The width of the water is

nearly as great as formerly, or about 100 or 120 yards. As passing down it in a canoe, the water was much less muddy than that of the river; but was covered by a floating pellicle, or scum, & was quite offensive to the scent. No doubt it is one of the very surest sources of malaria.

Our objects were to examine a marl deposite on the landing, & a bed of rock lying under Scott's lake. The former, as I anticipated before seeing it, was just another such heap as I had seen yesterday, brought to make lime for the indigo works. I earnestly hope that this may be the last of the many *indigo "marl" deposites* that I may be carried to, or have specimens sent me to analyze. Of both, there have been about a dozen—all of which could have had the true character & limited extent ascertained by the proprietors as well as by myself, if they would have given 10 minutes labor & attention to the examination.

The next object was interesting, though I fear likely to be of no direct practical utility. The bed of rock extending across the old bed of the river was slightly traced & found to go quite across, & to vary, irregularly, from 7 to 11 feet below the surface of the water. . . . I was able to bore enough into the rock (which is not hard) to bring up a specimen, & found it to be the white & rich marl of this region. . . .

The low position of this marl, & the total absence of it at any visible & accessible heights, satisfied me that all on this side of the river for a considerable distance down must be of a very low level compared to that on the opposite side. In the general upheaving, there must have been a dislocation, the line of which forms the line of the river, & the broken marl stratum was lifted up from 35 to 45 feet higher on the Orangeburgh side than in Sumter. This conclusion, added to my feeling some symptoms of indisposition, made me change my previous intention of proceeding lower to examine the borders of Potato creek, which empties not far from the Williamsburg line. Without going there, I have learned what I can rely upon as to the existence of the marl, & of its having formerly been dug out to burn for the indigo works. So says an old negro belonging to Mr. James McKnight; & the appearance of the old excavations, & the marl scattered around still are visible. If I were to give the time & labor necessary to examine the localities, I infer that I could learn nothing more, unless it were by boring to *sound* the depth of the marl below the surface of the earth. I have instructed Mr. McKnight how to proceed in the investigation, & to choose specimens of the marl for me.

[Ruffin inserted the following cutting later in his diary.]

For the Mercury
Instructions for marl-seekers, on minor points.
"Indigo-vat marl" formation.

Among the many sources or causes of misdirection which have embarrassed or impeded my progress in searching for marl in South-Carolina, there have occurred so many cases of one general class, that it is found necessary to publish the following remarks with the view of preventing future unnecessary trouble on this score, to all parties concerned. I allude to the many supposed discoveries of deposites of marl, of which specimens have been sent to me to be examined and analyzed, or which I have been induced to visit by erroneous reports, and which deposites a very little labor in digging could have shown to the discoverers to be the used lime of old and forgotten indigo vats or some other equally artificial deposites, not less limited in amount, and of which the true characters were equally obvious, or open to easy detection. In sundry cases I have been called upon to examine what proved to be such deposites, either *in situ,* at much waste of my time and labor, or otherwise much to the trouble of the proprietors in obtaining and forwarding specimens. Several new supplies of the "Indigo-vat marl" have very recently been received. And as there must be many thousands of such forgotten deposites yet undiscovered in South Carolina, (that is, one or more on every plantation on which indigo was formerly made.) it may be by no means a work of supererogation for me to suggest the easy and simple means by which all such artificial and limited deposites may be distinguished from the natural and abundant, and also both kinds from all non-calcareous earths.

. . . If the earth in question is calcareous, (as marl or limestone,) . . . application of acid will produce (by the disengagement of carbonic acid gas,) an immediate and violent effervescence, which it will be impossible to mistake. A mistake however, is frequently made by new experimenters in this manner. If the acid, (as indeed would water) be poured on dry clay, or earth not calcareous, the atmospheric air enclosed in the pores or crevices will give place to the liquid, and escape in air-bubbles, which may be mistaken for a very slight disengagement of carbonic acid gas. The moistening of the earth with water, before touching with acid, will guard against this slight source of error.

This testing by acid decides as to what is calcareous or not, but of course does not distinguish between natural and artificial formations. . . . Nearly all the artificial deposites (indigo vats, transported lime-stone, or other kinds,) which I have been called to decide upon, might have been dug through and exhausted in half an hour, and the most extensive of them would not have supplied a single cart for three days. Therefore, in all such cases of doubt, after ascertaining the earth to be calcareous, let the proprietor put a grubbing hoe and spade to work, and a cart if need be. . . .

Even in regard to natural deposits of *bona fide* marl, it is desirable and necessary that the specimens sent for my examination should be correctly designated and labelled, so that the precise locality and important circumstances shall appear, and also the person be known who selected the specimen, and who is responsible for the accuracy of the description. . . . Very different from the course here recommended, specimens of earths and minerals have been sent to me for examination, without any proper designation of the land, locality, or position in the beds, and sometimes even without the name of the sender. Of course it would be throwing away any labor of investigation to bestow it on such indefinite subjects. . . .

EDMUND RUFFIN

Agricultural Surveyor of South-Carolina
(The Editors of the Courier and the Agricultural Journals of South-Carolina are requested to copy the above)

50. In and around the Sandhills of Sumter (27 June–30 June)

After an early dinner, left Dr. Felder's plantation on my return. After first waiting for the close of one heavy rain, of short duration, hoped to escape any more. But four others occurred before I reached Mrs. Manning's residence in the sand-hills, where as before I found Gov. Richardson. As usual, having lost my way, (in the sand-hill settlement,) I added 3 miles of this to the necessary length of my journey of the day. In all 36 miles—including 3 of walking & in the canoe at Scott's lake, without my umbrella, from which I felt more oppressed by heat & fatigue than at any time this summer.—Heavy rain at night.

28 JUNE. While lost yesterday evening, & driving about the sand-hill settlement, I came upon more very handsome & costly mansions scattered about in the woods. But this morning I saw what surpassed them all so far, as to be a wonder for the location. Within a few hundred yards of Mrs. Manning's residence, is the mansion of Col. John Manning, which is the sand-hill wonder referred to. The house is indeed a palace—of large size, & beautiful in its plan & construction. Every other out-building & structure of every kind is made a part of the general plan, & serves to add to the general effect. The fountain, in a valley, is covered by a miniature copy of an ancient gothic church. The almost only natural beauty of the sand-hills, the trees, are made the most of as ornaments to the grounds. And as the mansion, for the sake of beauty, has been placed near steep slopes descending to streams, the land is better & the trees much larger than ordinary. Indeed, there is as much of beauty in the location as can well be where there is no prospect, nothing seen but trees, & where even grass cannot grow, except at great cost, (including good liming, which *must* be done for this end,) & as yet very imperfectly. The interior of the mansion, which I did not see, (there being no one at home, & the house locked up,) is as elegantly furnished as the exterior. And no where have I seen a private edifice so beautiful, & so magnificent. Yet its grandeur is combined with simplicity, & the erecting for display seems as little apparent in the building itself, as in the hiding it in these woods. Yet with all my admiration of its beauty, I could not but condemn the building, in its location, as a waste which even the great wealth of the proprietor cannot serve to justify.

I continued to feel head-ache, with other bilious symptoms; & this, added to the still threatening appearance of the sky, made me change my previous intention of proceeding on my journey, & to wait a day, to which I had been urgently pressed by my hospitable hosts. I shall also take some mild medicine.

The lands next the river swamp below the sandhills, & stretching some 10 miles along the river road, are of the peculiar red soil before described. They improve in their peculiar quality below, in the neighborhood embracing Gov. Richardson's extensive & valuable plantation. The general surface of this region is level, the soil reddish brown, & the subsoil the same. The original quality was excellent, & the durability great. The production is still better than any highlands I have seen latterly. The culture has been almost continual, since the first clearing. In all this land there has been no indication of visible or accessible marl, except those which my examination served to explode.

A small creek called Half-way swamp enters the Santee just opposite to the other "Half-way swamp" in Orangeburgh. Above this stream is found plenty of the hard rock containing silicified shells, but which are not found below that limit.

In the sand-hills, there are plenty of good & bold springs, & streams, which are very slightly affected by either dry or wet weather. There are mill-ponds in plenty around & they are not believed by the residents to be injurious to health. And if any can be harmless these should be, from their scarcely varying level, & the poverty of the land covered.

I have never seen any persons more anxious than the planters here to obtain marl or lime, & more willing, as some of them have proved, to incur great expense or labor for the purpose. And this is in the full confidence of remunerating profits from the application, inferred from the results already obtained, or which they believe have been obtained. Peter Mellett, a judicious & successful planter & manager, commenced this peculiar practice of very light limings some 6 or 8 years ago. He was not an educated, but yet a reading man—& especially accustomed to read agricultural works, & took the Farmers' Register. I am sure that it was from my translation of Puvis' Essay on Lime, in the F.R.[1] that Mr. Mellett acquired his views, & took his practice. Living 90 miles from Charleston, he was in the habit of sending his cotton there by his wagons to market; & not then knowing of any cheaper plan, he brought back return loads of Maine[2] stone lime, in casks, for which he had usually to pay in Ch.n $1.50, or nearly 50 cents the bushel of the unslaked stone. He made compost of leaves, & perhaps some little of other manure, the layers being sprinkled by a very thin cover of slaked lime. The heaps after remaining long enough to be in good state, were cut down & the compost applied to the crops. The quantity of pure slaked lime thus given to the acre was astonishingly small—only 2 ½ bushels. And though repeated for every crop, (which was generally every year,) I should myself scarcely have believed that any visible & appreciable increase could have been produced from so small a quantity within a few years. But whether such effects were produced, or whether all opinions thereupon have been mistaken, it is certain that Mellett did improve his land fast, & greatly, before his death; & that he, & all whom I have heard speak of his practice, ascribed his superior products solely to the lime & its action on the other manures. Col. James Richardson has tried the same practice, & with still smaller proportions of lime; & is as well satisfied of the beneficial results. He obtained his limestone from the other side of the Santee, & then hauled it by land 8 miles from Wright's Bluff & burnt the stone. And if he cannot get it nearer or cheaper, is willing to continue to get it by the same laborious route. Now with all my confidence in the value of the operation of marl & lime, I cannot go to this extent. But all respect is due to the opinions of these practical & judicious planters—who have no theory to sustain, & knew & cared nothing about the *mode of operation,* & who merely adopted the mode of application cheapest in their difficult position, & judged of profits & values by the effects they saw, or supposed they saw. All their neighbors whom I have heard speak of the facts, fully believe in the benefits, & also of the cause being the lime.

But a slight sprinkling, through threatening rain all day. Late in the evening went to the site of Dyson's cotton factory, to examine the fuller's earth[3] for marks of shells, which were said to be in it. This fuller's earth is very abundant hereabout, forming the lowest stratum & of great & unknown thickness. . . . It resists the washing of water so well, that it forms cascades like granite. . . . The earth when dry, appears not a soft stone. I am sure that it is the same that is found truly a stone in a thin layer, & in such abundance, in the waters & in the land, about Ashepoo & Huspa, & many other parts of the low country—& which is there believed to be petrified wood. As it lies below the silicified shell-rock (& the same was seen at Col. McCord's) which formerly was marl, it would seem that this fuller's earth must be the stratum under the great body of marl. It shows however in a stream flowing into the Santee between Dr. Palmers & Coutorier's [sic], in the marl region. Took blue pills at night.—3 miles.

1. M. Puvis in *Annales de l'agriculture francaise* (1835), extracts translated in *Farmers' Register* 3 (March 1836): 690–91; (April 1836): 705.

2. Maine was the prime source of stone and lime for the Southeast; its principal kilns were in the town of Thomaston.

3. Fuller's earth is soft, absorbent bisilicate of alumina, used for fulling cloth. It is found in strata with chalk or limestone.

29 JUNE. Proceeded after breakfast on my route.

The "fuller's earth" is seen 30 feet thick at the next valley crossed above Mrs Manning's house; & there & elsewhere I saw the silicified shell rock, though in small & separate masses, lying 3 to 4 feet above, on a loose & coarse sandy clay, which lies immediately upon the fullers earth. The latter has a very uneven upper surface. At a deep washed gulley between Manchester & the river swamp, which I went this morning to see, the fullers earth is near the top of the hill, & below it some 40 or 50 feet of numerous & various strata of clay, sand, ochre, & thin iron ore. . . .

The sandhills furnish a number of springs & streams, & a quantity of water, as may be seen even on the map. The temperature of the spring at Mrs. Manning's I found to be 66°, which is supposed to be like the springs generally. But at Col. Manning's fountain, it is 61 ½°. This source however is not from the sand, but from under clay strata, & comes from a different direction. [This passage was intended for another page not specified by Ruffin.]

By stopping at Manchester to dine & feed my horse, & because of the threatening appearance of the clouds, I escaped a very heavy rain. After it was over, proceeded to Dr. Anderson's at Statesburgh. He was anxious for me to stay through the next day, & to invite some of the neighboring planters to dine with me; to which I readily consented, as well as for the need of rest, as for the gratification of the meeting.

Statesburgh is on the elevated range known as the "High Hills of Santee", which range extends 3 or 4 miles lower, & joins the fine level red lands at Messrs. Macrae & Singletons, & stretches some 20 to 25 miles above Statesburgh. This high country is healthy.

Along the road passing through the adjoining plantations of Messrs Macrae & Singleton, there is for more than a mile a beautiful & well-grown thorn hedge. But like all other hedges I ever saw, it requires a good fence to make it a secure barrier. This hedge has been trimmed improperly. The top is wide, & consequently the bottom is open. The enclosure is otherwise made by a low & good straight fence on a bank & ditch; but there is an unusual feature deserving notice & imitation. The ditch (next the road) has its outer side cut away, & the earth thrown on the road. . . . This permits a much better shape to the road, gives greater width, & allows carriages to run safely even to the bottom of the ditch. The fencing in this state, especially along public roads & elsewhere if fencing timber be scarce, is generally of ditch & bank, with a low straight fence. The banks stand better than I had inferred; & when the plan is well executed, it is much superior to most of our fencing in Va, & offers a good example for imitation, by which I hope to profit.

30 JUNE. Had an early breakfast, & rode with Dr. Anderson to see his plantation on the Wateree, 7 or 8 miles above, & especially his river low-grounds. He has commenced embanking, & has already secured nearly 200 acres, which will be increased to 400 by the completion of the plan. The land seems as rich as could be desired.

I am more fully convinced by larger general observation, of the practicability of reclaiming much the greater part of the immense expanse of river swamplands on these rivers & others in S.C. But my plan, for the large rivers, would not be like any yet adopted. Neither like Col. Hampton's, which takes in all the low ground, & connects it by his bank with the highland—nor Major Porcher's, which excludes the freshes only above, & shields his land in part, by giving the current another direction. The last, in its results is necessarily very imperfect; & the swamp left outside of the embankment next the river, is always the highest, would require the least height of bank, & is much the most valuable & safest land for cultivation. The latter plan is doubtless the safer of the two. For if all the swamp lands were embanked on the former plan, the present bed of the river would be the only part left for the passage, & the freshes being thus confined between dikes, would rise so much higher as certainly to overtop & break them. My plan would be to run the embankment near to the river, leaving however sufficient margin for its protection, & *never* joining to the high land, but leaving next to it an open passage of 200 to 400 yards wide of swamp, which is *lowest* next the highland; & to supply earth for the rear dike, digging a straight & clear canal of 25 or 30 feet wide. . . .

The Santee swamps are generally 3 miles wide, & those of the lower Congaree & Wateree not much less. If the work were done upon this plan, the side passages left of low swamp (kept clear of fallen trees) & the clear & nearly strait canals therein, would furnish as clear a passage for the high waters as the whole width of swamp obstructed as it is now. Indeed, the river would soon choose these shorter passages, making them deeper & wide enough, & perhaps desert & fill up the present crooked bed.

No rain today to prevent the arrival of the visitors, with whom I spent an agreeable time. A short but heavy rain at night.—17 miles.

Mr. Nelson, one of the company, owns the land on the lower part of Potato Creek. He brought me a specimen of the marl, & also better information of its position & extent. It rises (as visible) 3 or 4 feet above the level of the water, besides showing in the bottom—& is seen for considerable distance along the creek. The outlet is at Black Oak Island, which lies just opposite the entrance into the Santee Canal.

51. Unwell at Columbia and the Monticello Convention (1 July–8 July)

1 JULY. To avoid as much as possible the risk of rain, & the heat of the sun, had an early breakfast. This enabled me to reach Columbia before 2 o'clock; & it was well that I did so—for in an hour after, there fell one of the most sudden & yet heavy rains I ever witnessed, & in which I should have been completely drenched, if on the road, in spite of umbrella & great coat. I have thus, in 4 days of the last 5, barely escaped heavy rains; & being thus made wet is deemed dangerous & would be especially to me, unwell as I have been & still am, though not much so. I have had to continue to take more & stronger medicine, for bilious symptoms, tonight.—33 miles.

This day excessively hot—& continued so even after the rain.

2 JULY. Sunday. Another very hot day—& the first since yesterday week in which it has not rained. Remained in the house nearly all day. Bilious symptoms continue, & took more medicine.

3 JULY. Feel much better, but weak. Was obliged to go to my collection of specimens in the State House, to select the fossils designed to be carried home for more full examination.

4 JULY. Left Columbia early, & reached Monticello at 2 P.M. to meet my appointment tomorrow with the Agrl. Convention. At night, the questions of the Committee (just completed) were put into my hands, & for the first time I had some definite idea of what was required of me, or expected. The questions would puzzle a much wiser man to

answer—nevertheless, as I am not afraid to confess my ignorance, I am very well satisfied with the manner of procedure which the Society have chosen. The questions are mostly very illy adapted to their end of gaining such information as I could give them.

Except a little headache, I have no remaining feeling of sickness. But am very weak.—32 miles.

The surface travelled over today is very hilly, & the lands not generally good. At & near this place, much improvement both of soil & culture. The land mostly red near to Monticello, & much like the upland of Goochland & Buckingham in Va. Slate showed in the lowest ground within 4 or 5 miles of Columbia, & more considerably at intervals afterwards. But the stone, which is plentiful, is mostly quartz, in loose fragments. A few miles below Monticello, granite shows in huge boulders, some 25 feet across, as seen above ground—& these lands show granite every where some 10 feet under the surface—& it sometimes appears, more or less disintegrated at the surface. Corn more than cotton cultivated, & the crops of the former better.

Dr. James B. Davis brought me from the tavern forthwith to his house which is within a short distance.

In the evening, paper of the formal questions to be asked, which had been concocted by a large committee of the Society of Monticello yesterday & today, was placed in my hands. There are 19 questions; & different from what I would have supposed in this region. They mostly are in relation to calcareous manures & kindred matters of agricultural chemistry. Some of them would puzzle Liebig[1] himself, & to which I can reply only by admitting entire ignorance. However—though I would have much preferred questions of different form, in the general, yet if these suit the inquirers, I have no objection to answer them or any others.

1. Justus von Liebig, later Baron (1803–73), German scientist and founding father of agricultural chemistry, was interested in plant nutrition and stressed the fertilizing value of phosphates.

5 JULY. The society & visitors assembled, making as I supposed at least 600 in number; mostly plain respectable planters of this & the neighboring districts. I did not see one who appeared to be of the lower class, of mere spongers & loafers, though such might be expected to

be attracted by the plentiful barbacue [sic], (though without any liquor than water,) & the freedom of such occasions which generally admits as many of such undesired guests as choose to intrude. After some preliminary business of the Society, & debate, my *questioning* was begun. It was not confined to the prepared bit; but many off-hand questions, suggested by what was said, were asked by sundry of the company. The conference lasted for more than two hours, divided by the recess for taking dinner. My replies extended to a wide range of agricultural views. I have rarely seen more interest & attention shown by an auditory for so long a session. And though I thought & feared that my words in answer were perplexed & obscure, as well as formal questions being far from judicious, still, if I may believe the subsequent expressions of many individuals to me, the whole gave great satisfaction & pleasure, & as was said valuable & new instruction. This is much better than I had conceived possible from such a meeting, in this place. I came to attend it with the greatest reluctance, because I thought the people here would take no interest in the subjects of my investigations elsewhere, & that I knew nothing to tell them of what they might expect to hear—& that the whole affair would end in great disappointment, of which I should bear the blame, & be made the scapegoat. My weakness of body made the day a severe physical trial—but I suffered less than I had expected.

[Ruffin later inserted the following published report of the questions and the proceedings.]

For the Planter
AGRICULTURAL CONVENTION
Monticello, Fairfield Dist. July 5

The convention was organized, by calling Dr. Thomas Smith, of Darlington to the Chair, and appointing J. J. DuBose, Secretary—after which, Mr. W. J. Allton[1] offered the following resolutions.

Resolved, That in the opinion of this Convention, an agricultural survey of the State, may be of much benefit to the planters, individually, and to the State at large—by "collecting accurate information on the condition of its agriculture, and every subject connected with it; and in pointing out the means of improvement."

Resolved, That it is the true policy of the cotton planters of the

State, to curtail the cotton crop, and increase the provision crop—so as to supply all the bread stuffs, and raise all the different kinds of stock, to wit:—horses, mules, cattle, sheep and hogs—which may be necessary for family and plantation use, and profitable for market.

Resolved, That in order to accomplish this, it is desirable to ascertain what kinds of grain and root crops are best adapted to particular soils and sections of the State, and the most approved modes of cultivating the same, also what are the best breeds of cattle, sheep and hogs, suitable to the lower and upper countries,

Resolved, That information is required, as to the best mode of reclaiming worn out land—embracing the most improved rotation of crops, the different kinds of manure, and their application; also, how far the native or foreign perennial grasses may be profitably cultivated. And whether it would not be a valuable improvement in the agriculture of the middle and upper countries, to curtail the number of acres to the hand, and, consequently, the number of horses and mules necessary to cultivate them. "And by manuring and resting the soil, and superior cultivation, to produce a given result, from the smallest number of acres?" And whether, in some sections, live fences or hedges, would not be the cheapest, and what kind of hedge is best adapted to our State? and how planted?—and whether with or without ditch?

Resolved, That in collecting accurate information on these, and other subjects of agricultural improvement, the zealous co-operation of the Agricultural Societies, and of individual planters, will be of the utmost importance; and, therefore, this Convention do hereby earnestly recommend to every local Agricultural Society, and to the planters in such districts as may not have such societies, to collect and report to the Agricultural Surveyor of the State, in as condensed a form as may be practicable, all such information as they may be able to acquire upon the various subjects indicated in his circular, and the foregoing resolutions—and upon such other subjects connected with the agriculture of their respective districts, as they may deem valuable.

Mr. J. B. Davis stated, that a committee had been appointed by the Monticello Planter's Society, to prepare questions on subjects respecting which, they desired to obtain Mr. Ruffin's views. The Rev. J. Davis, chairman of that committee, submitted the following, as the questions prepared.

"The committee appointed by the Planter's Society, at Monticello, to prepare such questions as may be most expedient to propound to Mr. Edmund Ruffin, Agricultural Surveyor of the State of South Carolina, at the convention, to be held July 5th at Monticello, relative to the constituents of marls, and their effects, as a manure, together with other subjects connected with agriculture—report the expediency of the following questions:—

1st. In your analyses and experiments in the different qualities of marls, have you ever been careful to designate the presence or absence of potash, magnesia, manganese, soda, *gypsum, silex,* common salt, &c. as well as the lime, and to mark their results as a manure, so as to determine which of those constituents, or what proportionate combination of them, rendered the marl most valuable?

2d. Have you ever analyzed the different soils and subsoils of lands where marls were applied, previous to their application, and observed the results, so as to prove on what qualities of lands, certain different qualities of marls proved most successful?

3d. Have you ever tested or witnessed effects of marls, not formed by the decomposition of shells, or alluvial deposits, but such as where steatites or soap-earth, chalk or other argillaceous[2] substances, &c. abounded, when used upon the different soils?

4th. What have been the results of your experiments and observations, relating to the best proportionate admixture of silicious and argillaceous matter, or sand and clay, for the different plants most generally cultivated in our climate? Or have you ever witnessed the effects of sand applied to stiff clay lands, and *vice versa;* and if so, what were the proportions applied, with the results?

5th. Have you ever witnessed the effects of lime and gypsum, as a manure, which did not contain potash, magnesia, soda, common salt, &c. or where only a trace of those constituents were present, so as to determine whether the chief value of lime and gypsum may not be justly ascribed to the presence of such constituents?

6th. Have you ever witnessed the effects of common salt, potash, magnesia, nitre, soda, alumina, or clays abounding in those properties, or any one of them, on soils where these properties were not present, so as to determine their value as a manure?

7th. Have you any experience in charcoal, as a manure; and if so, what were the results?

8th. What do you consider the best plan to preserve fertilizing properties of compost manures, until applied for use?

9th. What do you consider the best manure for wheat?

10th. In your travels in South Carolina, have you observed any specimens of red clover, which would compare with those you have observed in Virginia; and if so, did you ascertain what proportion of lime (if any) was present in the soil where they were produced?

11th. What have been the results of your experience and observations, upon the relative expenses of manuring lands, by compost manures, of ploughing or listing in the vegetable productions of the soil, and of using marls; and if a combination of those plans of improvement, would nett a higher profit, according to the expenses?

12th. Have you ever made any discoveries of the cause, or for a preventive, of rust in wheat;[3] and if so, what are the facts on which they are founded?

13th. What are the conclusions you have come to, from your observations, of the best season of the year for breaking up land, preparatory to cultivation?

14th. What are your views relative to the depth of ploughing, for the preparation of lands for planting—and during the cultivation of a crop?

15th. Have you ever discovered whether the colours and tenacity of soils were any certain index of the presence of alkaline, calcareous or mineral properties in the soils; and if so, what, particularly, are the peculiarities of color.

16th. Have you ever observed, where those properties abounded in the subsoil, any peculiarities in their changes of colour, when exposed for some time to the sun and atmosphere?

17th. Have you ever observed any peculiarities of colour, in compost manures, when in their most favourable state of putrefaction for use as manure; and if so, what are the characteristics of color?

18th. Have you, in your survey of this State, discovered any of the green sand-stone, and to what do you attribute its virtues as a manure?

19th. From your experience in the properties of lime or calcareous manure, and of ashes, or alkaline stimulants, would you advise the practice of deep ploughing on all soils, and of spreading the brush and underwood, and burning them over the surface, or would you prefer the system pursued by some of our distinguished and experimental planters, of collecting and piling them, (the brush and underwood, logs, &c.) in rows, at convenient intervals, and removing them from year to year?"

(Signed)

JONATHAN DAVIS
Chairman of Committee

Wm. J. Allston Secretary

Dr. Davis moved, that Mr. Ruffin be not called for an answer to any of these, until tomorrow. After some discussion, it was agreed that he should proceed with his replies at once, as he felt willing to do so. Mr. R. commenced in a conversational style, to give his views on most of the various topics suggested by the committee, (some of them, he remarked he was wholly unable to discuss properly) and on others suggested, from time to time, by members of the convention, until the hour for dinner arrived, when on motion, the convention had a recess until 4 o'clock. . . .

Mr. Ruffin was then invited to proceed with his replies to the questions propounded. He accordingly resumed his explanations; after which other questions of interest were proposed, and a conversational discussion carried on for some time. . . .

Mr. W. J. Alston introduced the following resolution, which was unanimously adopted.

Resolved, That the thanks of the Convention be returned to Mr. Ruffin, for his very instructive and satisfactory answers, to the various questions which have been propounded to him.

On motion of Dr. Davis, it was agreed that the Convention stand adjourned until to-morrow, 11 o'clock.

1. Though the name was spelled three different ways in the report, William J. Allston was probably meant.

2. Pertaining to (or compound of) clay or material of that texture.

3. This rust, consisting of orange-brown blisters on leaf blades and sheaths, was caused by the fungus *Puccinia rubigo-vera tritici.*

6 JULY. The meeting continued & adjourned. Nothing of much interest in the formal proceedings. A number of the principal members invited with me to dine at Dr. Davis', where we spent a pleasant afternoon.

7 JULY. Dined at Gen. J. Means', with a number of gentlemen, 14 miles above.

The lands of this part of Fairfield are all on granite, red, hilly, & of good quality. The culture is among the best, & the proprietors among the most successful in the state. There is some effort at a rotation of crops. Col. Jonathan Davis, one of the oldest & best farmers, has the succession of 1. cotton, 2. corn, 3. wheat & oats, then beginning again with cotton. They make much manure, mostly of leaves, used as litter for stables & cattle pens. Cotton seed, for manure, is higher prized here than any where. It has sold, formerly, as high as 18 cents the bushel, for this use. Now, no planter will sell cotton seed; but at public sales of estates, it usually commands 10 cents. The germination is prevented by heating the seed 5 or 6 days only before applying them. This is done by pouring hot water into the heap, which produces fermentation immediately.

The great defect of tillage here, as everywhere in S.C. is the planting all the land, & resting none, until it is worn down so low as to require turning out for a long time.

8 JULY. Returned to Columbia.—32 miles. Granite, in large masses, appears at the first stream just above Columbia—but manifestly not *in situ*. . . . Four miles above Columbia, slate shows nearly the whole length of a long & steep hill through which the road exposes a section; & it shows for some miles farther up. . . . The layers of slate[1] *cross* the direction of the road, & their dip is very steep, & not regular, & sometimes almost perpendicular. . . . Next, say 15 to 20 miles above Columbia, the stone abundant on the surface . . . forms a stiff adhesive poor clay, on which small black-jack oaks form nearly the exclusive growth. Next to this up the country, & beginning some 8 to 10 miles below Monticello, granite . . . is the underlying rock, & the soil formed, as I infer, mostly from its decomposition. It is much the best land—still it varies much in value, & I think the *best* soils are caused by the greater

quantity of granite containing hornblende[2] in place of mica, which kind is abundant.

Not well yet, but much better.

1. Ruffin refers to highly metamorphosed clay, easily split into thin plates.

2. Hornblende is a mineral compound with lime and also, commonly, magnesia and silica.

52. To Charleston, Indisposed, and Home to Virginia (9 July–11 July)

9 JULY. Leaving my horse at the tavern stable, & my heavy baggage to be kept to my return, set out at 6 A.M. in the railway train for Charleston, & arrived at 2 P.M. After dinner, though more indisposed called to see Mr. & Mrs. Roper, & spent the remainder of the evening with Dr. Bachman & Dr. Ramsay. At night took more medicine.—125 miles of railway.

10 JULY. Packed up all specimens necessary to be taken home, & attended to other arrangements. Had sundry friends to call at my room in the Fire-Proof Building, so as to take up very agreeably most of the forenoon. At 4 P.M. took my place in the Steamer for Wilmington, & thence, the next morning, the railway for home—where I arrived on the 11th, at 2 A.M. more indisposed than recently, & needing medical treatment, which, now having opportunity for, I shall forthwith resort to more properly. Still I have not had a fever yet—only head-ache & other symptoms of acculation [sic] of bile. Found my family well, & well-doing.

III

Late Summer & Early Fall

TOWARD THE MOUNTAINS

53. Return to Columbia and Buggy to Fairfield (22 August–27 August)

22 AUG. Left Petersburg, Va. at 7 P.M. & reached Columbia, S.C. on the 24th, at 5 P.M.

My health has continued slightly affected by bilious symptoms for about 4 weeks, & it was only after resorting to a course of medicine, that I had recovered, & had about 10 days of freedom from indisposition to attend to business before returning. Previously, I had written as much of my report as materials already acquired permitted, & made other arrangements deemed necessary to the same end.

25 AUG. Selecting & assorting & putting up specimens of fossils at the State House, during the forenoon; & after dinner set out in my buggy for Fairfield. Reached the house of the Rev. J. Cook, above Cedar Creek.—15 miles.

26 AUG. Went, guided by Mr. Cook to see what had been reported to be *limestone,* on the lands of T. Center & Mrs. Leitner, near Cedar Creek, & near the line of Richland & Fairfield. It turned out, as I had suspected, to be scattered lumps or nodules, or stony calcareous concretions, in a mass of earth. . . . They are too few to be separated & collected for burning (even if rich enough) & the matrix has so little of calcareous matter, that the whole mass would be scarcely worth using as manure. . . .

Rode to a cavalry muster & barbacue 10 miles distant, where there was a numerous assemblage of respectable people, including many ladies in their carriages, & who partook of the barbacue. Here, as at the much larger meeting on the 5th of July, there was nothing to drink but water. These, added to other facts, afford gratifying proof of the great progress of the temperance reform in Fairfield. There is but one whiskey shop in the district. At night, to Col. Jonathan Davis' house, near Monticello. He was not at home much to my regret, though I was kindly received & entertained by his family. This old gentleman is the father of agricultural improvement in this neighborhood, as well as of several good & improving planters.—22 miles.

27 AUG. Sunday. Spent the day with Dr. J. B. Davis & the evening & night with Mr. Wm. Davis, both close to Monticello. Graduated ditching is practised by all these gentlemen on their hill-sides, & with a greater descent than I have ever seen, which seems a valuable improvement.

54. Union and Spartanburgh: Limestone Springs and Gold Mines (28 August–1 September)

28 AUG. To Unionville—39 miles.

The season has been very favorable to vegetation, & perhaps too much rain since I was before in Fairfield, & the crops appear to great advantage. The surface above Monticello, as of all the other land seen, very hilly, red, & much washed where not in careful hands. 14 miles above, crossed the Broad River at Shelton's ferry. Afterwards scarcely any pines seen in original forest growth, it being almost entirely of oak. The second growth of exhausted old fields, of pine mostly, but very scattering, scrubby & stinted. All of short leaf kinds, & not the old-field pine of Va.

The road on the middle ground between the waters of the Tyger & Broad rivers—& the latter part of the days journey on a level though narrow ridge, gray soil mostly, & sometimes the road even sandy. Fine land when new, but soon washed & worn out.

At Unionville by rare good luck I met with Capt. James A. Black, whom I was to meet by appointment, but whom I expected to have to seek at his home, & that after writing to him. He had offered to accompany me through the limestone region. No one can be a better guide, or give me more advantages by his company. He is the member of congress from this district; &, from being long engaged in iron-works, is well acquainted with the limestone localities,[1] & the little that has been done in quarrying & burning the stone.

1. Limestone was used in blast furnaces to keep the slag (scum of impurities over molten iron) runny and to remove sulfur from it.

29 AUG. With Capt. Black, went to Glenns Spring, in Spartanburgh, 16 miles, where we propose to stay several days. The spring is impregnated with sulphur & (as said) with magnesia—& has cathartic effects. There are here now, as usual at this season, many visitors for health—though not as many as I had hoped to see of planters. Showery.

The land on todays road hilly, though much less so than in Fairfield. About the junction of Union & Spartanburgh, & thereafter, the land becomes redder, richer, & chestnut trees[1] begin, & become common.

1. *Castanea dentata* grows often as small, shrubby trees, with oak in the Blue Ridge; it provides nuts and durable wood.

30, 31 AUG. This is the gold region of S.C. & within a few miles of this place there are 4 mines now worked. Visited two of them Dr. Nott's, & Mr. Black's. At the latter the vein of ore is principally of a very silicious rock lying between slaty clay. The vein is nearly perpendicular, & the dip westward, & from 2 to 4 feet thick. It shows at the surface of the earth, & is worked to 70 feet deep, where the further progress is stopped by the too rapid rise of water. It is richer at bottom; but the strength generally very regular. The gold in the richest ore is rarely visible to the eye, or even by a magnifying glass, & then in very minute & very few specks. But though the ore is poor, it is easy to work, & the product very regular, & the proceeds of this month are about $2400 of gold, with the labor of 6 hands.

The ore is drawn up through shafts by buckets & windlass, & when got up much of the rock is reduced to sand. It is then taken to the mill close by to be ground fine beneath a stream of water. The mill turns 4 heavy cast-iron wheels running on edge, which follow each other in a circular iron trough, in which the ore is gradually thrown, & into which a stream of clean water is constantly entering, & running out at an aperture carrying within it all the finely ground ore. The 4 wheels weigh lbs. 1500 each, & the machinery which turns them is worked by 6 mules.

The water passing out, & conveying the ground ore, flows first through a wooden trough in the bottom of which is placed a thin layer of mercury. The strong attraction of the gold for mercury causes the former to combine with & be retained by the latter, forming a solid amalgam. Still much of the gold passes off with the water; & the sand (though reduced very fine,) is again collected & ground over in like manner, not only once, but twice; & each time produces valuable returns of gold.

Though the ore is generally poor, as above mentioned, there is one small vein passing through the general body of ore, of extreme richness. It is only about a quarter of an inch thick. Of this part about a quart of ore was washed in a pan, to show us the product, which seemed about a table-spoon-full of the pure & brilliant metal, supposed to be worth about $15.

At Dr. Nott's mine, the ore is a very hard quartz rock, mixed with iron. It requires to be burnt, before being ground, & the labor is much greater than at Black's mine.—15 miles.

1 SEPT. Left Glenn's Springs more indisposed than when arriving there, for the Limestone Springs. This another extremely hot day.—27 miles.

I had thought my sickness entirely removed, some 8 or 10 days before leaving home. But whether it be owing to my passing through the low country of N. & S.C. on my rapid course to Columbia, or to the oppressive heat of the weather every day of my travelling since, my previous bilious symptoms have returned, & have now been increasing for 4 days, though still not worse than constipated bowels (notwithstanding the water of Glenn's Spring,) slight headache in the mornings, & yellow eyes.

55. Legume Culture and a Company Hotel in Spartanburgh (1 September)

I was accompanied & guided by Mr. Black, & our ultimate destination was his residence. The road entered Union district very soon after leaving Glenn's, & continued therein for the greater part of the day's ride, before re-entering Spartanburgh. The country continues hilly,

& much of it scattered over thinly with small stone—rarely any large. The soil more of gray than red, & but of moderate quality. Cotton culture has ceased to be extensive or general; corn, wheat & oats the principal crops; & these always in the succession of 1 corn, 2 wheat or oats—& in many cases followed by a year of rest, forming the 3 shift rotation of Va.

After crossing Pacolet river, at a ford, we passed by the plantation of Major Littlejohn, whose general good farming & peculiar pea culture would have made us call, but for his sickness, which renders it improper. We went a mile out of our way to see his field from which wheat had been reaped this summer. It had been last year in corn, with peas,[1] (as is almost the universal practise [sic] in S.C.) & the peas gathered, or the vines removed with the corn, for sowing wheat. The peas were of the black (or "tory") variety, which is so hardy that the seeds will lie on or in the ground all winter, & come up in spring. From the scattered peas left, there came up in the spring enough to make a tolerable cover—in some spots plenty thick, though generally too thin. On the richest spots, the pea vines now stand more than knee high, & a few pods were fully ripe. But most of the land is poor, & the growth low in proportion. But though not usually leaving the land *without* seeding in peas, Major Littlejohn has for 5 or 6 years sown this kind of pea at the same time with his crops of wheat & oats—& always succeeded in obtaining secondary crops after reaping the wheat or oats. It was also said that he sowed as little as a peck of peas to the acre. Before learning these facts, I had heard from Mr. Wm. K. Davis that he had for 3 years sown the hardy red pea (believed to be the same as the black, except in the color of the seed,) with oats—& obtained good succeeding crops. The peas grow very slowly & look badly until the grain crop has been reaped, & therefore cannot injure its product. After removing the grain crop, the peas grow rapidly; & as in September is their principal growth, the crop I saw today has the greater bulk yet to gain.

If this use of broad-cast peas can be regularly made in Va, & also, if it would not injure clover sown the same season, this practise [sic] would be an admirable adjunct to our (or indeed any) rotation. For at no other cost than the sowing say half a bushel of peas to the acre, with wheat in autumn & oats in spring, there would be gained an excellent green manuring crop to follow. It is said that on Major L.'s lands so treated, the peas will fatten a hog to the acre. But I would be content with the *manuring* operation only of the crop, even if no crop of peas were made, or secured.

On approaching near to the Limestone Springs, I saw the first appearance of the limestone which was the object of my present route. It rises above the surface of a valley, & has been quarried to considerable extent for lime for building.

I had before seen the sites or the remains of deserted villages in lower S.C. & of a closed & vacant palace on the barren sand hills. But here seemed to be a palace in extent, though not in splendor, which was open & ready for hundreds of residents, & yet contained none. We walked into a spacious & well furnished parlor, where after sitting until I was tired of it, & seeing no one belonging to the house, I sought in vain for a servant, or for any means of obtaining one. I at last got a drink of water by going to the spring for it—being directed by a person who was also a visitor. The only attendant on guests had taken our horses to feed them, & the landlord had gone away. When he returned, I found him very attentive, & his accommodations very neat & comfortable. This was but a few years ago, a newly established summer resort, at which met numerous visitors of the best class in S.C.: & the company, in expectation of the permanency & increase of the support, spent some $30,000 of their money, & went in debt $60,000 more in buildings & other improvements. But a mere change of fashion has turned all the former visitors to other directions, & this immense establishment is now merely an ordinary country inn, where a few travellers call for the entertainment of a single meal, or to stay a single night. Still the place is beautiful, & the accommodations excellent for 200 guests. The principal building is of brick, 4 storeys high, 250 feet long (besides a 2 storey wing attached) & 40 feet wide, with broad piazzas to all the floors extending the whole length of the main building, & on both sides of the house. The style of construction is plain, but neat & abundantly good for the object. There are besides 8 or 10 smaller houses for lodgers. Of course the company became bankrupt, & the establishment has been sold for its bank debt alone, at $20,000. There is now on foot a scheme to obtain a grant of it from the state (which owns the bank & the debt) & also a sufficient annuity to endow a high school. In this scheme Mr. Black is zealously & ardently engaged, & for his gratification, as well as for the real value to the public, I hope it will be carried through.

1. Like clover, these peas were a valuable double-function legume, yielding fodder to livestock and fixing nitrogen in soil; they are intolerant of acidity.

56. Iron and Lime Production in York (2 September–3 September)

2 SEPT. After breakfast set out for the Nesbitt Iron works, on Broad River, at the head of the 99 islands falls, the foot of which is the head of continued navigation from Columbia. The works are very extensive. The woods are cleared for some miles to get wood & charcoal for the furnaces. The ore is smelted, & the pig iron afterwards worked into bar iron in the rolling mill. There is also an extensive nail factory, a foundry for casting, workshops of every kind necessary, & also a large flour mill. There is a wooden railway about 6 miles long, to bring fuel, lime-stone, & ore from that side of the river—& ore & lime-stone are also boated down the river. Several hundred thousand dollars have been embarked in this great enterprise; & it is to be regretted that half of the capital is already gone, & it is very doubtful whether the balance will yield any returns of profit. This is but one of the thousands of operations which have been caused by the intoxication of paper-money banking—& by which almost every individual of the people has suffered more or less, unless he be one of the "knowing ones"—a cunning stock-jobber who knew when to sell out, or otherwise a fortunate gamester, who won by luck, when by reason he ought to have lost.

After dining at the Iron Works, where I was kindly attended to by Mr. Clarke, the intelligent superintendent, & Col. Nesbitt, one of the proprietors, we rode on to Mr. Black's house, on the border of King's Mountain, & about 3 miles from the N.C. line. On the days route viewed sundry exposures of lime-stone. It shows only in the lower places, & there barely juts above ground. It is doubtless the continuation of the same narrow exposure of similar limestone which is seen in Va. passing through Campbell, Nelson & Albemarle, east of the most eastern range of mountains.

The iron ore used by this neighborhood is of different kinds. The best is the gray ore, in which the iron is intermixed with talc, & the vein of ore lies between talc. It generally contains about 50 per cent of iron, which is as rich as desirable, being harder to work when richer.

Some specimens contain 80, but neither would be suspected to be a metallic ore, but for its weight. The iron made from the gray ore is said to be the best in the world. Another large company, the King's Mountain, have their principal works some miles down the river. The whole neighborhood is rich in veins of iron ore.

3 SEPT. Sunday. Two very intelligent neighbors of Mr. Black's came by his previous invitation to see me—both mechanics & plain men, but who gave me much useful information in regard to lime stone, & its burning etc. Both are from the limestone region of Pa. Lime sells in this neighborhood usually at 25 cents (unslaked) from the Kiln. Mr. Janney, keeper of the Limestone Springs, is selling regularly, but at a slow rate, at 33 cents, & hiring all his labor for the purpose. He says, that even with the present very limited & badly contrived means for raising the limestone & burning, he could make money at 10 cents for a large & certain demand. Mr. Graham, one of the persons above referred to, & the best informed & most judicious calculator hereabout, thinks that 8 cents would pay, with proper operations & on a large scale.

I was disappointed at the little resemblance to mountain scenery, & mountain elevation seen here, though on the outskirts of the celebrated King's Mountain, & within two miles of the battlefield.[1] The highest ranges in view rise not much above the general horizon formed by the forests, & would be deemed very small mountains at any rate. The highest part of King's Mountain is in N.C. & not in sight from this place.

The soil of this neighborhood is not very stony, & of moderate fertility. It is supposed capable of producing 15 bushels of corn to the acre when new, & in its best state. All hilly, but not as much so as most of the land in Union & Fairfield.

My indisposition seems to increase, & I have returned to blue pills.

1. At Kings Mountain a force of "overmountain men" and militia attacked Colonel Patrick Ferguson's eleven hundred loyalists in October 1780, killing or wounding nearly three hundred and capturing most of the rest. The battle was heralded as the end of the British southern campaign.

57. Illness in Limestone Springs and an Address to the Planters of Spartanburgh (4 September–11 September)

4 SEPT. Returned to Limestone Springs, intending to stay there a few days to nurse myself in quiet, & to write letters, & bring up my journal, which I find it almost impossible to do in any private house. The very extent of hospitality, & kind attentions shown to me as a guest, deprive me of all command of my own time for writing, or for anything requiring me to be alone & undisturbed.

By the time I reached the Springs, my headache had become much more troublesome, & to which was added the toothache. Took immediately a Seidlitz powder,[1] & to my bed.—16 miles.

1. Tartrate of soda and potash, bicarbonate of soda, and powdered tartaric acid was dissolved in water to give an effervescent laxative (named after Seidlitz springs in Bohemia).

5 SEPT. Passed an almost sleepless night, owing partly to the pain which has spread from the temple to the lower jaw—but more to two cups of strong tea which I imprudently drank, & which affected my nerves all night. The pain is much less this morning—but I cannot talk without increasing it; & that is difficult to avoid, as several gentlemen have called upon me. Sent for Dr. Nott, & had a tooth extracted, & took his advice on my more general indisposition.

9 SEPT. After being confined 5 days, & still much indisposed, but better, left the Limestone Springs, early in the morning & reached the village of Spartanburgh (the C.H.) by 11.—21 miles.

The road passed near to Watkins' quarry, 7½ miles from the Limestone Springs, which is the most southern exposure of the limestone in this district. But the stratum has been reached by the digging of a well within 200 yards of Pacolet river, & 7 miles from Spartanburgh C.H.

Being very unwell still, headache & bilious symptoms continuing, I have to continue taking blue pill & aloes[1] every night. Determined, from accounts heard here to proceed on Monday to the Blue Ridge in N.C. for the benefit of health. But before bed-time, upon the invitation of some of the most respectable gentlemen of the place I agreed to stop through Monday & to meet the people on that day.

1. *Agave americana,* the fermented, ill-smelling sap of which produces, in quantity, intoxication and sleep.

10 SEPT. Sunday. Attended the Baptist church in the forenoon & heard an episcopal minister in the afternoon.

11 SEPT. Rode out with Col. Leitner to see something of the land of the neighborhood, but feared to be long in the sun. My head aching worse, owing to having omitted taking medicine last night for the first time in the last 10 days.—6 miles.

At 3 P.M. attended the meeting, for which notice had been given yesterday. There were assembled from 50 to 70 persons, mostly of the plain & respectable planters of the adjacent country. For the first time I ventured to address the meeting without interrogations, or rather in the form of an extemporaneous lecture, instead of that of conversation as on all former occasions. After speaking thus for an hour or more, I then had proposed & answered as many questions as persons thought proper to put, & which conversation continued for an hour longer, when I thought it time to close the meeting, as many persons present had to ride home a considerable distance. Still, a smaller number remained, & accompanied me to the tavern, where our conversation in continuation of the same subjects lasted until near dark. I had aimed to adapt my voluntary remarks to the peculiar circumstances of Spartanburgh. A large proportion of my auditors from the country were past middle age. I had every reason to think that the meeting was gratifying & satisfactory to them; which was the more satisfactory to me, as in this region there had been much prejudice against the agricultural survey, & Major Henry, the representative, had incurred some blame for giving it his support in the legislature.

58. Over the Line into North Carolina (12 September–13 September)

12 SEPT. Thinking that it was necessary for me to attend better to my health, & to seek a perfectly healthy climate for recuperating for a week or two, set out early in the morning for Flat Rock in N.C. The road passed through a still poorer country than lower in Spartanburgh,

until reaching the northern line of that district, & of S.C. Thence it is 2 miles to the first range of mountains, which seems to be an extension of the S.W. mountains in Va. In the valley, between the mountains, some fertile land—but mostly poor. Granite, the prevailing rock, through the whole, & mostly disintegrated. Stopped at Foster's, a new & open log house, but which in other respects furnished pretty good fare.—33 miles.

13 SEPT. The night was rainy, & there was every appearance of a long bad season, to avoid which I set out in a driving mist, which was full as bad as a *down* rain. The mountain & bad road continued 9 miles farther, & then almost level land (compared to any seen above Columbia,) to Flatrock—12 miles in all. The summit of Blue Ridge was passed in part of this seeming level, without my knowing it. In an hour after my arrival, the rain poured, & I was thankful to be in good quarters, & in agreeable company. There are about 30 persons here, mostly from the low country of S.C. & Ga. on excursions for health. I feel considerably better, & shall try a cessation of my nightly medicine for such short intervals as it may be dispensed with.

These notes of late have been so little agricultural in their character, & being of no other value to me—& of none whatever to any one else—that I shall cease taking the trouble to continue them.

Biographical Supplement

AIKEN, William, owned all of Jehossee Island and ran one of the most successful rice plantations in the low country. He owned 904 slaves in 1850, making him one of the biggest slaveholders in the state. He also had major railroad interests. When Ruffin visited he may have been absent in Charleston, where he owned extensive real estate, or in Columbia, where he represented the parishes of St. Philip's and St. Michael's in the senate. Then thirty-seven, he owed his wealth and position primarily to the commercial successes of his Ulster-born father. Aiken succeeded James Hammond as governor in 1844 and served three terms in the U.S. Congress in the 1850s. He held broadly Unionist convictions. He died in 1887. [Davidson, *Last Foray;* Scarborough, private communication to the editor; MacDowell, *Gaillard Genealogy; Reports of the General Assembly, 1843; Journal of the Senate, 1844*]

ALLSTON, Captain John Hays, of Breakwater, Georgetown District, was not one of the mainstream Alstons/Allstons (q.v.). Born in the Darlington District of a North Carolina family, he moved down the coast around 1813. At the time of the survey, in the mid-1850s, he was politically anti-Union. In 1832 the Georgetown State Rights and Free Trade Association chose him as its candidate for the state legislature; he lost the election. Ruffin approached him for his agricultural expertise and later thanked him in the report for "verbal information" supplied, describing him as "one of the most judicious and successful rice planters of the State . . . who, to long experience, adds the further advantage of having cultivated in several different parts of the State, and on all varieties of soil."

In the mid-1820s Allston introduced to the Georgetown area the process of "claying" rice, whereby seeds soaked in clayed water and subsequently dried were sown in open trenches without an earth covering. By the time of Ruffin's journey, this method had become very popular locally. "The advantage," according to R. F. W. Allston (q.v.), "is that the grain, furnished on its exterior husk with a hairy fuz, retains about it . . . particles of the tenacious clay: this, on the grain in the trenches being reached by the water with which the field is flooded immediately after sowing, makes it adhere to the earth, and of course prevents the seed from floating." Allston also adopted "extended flowing," a method by which water was left on the field until the crop was fully grown (with short breaks for hoeing), thus reducing labor requirements, preventing rust, and, allegedly, improving the quality of the rice. He was a founding member of the Pee Dee Planters' Club in 1839. He died 1849. [Allston, *Memoir of Rice in South-Carolina;* Ruffin, *Report of Survey*]

ALLSTON, Colonel Robert Francis Withers, a descendant of a seventeenth-century English immigrant, was then aged forty-two. Despite his unavailability, he was able to make major contributions to the survey. Over the previous twenty years, Allston had been, successively, surveyor general, representative, senator, and candidate for governor, just failing to defeat James Hammond (q.v.) the previous year and losing to William Aiken (q.v.) in 1844. In 1847 he became president of the state senate and finally won the governorship in 1856. He was always a strong advocate of states' rights, though a wary secessionist. Ruffin was in touch with him at White Sulphur Springs in Virginia and in Charleston in the weeks preceding the war.

Allston was one of the largest and best-informed antebellum rice barons, thereby attracting Ruffin's attention. He owned Chicora Wood on the Pee Dee and several other plantations in the Georgetown District. In 1850 he owned more than four hundred slaves, and in 1860 he marketed an impressive 1.5 million pounds of rice. He was very active in the State Agricultural Society and was elected vice-president of the Agricultural Convention in 1839. Because he was a leading authority on rice cultivation, in summer of 1843, Ruffin asked him to prepare an account of rice farming for the benefit of the survey. Allston obliged, submitting the "best compilation which I could make" in November. It was placed in the report's Appendix and later published separately, acquiring some fame as *Memoir of the Introduction and Planting of Rice in South-Carolina.* Some years later he published *Essay on Sea Coast Crops.* In the mid-1850s he was twice awarded medals by Paris expositions for rice culture.

In 1846 James Hammond sent him a paper on marling, which Allston read "with great pleasure and much profit," but there is no evidence that he resorted to the practice or that the profit came as money. The *Southern Agriculturist* did report in 1844, though, that Allston and some others were leading the way out of rice monoculture by "rest, manuring and rotation." Allston supported the survey, cooperating as mentioned above and declaring in 1854 that the identification and evaluation of marls by "the venerable Ruffin, together with his clear explanation of and rules for the use of them, are a boon to the inhabitants of the respective localities, and to the productive wealth of the State." He died in 1864 at the age of sixty-two. [Allston, *Memoir of Rice in South-Carolina;* Allston, *Essay on Sea Coast Crops;* Barnwell, *Love of Order;* Davidson, *Last Foray;* Easterby, ed., *South Carolina Rice Plantation;* Faust, *James Henry Hammond;* Joyner, *Down by the Riverside;* Reynolds and Faunt, *Biographical Directory of the Senate;* Rogers, *History of Georgetown County;* Ruffin, *Report of Survey;* Ruffin, *Diary,* vol. 1; Smith, *Economic Readjustment of an Old Cotton State; Catalogue of South Carolina College,* 1843; 1844; *Proceedings of the Agricultural Convention; Southern Agriculturist,* 1830, 1844]

ALSTON, Colonel John Ashe (ca. 1806–56) could do little for Ruffin other than show him around and arrange introductions. He probably issued Ruffin an invitation in Charleston, where Alston and his wife, Fanny Buford Fraser, lived in some style and owned a notable art collection. He seems to have had no plantation of his own though he sometimes lived at one of his father's estates, Strawberry Hill. Later, though, he purchased the nearby Hagley property. Rogers suggests that he may have become the owner of over 150 slaves. He was a first cousin of Colonel R. F. W. Allston (q.v.), both being founding members of the Pee Dee Planters' Club in 1839. His father, William Algernon Alston, was then the grandest planter on Waccamaw neck. [Allston, *Memoir of Rice in South-Carolina;* Rogers, *History of Georgetown County;* Ruffin, *Report of Survey;* Walker, *History of the Agricultural Society*]

ANDERSON, Dr. William Wallace, was born in Maryland in 1789 of Scottish families. His father, Robert, had fought at Cowpens, Green Swamp, and Camden and knew South Carolina well enough to encourage William, on his graduation from the University of Pennsylvania, to move south and practice medicine in the High Hills of Santee. This he did in 1810. His plantation in Sumter was known as Red Hill; his mansion, close by the village of Statesburg, was called Borough House. He conducted extensive farming operations, using the labor of about 150 slaves.

To Ruffin's taste, he also had a reputation as a scientist. When he first moved south he served as professor of surgery at the state Medical College in Charleston. In 1832 he published a paper in the *American Journal of Medical Science* on the successful treatment of jaw cancer. He was a member of the Philadelphia Linnaean and Medical societies. Anderson enjoyed some local fame for the construction work mentioned in Ruffin's diary, using the newly learned technique to add wings to his house. (*Pisé de terre* was earth that had been kneaded with gravel and rammed between boards, which were later removed as drying took place.) He also had botanical interests, which led to a friendship with Joel Poinsett (who died during a visit to Borough House).

With the outbreak of war the émigré declared, "I will risk everything I am worth in support of my adopted state, through every trial and every danger." He sent slaves down to help build fortifications around Charleston. His son Richard Heron Anderson became one of the military leaders of the Confederacy. Anderson died in 1864, aged seventy-five. [Davidson, *Last Foray;* Nicholes, *Historical Sketches of Sumter County,* vol. 1; Waring, *History of Medicine in South Carolina*]

ASHE, Colonel John Algernon Sidney, then aged forty-seven, came from a family that had been in the low country since the seventeenth century. He owned Roxbury, Hywassee (or Rotterdam), and other plantations in St. Paul's Parish, as well as 26 South Battery, Charleston. Educated initially in New York, he later studied law in South Carolina under Langdon Cheves. Ashe served as senator for St. Paul's, 1830–35, and for St. Philip's and St. Michael's, 1844–49. In 1843 he was director of the Bank of South Carolina. He remained a bachelor and was an experienced dueler.

Ashe started planting in 1818, acquiring some reputation as an improver. He had been appointed to the State Agricultural Society's three-man standing Committee on Cotton in 1843 (along with F. A. Porcher, q.v.) and on another, in the same year, to look into corn cultivation (along with John H. Tucker, q.v.). He died in 1868, aged seventy-one. [Heitzler, *Historic Goose Creek;* Reynolds and Faunt, *Biographical Directory of the Senate;* Walker, *History of the Agricultural Society*]

BACHMAN, Rev. Dr. John, had first met Ruffin during the latter's visit to Charleston in 1840. It was the start of an enduring relationship. Ruffin wrote in 1860 of "my old friend Dr. Bachman, who is a great man, and one of the best of all the good men whom I have known." His final extended visit occurred in 1863, during the federal bombardments around Charleston Harbor. Born in New York State in 1790, of Swiss and German ancestry, Bachman moved to Charleston in 1815 and lived there until 1874. He befriended John James Audubon as a result of their common interest in natural history. In the late 1830s two of Bachman's daughters, Maria and Eliza, married Audubon's sons John and Victor. The deaths of the young women, in 1840 and 1841, were the "heavy afflictions" referred to in the diary.

Bachman had been active in the state Horticultural Society and had written editorial copy for the *Southern Agriculturist.* At the time of Ruffin's arrival in Charleston, he published a skeptical *Inquiry into the Nature and Benefits of an Agricultural Survey.* He was not averse to reform as such, however, insisting that if "South-Carolina ever recovers her proud pre-eminence among her sister States, it will be through the means of Agricultural knowledge." Bachman had been a strong Unionist during the nullification crisis and, although selected to deliver a prayer at the opening of the Secession Convention in 1860, had continued to caution against reckless action and violent separation.

He enjoyed some scientific reputation, having published in national journals, met Alexander von Humboldt in Philadelphia and Berlin, argued with Louis Agassiz, conversed with Charles Lyell, and co-authored a study of vivaparous quadrupeds with Audubon. He belonged to the ambivalent intellectual category of natural theologian, evading trouble for the most part by his emphasis on mere observation and classification and by resort to tautology in analysis. William H. Longton describes some of his ideas as "almost mediaeval."

Science and Scripture contributed to a potentially liberating view of race, Bachman contending that the whole of mankind was descended from Adam and Eve and the subsequent variety of people resulted from God-given capacities for environmental adaptation. Such antipluralism, however, was not antiracist. Human adjustment in Africa had produced "the permanent marks of inferiority." This was all to the good because "the negro by his constitutional adaptation to labour in situations where we would only find disease and degeneracy, contributes to our wealth and comfort." Bachman died in 1874, in his mid-eighties. [C. Bachman, *John Bachman;* J. Bachman, *Address before the Horticultural Society;* J. Bachman, *Inquiry into the Agricultural Survey;* J. Bachman, *Discourse on the Anniversary of His Ministry;* J. Bachman, *Doctrine of the Human Race;* Longton, "Some Aspects of Intellectual Activity in South Carolina"; Ruffin, *Diary*, vols. 1, 2, 3]

BAILEY, Dr. S. J., may have been the owner of "Dr. Bailey's Mill" shown in Mills's 1826 map, on Four Mile Branch tributary of the Savannah, about six miles downriver from Shell Bluff. He was a Virginian and therefore probably no relation of the old Bailey family of Edisto. [Mills, *Atlas of South Carolina*]

BARKER, Dr. Sanford William, of South Mulberry, St. John's Parish, was from an English family, latterly with Rhode Island connections. He was a Gaillard on his mother's side and brother-in-law of Colonel James Ferguson (q.v.). He married, in 1835, Christiana Broughton, to whose family the plantation had formerly belonged.

Then thirty-five years old, with a reputation as a physician, botanist, and book collector, he was one of the tiny body of planters who experimented with marl for a year or two. Later he served as senator for St. John's, Berkeley (1850–63), and vice-president of the State Agricultural Society. Barker was an active secessionist, meeting and corresponding again with Ruffin. [Heitzler, *Historic Goose Creek;* Irving, *Day on Cooper River;* MacDowell, *Gaillard Genealogy;* Ruffin, *Diary,* vols. 1, 3; Shaffer, *Carolina Gardens;* Walker, *History of the Agricultural Society*]

BARNES, John J., despite his knowledge and experience as an overseer and his being "the best I have ever had if left to himself," was sacked by James Hammond in 1845. Hammond explained to Ruffin that Barnes would not "attend properly to the details of my affairs—experiments particularly." Considering "the dozens of nice experiments" going on at Silver Bluff, "he worried me beyond endurance & continually thwarted all my views in detail." [Faust, *James Henry Hammond*]

BARNWELL, Colonel Robert Woodward, had met Ruffin earlier, possibly at the Beaufort meeting or at some dinner table. Then forty-one years old, he was the son of a U.S. senator and revolutionary war veteran and a native of the Beaufort area. He owned a large house in town and extensive plantation properties on the mainland and islands. Barnwell had signed the Ordinance of Nullification in 1832 and was a delegate to the Southern Rights Convention in 1852. Despite some moderation on sectional complaints, he articulated strong proslavery views and engaged in active secessionism in the months preceding the outbreak of war. He helped draft the Confederate constitution but turned down Jefferson Davis's offer of the secretaryship of state.

Since his student days at Harvard College (and friendship with Ralph Waldo Emerson) Barnwell had enjoyed a scholarly reputation, generally preferring life in education and country pursuits to one in politics (though he did serve in Congress periodically between 1829 and 1850). He had been president of South Carolina College for some years before Ruffin's visit and returned there after the war, as president and professor of political science, when it was reconstituted as the University of South Carolina. He withdrew for some years in the 1870s in protest against admission of black students.

Ruffin met him again in Richmond when Barnwell was serving in the Confederate Senate. Barnwell spent his last years as a virtual recluse in Columbia, taking charge of the university library. He died in 1882. [Barnwell, *Story of an American Family;* Davidson, *Last Foray;* Ruffin, *Diary,* vols. 1, 2; Wakelyn, *Biographical Dictionary of the Confederacy; Catalogue of South Carolina College,* 1843, 1844; *Journal of the State Convention of South Carolina; Garnet and Black,* vol. 1]

BELIN, Colonel Allard Henry, of Sandy Knowl, was then thirty-nine years old. He was a Harvard graduate and came from a long-established Huguenot family (the original pronunciation of the name was corrupted to "Blane"). A states'-rights activist with Robert Allston, he was a delegate to the 1832 Nullification Convention and a state representative since 1838 (voting for the survey). A man of notable argumentative and conversational powers, he was a successful, improving planter. Belin died in 1871. [Davidson, *Last Foray;* Joyner, *Down by the Riverside;* Rogers, *History of Georgetown County;* Ruffin, *Report of Survey*]

BIGHAM, Mr., was an experimental marler, possibly Leonard Bigham of Lepre Creek (father of Leonard Smiley Bigham, future state representative and senator). [Ruffin, *Report of Survey;* Reynolds and Faunt, *Biographical Directory of the Senate*]

BLACK, Mr., would be William, a mine owner and planter who conducted some successful experiments with lime. [Tuomey, *Report on the Survey of South Carolina*]

BRISBANE, John Stanyarne, was then around seventy and had been farming behind Charleston (Otranto, Malona, and other places) since the turn of the century. A lively and opinionated man, he was much stimulated by Ruffin's company. His grandfather came over from Glasgow in the 1730s, practicing medicine in Charleston and planting on the Ashley. His father was banished by U.S. authorities in the early 1780s, but the young Brisbane jumped ship in Charleston Harbor and found refuge with an aunt. As a planter he held office in farming clubs, succeeded Robert Roper as orator to the State Agricultural Society in 1835, and became first president of St. Andrew's, Ashley, and Stono Agricultural Society after its foundation in 1842. It was in this latter rank, presumably, that he received Ruffin in 1843. He had been mentioned by James Johnson (q.v.) in an 1840 issue of the *Farmers' Register* as a source of marl specimens for analysis.

The report of his 1844 presidential address in St. Andrew's described his remarks as "neat and happy." Among them were some Old-World speculations about whether bookish people could be good planters. His conclusion was that a man who took pleasure in Byron and Scott was unlikely to care for the "minute attention to objects" needed in successful agriculture or be capable of dealing with slaves. (The matter had probably been discussed with Ruffin, a man of wide literary tastes and practical inefficiency.) Brisbane revealed some distaste for James Hammond, praising him for beginning the survey and advising his audience not to worry themselves overmuch as to his motives: "A good act is done by different agents." Ruffin, however, was a much purer soul and had, through his labors in the state, been acting only for the well-being of the community. The Virginian was particularly to be remembered for his conversations, "for they not only contained much information, and that imparted in the most agreeable manner, but they excited to action the powers of our own minds hitherto dormant." The visit, Brisbane hoped, would be "recollected as an era in agriculture." These were not entirely empty words, Brisbane being one of a handful of planters who made some brief effort to experiment with marl and publicize his results. He died around 1850. [Mathew, *Edmund Ruffin;* Walker, *History of the Agricultural Society; Farmers' Register,* 1840; *Proceedings of the Agricultural Convention; Southern Agriculturist,* 1844]

BROWN, Major J. G., of Barnwell village, also planted on the Lower Three Runs tributary of the Savannah, about half a dozen miles to the east. Possibly he was the J. Brown appointed to the State Agricultural Society committee in 1843 to determine measures to make that body more useful. [Ruffin, *Report of Survey; Proceedings of the Agricultural Convention*]

BUFORD, Dr. William June, a physician and planter of Westee on the Williamsburg side of the Santee swamps, was then in his late fifties. He was descended from William June Buford, who moved into the area from Virginia in the 1700s. Formerly a state representative, he was then a senator and also represented Williamsburg in the State Agricultural Society. He died in 1845. [Nicholes, *Historical Sketches of Sumter County,* vol. 2; Reynolds and Faunt, *Biographical Directory of the Senate; Journal of the Senate, 1844; Proceedings of the Agricultural Convention; Southern Agriculturist,* 1844]

CAIN, William, was important to Ruffin and his pursuits. Then fifty years old and senator for his native parish of St. John's, Berkeley, he had earlier served twice as representative and later as lieutenant governor. Cain was excellently connected, being married first to a DuBose and then to a Palmer (this second wife also being the daughter of a Porcher). He attended local Pineville Academy and South Carolina College, Columbia, graduating in 1812. He farmed Somerset plantation in Black Oak, dying at Pinopolis in his mid-eighties.

Cain appealed to Ruffin because of his seigneurial status and involvement with the State Agricultural Society when it was pressing for the survey. They probably agreed politically as well. In 1832 Cain had joined his local Nullification Association. His second son, born in 1835, was named John Calhoun Cain. A member of the Secession Convention of 1860, he subsequently signed the ordinance detaching the state from the Union.

Ruffin, visiting him again in 1863, observed that no one had made "a single application" of the rich and accessible marl in the vicinity, though Cain's son had said he hoped to give it a try. [Davidson, *Last Foray;* Reynolds and Faunt, *Biographical Directory of the Senate;* Ruffin, *Report of Survey; Journal of the Senate, 1844;* Ruffin, *Diary,* vol. 3; *Proceedings of the Agricultural Convention*]

CAPERS, Charles Gabriel, was a member of the St. Helena branch of a prominent agricultural and ecclesiastical family that had been in the area since the end of the seventeenth century. Capers was born in 1775 and therefore was in the oldest generation of active planters. His father, Charles Capers, had been deputed by St. Helena planters to receive and entertain Napoleon Bonaparte in exile (as they anticipated) on their island. The son owned several plantations, including one in Charleston District, but lived principally on a five-hundred-acre estate on Chowan Creek and in a Beaufort town house. He was a near neighbor of planter and diarist Thomas B. Chaplin of Tombee and in 1843 was in a dispute with Henry McKee (q.v.) over plantation road rights.

Capers was unlikely to have won Ruffin's admiration. Reputedly he was a "devil in cruelty." Theodore Rosengarten calls him "a wealthy old man who kept close watch over his estate while his sons killed time waiting to inherit plantations of their own." Of a St. Helena agricultural society meeting in December 1853 at which he and Capers were present, Chaplin wrote: "There was nothing done, but eating & drinking, & a great dog fight. Some men came near making up a horse race, & some trying to swap horses." Such lack of attention to farming matters was normal. The "principal &

most important business" of such meetings, Chaplin later noted, was "eating & drinking," the planters taking turns to provide. Capers died in 1857. [Johnson, *Social History of the Sea Islands;* Reynolds and Faunt, *Biographical Directory of the Senate;* Rosengarten, *Tombee; South Carolina Genealogies,* vol. 1]

CARSON, Colonel William Augustus, of Dean Hall (later called Cypress Gardens), had purchased the place from the Nesbitt family in 1821. He married a daughter of Unionist James Louis Petigru. The plantation, wrote John B. Irving in 1842, "resembles a well ordered village more than that of a single plantation. The residence of the proprietor—the condition of the fields—the banks—the white and cleanly appearance of the negro houses—the mill and thrashing machine in complete order, all excite a strong feeling of admiration, and stamp at once the proprietor as an experienced and skilful planter."

Carson was, however, currently running into financial trouble in consequence of the improvements noted by Ruffin. "Our friend Carson," wrote his father-in-law to Robert Allston (q.v.) in June 1843, "has met I am told with a great loss. He hired Irishmen and had some beautiful embankment on an obscure stream which is called Back River, and when he settled with them the work came to 6000 Dollars. He was not surprised at this for he supposed he had 300 acres. Last week he got Quash Pinckney to survey it, and it turns out only 130 acres." Further difficulties led one of his sons to describe him as a man who "wore out his life watching a salty river." He died in 1856. [Easterby, ed., *South Carolina Rice Plantation;* Irving, *Day on Cooper River*]

CHEVES, Dr. John Richardson, planter of Cavehall by the Santee in Orangeburg District, was twenty-seven years old. He was descended from eighteenth-century Scottish immigrant Alexander Cheves and was the seventh of Langdon Cheves's and Mary Elizabeth Dulles's fourteen children. Like his brother Langdon Cheves, Jr., he attended West Point and failed to graduate. He was the younger brother of writer Louisa Cheves McCord and brother-in-law of Colonel David McCord (q.v.). [Ruffin, *Report of Survey; South Carolina Genealogies,* vol. 1]

DAVIS, Henry, lived on the Pee Dee in Marion District and was a state representative and supporter of the survey. To Ruffin he was of interest as the owner of extensive marl deposits. In consequence of Ruffin's visit, he experimented by putting three hundred bushels on an acre of fairly new ground. "The season," Ruffin reported, "was very unfavorable and the whole crop of cotton bad. The product of the marled acre was 592 lbs. of seed cotton, and of the unmarled, 336 lbs." Such good results could, he suggested, be sustained for another ten to fifteen years, but there is no record of Davis continuing the work. [Ruffin, *Report of Survey; Journal of the House of Representatives, 1843*]

DAVIS, Dr. James B., should not be confused with the recently deceased Senator James Davis of Union. He was probably the James Davis who taught at South Carolina College and had been an early graduate of the state Medical College, in later years serving as vice-president of the Medical Society of South Carolina. He is likely, too, to have been the "J.D." who told readers of Ruffin's *Farmers' Register* in 1838 about the existence of rich marl beds along the Santee.

Davis was one of Fairfield's representatives at the 1839 Agricultural Convention, where he formally proposed that the State Agricultural Society be revived with headquarters in Columbia. He was the society's secretary when it decided to promote the agricultural survey, remaining in that position for a number of years thereafter and claiming that "no occupation in his life afforded him so much pleasure." He enjoyed a reputation for good management of his own farming affairs. [Waring, *History of Medicine in South Carolina; Proceedings of the Agricultural Convention; Farmers' Register,* 1838; *Southern Agriculturalist,* 1842, 1843, 1844]

DAVIS, Colonel Jonathan, Sr., an improving planter, had been a Fairfield delegate at the recent meeting of the State Agricultural Society in Columbia, which appointed him to serve on the Committee on Corn for 1843. [*Proceedings of the Agricultural Convention*]

DAVIS, William K., with Dr. J. B. Davis (q.v.), had been a Fairfield delegate to the 1839 Agricultural Convention. He was cited by Ruffin in the report as one of only two men he knew of in the state who had inserted pea breaks between grain crops, securing the advantage of leguminous nitrogen-fixing. [Ruffin, *Report of Survey; Proceedings of the Agricultural Convention*]

DWIGHT, Isaac Marion, descended from an English family that came to Carolina in the seventeenth century, was forty-four years old when Ruffin visited his Cedar Grove plantation on the Ashley below Dorchester. He was the son-in-law of Colonel Thomas Porcher of Ophir and Elizabeth Sinkler Dubose, thus a member of another of the grand family alliances of the state. His eldest son married a Ravenel and two others married Gaillard sisters. In deference to Ruffin's opinions, Dwight applied marl to a small portion of rice land shortly after the surveyor left. Ruffin later inspected stubble, recording that "never was superiority of growth more manifest." Additional benefit on such tidewater lands, he suggested, was the countering of salt damage. There are no indications, however, that Dwight persisted with the practice. He was a St. George's, Dorchester, delegate to the Southern Rights Convention in 1852 and died in 1873. [MacDowell, *Gaillard Genealogy;* Seabrook, *Memoir on Cotton; Journal of the State Convention of South Carolina*]

ELLETT, William Henry, had clearly already met Ruffin. Further exchanges would have been to the surveyor's benefit as Ellett had been suggested for the job. He was a senior scientist at the college, having been professor of chemistry, mineralogy, and geology there since 1835 (later leaving to take up a teaching post in New York). He had shown some abstract interest in acidity, publishing a paper in the *American Journal* on compounds of cyanogen in 1830. In the mid-1840s he was approached by a committee of the state legislature to explain his ideas for cheaper production of gun-cotton explosive. The residue of his process, he suggested, could be combined with marl or lime to form a valuable fertilizer.

His wife was a daughter of a wealthy Manhattan real estate speculator with a reputation as a biographer and literary critic. Robert F. W. Allston (q.v.), then senator, wrote home in December 1844 about "a very pleasant and well conducted" party at Elizabeth Ellett's. "W. G. Simms and Mr. Holmes the authors were distinguish'd as guests." [Easterby, ed., *South Carolina Rice Plantation;* Longton, "Some Aspects of Intellectual Activity in South Carolina"; O'Neall, "Agricultural Address"; *American Journal,* 1830; *Catalogue of South Carolina College,* 1843]

ELLIOTT, William, Jr., then fifty-five years old, owned extensive properties in Colleton and St. Helena and the grandest house in Beaufort town. Eighteen years later Ruffin declared that there were few men "better informed, & no one more agreeable in conversation." Elliott held appeal as a man of culture and wide reading, capable of powerful argument and vivid expression, with a name as a writer on sporting, farming, and political subjects. He belonged to one of the first families of the sea islands, dating back to a seventeenth-century English immigrant and subsequent land accumulators along the coast from Beaufort to Charleston.

Alfred Glaze Smith wrote of him as a planter who recommended expansion of cotton and abandonment of diversification though this referred only to the prized sea-island variety of cotton. People growing the short-staple crop had, by contrast, to guard against overproduction. The southeastern cotton states now stood under a "suspended . . . sentence of death." Elliott, like Ruffin, admired British agricultural methods.

His student years at Harvard contributed to his Unionist convictions. Refusing to go along with nullification sentiment in St. Helena, he resigned his state senate seat after the 1831 session and abandoned his political career. Ultimately, however, he was a secessionist. He and Ruffin met again in March 1861 on a steamer cruising around the forts in Charleston Harbor. Elliott died in 1863. [Elliott, *Address before the St. Paul's Agricultural Society; Carolina Sports;* Davidson, *Last Foray;* Johnson, *Social History of the Sea Islands;* Reynolds and Faunt, *Biographical Directory of the Senate;* Ruffin, *Diary,* vol. 1; Rosengarten, *Tombee;* Smith, *Economic Readjustment of an Old Cotton State;* Wauchope, *Writers of South Carolina; South Carolina Genealogies,* vol. 2; *Southern Agriculturist,* 1828]

ERWIN, General James D., owned a place on Boggy Gut, ten miles south of Barnwell village and otherwise known as Gillett's Mill. He had gone to Columbia in 1839, along with James Hammond and William Gilmore Simms, as a Barnwell District delegate to the state Agricultural Convention. He should not be mistaken for Colonel James F. Ervin, former senator for Marlborough District, thanked by Ruffin at the end of his report. [Mills, *Atlas of South Carolina;* Ruffin, *Report of Survey; Proceedings of the Agricultural Convention*]

FELDER, Dr., belonged to a family that farmed on the Orangeburg side of the Santee (and included the departed Adam Felder, mentioned earlier for his limekiln). It is not clear who this one was, as Ruffin gives neither Christian nor plantation name. Possibly it was Paul S. Felder, then representative for Orange Parish—or John Myers Felder, then senator for same—but neither had the title of doctor. [Reynolds and Faunt, *Biographical Directory of the Senate;* Rosengarten, *Tombee;* Ruffin, *Report of Survey;* Seabrook, *Memoir on Cotton; Journal of the Senate, 1844; Proceedings of the Agricultural Convention; Reports and Resolutions of the General Assembly, 1843*]

FERGUSON, Colonel James F., of Dockon Gardens, set back from the West Branch of the Cooper, also purchased the Farm, the Briars, Seaton, and Harry Hill nearby. He was the grandson of a Scottish immigrant of the same name who had worked as an engineer with James Oglethorpe in Georgia. His father, Thomas, had served during the revolutionary war as an aide to Lafayette. James in turn had been an aide to General Thomas Pinckney in the War of 1812. Dockon had come to him from his mother, Ann Wragg, a woman of part-Huguenot descent. He had settled there in the 1810s, abandoning his father's property on the Stono River. He married Abby Ann Barker, sister of Sanford Barker (q.v.).

At the time of Ruffin's visit, Ferguson, almost sixty years of age, had done some political work as a state representative. His main planting interest was rice, though he was also well known as a successful breeder and trainer of thoroughbred horses—activities Ruffin held in low esteem (q.v. Hampton, Wade, II). Ferguson lived in Charleston in the malarial summer months, riding fifty-odd miles to Dockon and back each day. He was a delegate from St. John's, Berkeley, to the 1842 session of the State Agricultural Society, being elected to serve on the Committee on Rice for 1842–43. He experimented with marl.

The two men met again later, in Carolina and Virginia. Ruffin revisited Dockon in March 1861 and noted: "Col. F. is a well preserved & hale old man, 77 years of age. He & his wife are both well educated & well informed, & very intelligent & agreeable." On his third day there he rode out—as he had in 1843—with Ferguson and Sanford Barker, noting, with pained incredulity, that no one around had developed any interest in his old calcareous prescriptions. Ruffin returned again in 1863, inspecting "the extensive & admirable fixtures & apparatus . . . for thrashing & cleaning his rice, by steam power." Ferguson survived the war, dying in his ninety-first year. [Irving, *Day on Cooper River;* Ruffin, *Diary,* vols. 1, 3; Walker, *History of the Agricultural Society of South Carolina; Proceedings of the Agricultural Convention*]

FORD, Stephen, then in his late fifties was the son of another Stephen, of English origin, who accumulated properties on the south side of the Black River. Davidson locates him at Combahee, and Rogers suggests his possible presence at Peru and Ford's Point. As a planter, Ford was both wealthy and able. With a force of some 140 slaves, he succeeded in raising 420,000 pounds of rice in 1850. Ruffin never wrote his promised diary note on rice but did insert a lengthy account in his report, thanking Ford (and his nephew, Stephen C.) for information given him on the "Leggett's" and "All-Water" methods of cultivation. The latter was much the same as the "extended flowing" system used by John Hays Allston (q.v.). The system began as an "accident and supposed disaster" but was found to be beneficial and thus was continued. It had, wrote Ruffin, "now been several years in use, and preferred, by several good and judicious planters on Black river." [Davidson, *Last Foray;* Rogers, *History of Georgetown County;* Ruffin, *Report of Survey*]

FRASER, Mr., probably was Frederick G. Fraser, member of a Scottish family resident in the low country since the beginning of the eighteenth century. He was a planter in Beaufort and a factor in Charleston. His brother was the artist Charles Fraser of Charleston. [Mills, *Atlas of South Carolina;* Moltke-Hansen, private communication to editor; Ruffin, *Report of Survey; South Carolina Genealogies,* vol. 2]

GAILLARD, James, presumably was James Gaillard, Sr., of Walnut Grove, St. John's, Berkeley, and not his son of the same name, who farmed at the Rocks (the former being fifty-four years old, the latter thirty years younger). The family was an extensive one of Huguenot descent, dating back in Carolina to the seventeenth century. Gaillard was a graduate of South Carolina College. He married first a Porcher, then the daughter of a Ravenel and a Gourdin. Despite serving for a time as state representative, he appears to have led a fairly quiet country life, taking an interest in church and school affairs and working in the local magistracy. Gaillard is cited in the *Farmers' Register* as one of the St. John's planters who had resorted to manuring. On the eve of the Civil War he owned around one hundred slaves. He died in 1871, aged eighty-three. [Davidson, *Last Foray;* MacDowell, *Gaillard Genealogy; Farmers' Register,* 1840]

GOURDIN, Dr. Robert Marion, aged forty-four, came from another of Carolina's wealthy and long-established Huguenot families. He farmed Mepu by the Santee, on the borders of Georgetown, Williamsburg, and Charleston districts, and owned a house at the corner of Meeting Street and South Battery, Charleston. His father, Theodore, had been a U.S. congressman; his mother, Elizabeth, was a Gaillard; his brother Theodore was a disunionist senator in Columbia. Sent north for higher education, Gourdin took degrees at Harvard College and at New York College of Physicians and Surgeons. He never married. He became an active, intense secessionist, chairing the executive committee of South Carolina's belligerent 1860 Association.

Gourdin's name appears occasionally in records of the State Agricultural Society, but his principal claim to Ruffin's attention lay in his experimentation with marl over the preceding three years. "A field of fifty acres of flat, sour pine land," it was later recorded, "was prepared for planting in 1840. . . . Upon this new list, he brought marl one mile and a half, and applied it at the rate of one hundred bushels to the acre—excepting six acres, running the whole length of the field, one half acre in width." Cotton was planted, but no benefit was discernible in the first year. In 1841, however, the marled parts looked "decidedly superior," yielding an average of 128 pounds of clean cotton per acre, compared with only 63 for unmarled land. A substantial difference was also evident in 1842. "He is so well convinced of the benefit to be derived from marl, that he has applied it to twenty acres more . . . and contemplates a more extensive use of it hereafter." There is no evidence, however, to show that such intentions were sustained. It is notable that Gourdin conducted exact measurement and comparison only in 1841. His evaluation was largely visual and impressionistic. He died in 1876, aged seventy-seven. [Channing, *Crisis of Fear;* Davidson, *Last Foray;* Irving, *Day on Cooper River;* MacDowell, *Gaillard Genealogy; Southern Agriculturist,* 1843]

GREGG, Dr., was almost certainly the son of either Alexander or John, both of whom served in the senate. The first died in 1823, the second in 1839. He was probably the son of John, who farmed close to Mars Bluff. [Reynolds and Faunt, *Biographical Directory of the Senate*]

HAMMOND, James Henry, governor of South Carolina, was sometimes referred to as "General," the rank he held in the Third South Carolina Militia. A recent arrival among the Carolina plantocracy, his father, Elisha, had come from Massachusetts in 1802, working as a farmer, merchant, barge operator, and school and college teacher. His mother, Catherine Fox Spann, was from a family of middling Edgefield farmers.

James Henry was born in Newberry, 15 November 1807, and brought up in Columbia. He acquired an extensive landed base to support his political aspirations through marriage to heiress Catherine Fitzsimons in 1831. By then, he was a practicing lawyer and had also taken charge of the antitariff *Southern Times,* gaining prominence, in Drew Gilpin Faust's words, as "the nullifiers' up-country editorial spokesman." The Hammond-Fitzsimons plantation lay at Silver Bluff on the Savannah, comprising more than ten thousand acres of poorly attended farmland and swamp.

Eager to take advantage of his growing radical reputation and to secure national renown, Hammond sought and won election to the U.S. House of Representatives in 1834, running on a free trade and states'-rights ticket. In 1840, aged thirty-two, he made his first bid for the ceremonial but prestigious office of state governor, in contest with John P. Richardson (q.v.). Richardson won through prevailing Unionist sentiment in the legislature and some anti-Hammond animus. Personal antagonisms almost cost him the next gubernatorial election as well, but he won in 1842 over the self-declared noncandidate R. F. W. Allston (q.v.) by a mere 83 to 76 votes. His governorship, 1842–44, was described by Bleser as "neither historically significant nor innovative." According to Faust, he undertook a series of military reviews around the state in early 1843, hoping "to impress a dispersed rural population with the meaning and majesty of state power." The tour culminated in a grand parade held in Charleston during the annual race week (and is alluded to by Ruffin). Allston had been supported in the 1842 election by large numbers of low-country planters, and Hammond experienced some consequent social discomfort in Charleston—possibly one reason for his interest in securing Ruffin's company.

More intense embarrassments were shortly to follow. In April 1843 Catherine Hampton, second daughter of Wade Hampton II (q.v.), reported to Catherine Hammond, her aunt, that the governor had enjoyed physically intimate relations with her and her three sisters during their visits to the Hammond household over the preceding two years. This was no imaginary dalliance; Hammond himself recalled that he and the Hampton girls had committed "every thing short of direct sexual intercourse." By the end of the year news was circulating in Columbia of a serious family rift, and in 1846 Hampton, by threat and innuendo, managed to prevent Hammond's election to the U.S. Senate. Hampton's vendetta lasted almost until his death in 1858. Only in 1857 did Hammond finally arrive in the Senate to rescue some of his broken career. He resigned in 1860 on Lincoln's election and died four years later at his Redcliffe plantation, two days short of his fifty-seventh birthday.

Hammond was an intelligent, restless, troubled man, his disposition much affected by pursuing, and then abandoning, *arriviste* ambitions among old grandee families of South Carolina. He lived in a grand, patrician style, traveling in Europe and assembling a noted library and picture gallery. Solon Robinson, visiting in 1850, judged him "one of the best-read men in the country." Ruffin, as noted, thought he had the most powerful intellect in the South, an opinion owing something to Hammond's agreement with Ruffin on states'-rights, banks, tariffs, and calcareous amelioration and to his notable capacities for scholarly appraisal in these and other matters.

The two friends, though, did drift apart on secession, Hammond finding his old radicalism tempered for a time by fear of Carolinian isolation and his conviction that the South, through cotton, still enjoyed great leverage within the Union. He was also generally more sensitive than Ruffin to the tragic dimensions of the conflict over slavery. He was as rigid as any, however, in his attachment to the institution, eventually owning over three hundred slaves. In 1844, while governor, he penned a widely circulated letter to the Free Church of Scotland in Glasgow arguing that Christianized Africans in South Carolina were more secure and better rewarded than factory workers in Britain. He delivered his famous "mud-sill" speech in the U.S. Senate in 1858, asserting the perennial need for a menial class to support elite sophistications.

Despite his skepticism about reform, there can be no doubting that from 1841 on he attached the greatest importance to Ruffin's propositions and that he became, by acreage, the grandest of all southern marlers. "I commenced in November last," he informed the *Farmers' Register* in 1842, "to marl my plantation, at Silver Bluff, on Savannah river. There is no marl on the place. I procured it from Shell Bluff, on the same river, and had to boat it up the stream. It requires eleven prime hands to man the boat I use, and when the river is not too high, they make two trips a week, loading and unloading themselves. They bring about 1100 bushels at a load." The work was pursued on his expanding properties for the remainder of the 1840s (over 2,300 acres were marled by 1847), with smaller exercises continuing into the 1850s. The intensity of his effort owed much to Hammond's forced retreat from public life. But despite his well-documented successes—in particular with cotton and on new swampland—no one followed his example. [Barnwell, *Love of Order;* Bleser, *Hammonds of Redcliffe;* Davidson, *Last Foray;* Eaton, *Growth of Southern Civilization;* Faust, *James Henry Hammond;* Freehling, *Prelude to Civil War;* Mathew, *Edmund Ruffin;* Ruffin, *Address on Exhausting and Fertilizing Systems;* Ruffin, *Report of Survey;* Ruffin, *Diary,* vol. 1; *Farmers' Register,* 1842; *Southern Cultivator,* 1842]

HAMPTON, Christopher Fitzsimons, was one of the sons of Wade Hampton II and a nephew by marriage of James Hammond. He was named after his maternal grandfather, purchaser of Silver Bluff in the 1810s. [Faust, *James Henry Hammond*]

HAMPTON, Colonel Wade, II, was the son of Wade Hampton I and father of General Wade Hampton III (1818–1902). He was shortly to learn of James Hammond's (q.v.) indiscretions with his four young daughters (Hammond's nieces by marriage). The subsequent vendetta kept Hammond out of politics almost until Hampton's death in 1858. Hammond was surprised that Hampton was prepared to publicize such private episodes in the lives of his children. None of the girls ever married.

Hampton was three years older than Ruffin. He had been on Andrew Jackson's staff at the Battle of New Orleans and was deputed to carry news of the victory back to Washington. After the War of 1812 and incomplete college studies in Columbia, he became a full-time planter at Millwood in Richland District, just outside the capital (the sequence and timing being somewhat similar to Ruffin's own move into farming). Hampton later acquired land in Louisiana and Mississippi, becoming reputedly the wealthiest man in the South. He served as state senator between 1826 and 1830.

His main fame, earning him Ruffin's near contempt, was his racing establishment—possibly the largest and finest in the state. He also had a reputation for his strictly agricultural stock breeding. "Col. Hampton has the Leicester Stock lately imported," wrote Dr. R. W. Gibbes to Colonel Robert Allston (q.v.) in 1837, "and has lost some of them, they are very fine looking large and round animals, and probably will do well." He was one of the state's largest producers of short-staple cotton, dealing directly with Liverpool. In 1842 he was appointed to the State Agricultural Society's Committee on Printing (to decide on publications of reports, letters, and the like). He was thanked for his assistance in Whitemarsh Seabrook's *Memoir on Cotton.* [Bleser, ed., *Hammonds of Redcliffe;* Easterby, ed., *South Carolina Rice Plantation;* Faust, *James Henry Hammond;* Ford, *Origins of Southern Radicalism;* Moltke-Hansen, private communication to the editor; Reynolds and Faunt, *Biographical Directory of the Senate;* Seabrook, *Memoir on Cotton; Catalogue of South Carolina College,* 1843, 1844; *Proceedings of the Agricultural Convention*]

HARLESTON, Mr., probably was John Moultrie Harleston of Elwood, a planter in his early twenties, from an English family resident in the low country since around 1700. [MacDowell, *Gaillard Genealogy;* Shaffer, *Carolina Gardens; South Carolina Genealogies,* vol. 3]

HARLLEE, Dr. Robert, thirty-five-year-old physician and large-scale planter of Scottish ancestry, farmed Melrose near Mars Bluff. He was the son of Thomas Harllee and Elizabeth Stuart, who moved to Marion District from Virginia in 1790, acquiring considerable landed and mercantile wealth, and the elder brother of soldier and lawyer William Wallace Harllee, lieutenant governor of South Carolina at secession. Harllee graduated from the Medical College in Charleston. He served as a state representative for two terms in the 1830s and as senator for one term in the late 1840s.

Harllee first came to Ruffin's attention in 1841, when the *Farmers' Register* reproduced his report on the calcareous resources of the Pee Dee. The *Register* refers to Harllee's use of stone on Myers's land and production of lime for sale at $1 per barrel. Clearly, he had used lime for agricultural purposes before Ruffin's arrival. He died in 1872. [Davidson, *Last Foray;* Reynolds and Faunt,

Biographical Directory of the Senate; Eminent and Representative Men of the Carolinas; Farmers' Register, 1841]

HASKELL, Mr., possibly was Charles Thomson Haskell (1802–74), of the large Home Place plantation in Abbeville, son of Colonel Elnathan Haskell of Orangeburg and son-in-law of Langdon Cheves. Haskell was a Harvard graduate. He lived at Zante near Fort Motte, Orangeburg, before buying up a number of emigrants' properties in Abbeville, forming them into a single estate. He served for a time as a representative in the state legislature. [Davidson, *Last Foray; South Carolina Genealogies,* vol. 1]

HAYWARD, Mr., from the location and scale was almost certainly Charles Heyward (1802–66), of Rose Hill in St. Bartholemew's Parish, Colleton. He owned between four and five hundred slaves. [Davidson, *Last Foray;* painting by unknown artist in the Charleston Museum: *Landscape of Rose Hill Plantation on the Combahee River, South Carolina*]

HAZEL, Joseph (1798–1885), was a notably successful producer of sea-island cotton. [Rosengarten, *Tombee*]

HENNEHAN, Mr., possibly was John Hanahan (1798–1856). His second wife was the widow of William Mikell Clark, daughter of Sarah Jenkins, and granddaughter of Martha Seabrook, indicating strong relationships with the Edisto plantocracy. [MacDowell, *Gaillard Genealogy*]

HENRY, Reverend Robert, had been president of South Carolina College since 1842 (holding the position pro tempore in 1834) and also professor of metaphysics and moral and political philosophy. He was born in Charleston in 1792, son of a Scottish merchant in the West Indies trade. He graduated from the University of Edinburgh in 1811, beginning his college teaching career in 1818 (numbering James Hammond [q.v.] among his students). Some of his religious notions gave offense to the orthodoxy in Columbia, causing a brief move into farming and banking during the 1830s. He published a eulogy on John C. Calhoun in 1850, writing therein of the 1832 tariff that it "seemed to announce the knell of freedom" and of the African that "he has invented nothing, he has improved nothing: the world owes him nothing. . . . His lowest are his strongest instincts." [Faust, *James Henry Hammond;* Henry, *Eulogy on Calhoun; Catalogue of South Carolina College,* 1843; *Garnet and Black,* vol. 1]

HERIOT, Dr. Edward Thomas, was a practicing physician, approaching the age of fifty. He came from a Scottish family that arrived in the late eighteenth century and achieved distinction in the revolutionary war and mercantile prominence in Georgetown. Heriot was a pioneer in manuring of rice fields, giving him a reputation for improvement which attracted Ruffin. He farmed Mont Arena on Sandy Island between the Pee Dee and Waccamaw. In 1847 he inherited Northampton on the Sampit Creek from his friend Francis

Withers and in 1854, just before his death, bought Richland on the Pee Dee, at once changing its name to Dirleton after his ancestral home in Scotland. (Oddly, this famous ruin near Edinburgh was also once occupied by Ruffin's ancestors.) Such accumulation of properties, added to his summer residence on Murrell's Inlet, made him a planter of formidable scale; he owned 369 slaves in 1854.

In 1843 Heriot had just completed two terms in the state senate, representing All Saints'. Eleven years earlier, he was the Unionist candidate for the House of Representatives and was defeated by John H. Allston (q.v.) and Robert F. W. Allston (q.v.). It was rumored, mistakenly, that he might run against the latter in the senate election of 1850. Despite political differences, he and Allston remained friends, communicating on planting and jointly hiring a New York tutor for their children. Heriot also performed medical services for the Allstons.

In 1839 he began a system of manuring rice swamps with refuse straw and old tailings. He supplied an account for inclusion in Robert Allston's report on rice prepared for Ruffin in 1843. No one had fertilized extensively before, the common assumption being that, even on the poorest lands, sediment from controlled river flooding would suffice as a source of nutrients. Heriot shrewdly added a little crop diversification to his manuring, sometimes inserting pea breaks between rice plantings (gaining thereby from the fodder and nitrogen bonuses such legumes could provide) and sometimes using oats and potatoes. He believed that his rice yields had more than doubled. He did not experiment with marl or lime.

In 1851, Heriot was elected as a delegate of Winyaw and All Saints Agricultural Society to attend the Crystal Palace Exhibition in London. He went with some of his family, taking the opportunity to visit relations in Scotland. At the exhibition he won an award for Carolina rice (as did Samuel Colt, Charles Goodyear, and Cyrus McCormick for revolvers, vulcanized rubber, and reapers respectively). [Allston, *Memoir of Rice;* Allston, *Essay on Sea Coast Crops;* Davidson, *Last Foray;* Joyner, *Down by the Riverside;* Lachicotte, *Georgetown Rice Plantations;* Reynolds and Faunt, *Biographical Directory of the Senate;* Rogers, *History of Georgetown County;* Ruffin, *Report of Survey*]

HOLLY, Mr., owned a plantation that ran up to the mainstream of the Lower Three Runs, a left bank tributary of the Savannah. [Ruffin, *Report of Survey*]

HUGER, Dr. Benjamin (1793–1874), was a cosmopolitan member of a prominent Charleston and Georgetown family and a graduate of Harvard College and the University of Pennsylvania. He owned the large Richmond plantation on the East Branch of the Cooper. He is not to be confused with his younger kinsman of Mexican War and Civil War fame or with the politically prominent Benjamin Huger, then deceased. [Davidson, *Last Foray;* Irving, *Day on Cooper River;* Reynolds and

Faunt, *Biographical Directory of the Senate*]

HUGUENIN, Captain Abraham, and HUGUENIN, Julius, belonged to a family that was well established in St. Luke's Parish, Beaufort, lending its name to the substantial neck of land west of the Broad River and south of Bees Creek. Robert Mills visited the plantation in the 1820s (when it probably was in the possession of an earlier Abraham, who died in 1829), finding fragments of marblelike rock composed of shells and abundant "remains of marine animals." This no doubt stimulated Ruffin's curiosity. In addition, the present Abraham Huguenin was clearly a planter of repute, being specially thanked in Whitemarsh Seabrook's *Memoir on Cotton* for information he had supplied.

The Julius mentioned was probably his son, Julius Gillison Huguenin (1806–62), later a St. Luke's delegate to the Southern Rights Convention of 1852. [Davidson, *Last Foray;* Mills, *Atlas of South Carolina;* Mills, *Statistics of South Carolina;* Seabrook, *Memoir on Cotton; Journal of the State Convention of South Carolina; South Carolina Genealogies,* vol. 1]

JAMISON, General David Flavel, despite his imposing militia title, was only thirty-two. He and Ruffin shared powerful secessionist beliefs in later years, but in 1843 Jamison probably attracted Ruffin's attention because he owned land of calcareous interest, resources of which had already been put to well-publicized use. In 1838 the geologist Lardner Vanuxem had written to Ruffin's *Farmers' Register* citing six places in South Carolina that boasted rich deposits of shell marl, one of them being "Dr. Jameson's near Orangeburgh Court House." This was Dr. Van de Vastine Jamison, David's father, named by Robert Mills in 1826 as an energetic quarrier, lime producer, and indigo grower.

In 1843 David Jamison was a representative for Orange Parishes in Columbia, taking special interest in military matters. His main business, since termination of a brief law career in 1832, was agriculture, in which he acquired some local reputation as an improver. He had some intellectual interest in philosophy and wrote papers for the *Southern Quarterly Review* and State Agricultural Society. He attended the state Agricultural Convention in 1839.

Jamison's increasing involvement with sectional politics culminated in his election to the presidency of the Secession Convention in December 1860. Ruffin was able to take advantage: "By favor of Gen. Jamison . . . I got a seat in the hall, which is so small that but few others than the members could find places to stand." When the deed was done, Jamison sailed Ruffin and others around Charleston Harbor to inspect the forts. They met again in 1863. Jamison died of yellow fever in 1864. [Barnwell, *Love of Order;* Mills, *Statistics of South Carolina;* O'Brien, ed., *All Clever Men;* Ruffin, *Diary,* vols. 1, 3; Smith, *Economic Readjustment of an Old Cotton State; Farmers' Register,* 1838; *Journal of the State Convention of South Carolina; Proceedings of the Agricultural Convention*]

JANNEY, J. C., was a Springs lime producer. Rock had long been burned there, according to Ruffin's report. "The kiln is badly constructed," he wrote, "and there is great waste of fuel, and of heat, and of time in burning, as well as labor in blasting the stone in the quarry." The hypothetical ten-cent figure is based on current costs for quarrying, hauling, kiln-filling, wood-collecting, burning, and rents. Janney appears to have been securing income by high unit returns and low sales. Ruffin clearly favored his strategy of improved efficiency and reduced prices. [Ruffin, *Report of Survey*]

JOHNSON, Dr. Joseph F., was the son of William Johnson, Goose Creek planter and local revolutionary war leader, brother of U.S. Supreme Court Justice William Johnson, (friend of Thomas Jefferson). Then aged sixty-six, he was the patriarch of a large family. He had a distinguished career as a physician and scientist in his native Charleston, contributing to professional journals, serving as past president of the state Medical Society, and ranking as an authority on causes and treatment of yellow fever.

Johnson owned a plantation behind Charleston and later built a large house, known as the Castle (still extant), in Craven Street, Beaufort. He was long interested in agricultural matters and since the mid-1830s had been advocating the importance of calcareous deposits, following his penetration of marl strata during a well sinking in Charleston. Two of his pieces on marl were published in the *Farmers' Register,* citing the value of Ruffin's writings, the slow spread of experimentation, and the possibility that marling might check further emigration.

In other respects, however, their opinions and experiences diverged. Johnson, with considerable business and property interests, had served as president of the Charleston branch of the Bank of the United States—thus being associated with Ruffin's most detested profession. A graduate of the University of Pennsylvania, he also held strong Unionist sentiments and had been one of Charleston's most outspoken critics of nullification. He died in 1862, aged eighty-six. [Graydon, *Tales of Beaufort;* Heitzler, *Historic Goose Creek;* Moltke-Hansen, "Expansion of Intellectual Life"; Pease and Pease, "Charleston's Nullification Crisis"; Walker, *History of the Agricultural Society;* Waring, *History of Medicine in South Carolina; Charleston Medical Journal,* 1848, 1850; *Eminent and Representative Men of the Carolinas; Farmers' Register,* 1838, 1840]

JOLLY, Captain, probably was the Joseph Jolly who sat in the House of Representatives in Columbia in 1843. He was of most interest as the owner of land containing cretaceous marl. His plantation lay at the end of the neck between Jeffrey's Creek and the Pee Dee and very close to the "Myers' land" mentioned in the diary (and shown in Mills's Marion map).

[Mills, *Atlas of South Carolina;* Ruffin, *Report of Survey; Journal of the House of Representatives, 1843*]

KING, Judge Mitchell, a city court judge, was aged about sixty in 1843. He was born and educated in Scotland, arrived in Charleston in 1805, and was admitted to the South Carolina bar in 1810. He earned well as a lawyer but acquired most of his considerable wealth through astute marriages, successively, to sisters Susanna and Margaret Campbell. By 1830, according to Frederick A. Porcher (q.v.), he "was living in one of the most elegant mansions in the city." William Makepeace Thackeray later described him as "a bigwig of the place." A large, beetle-browed man of some intellectual substance, King was a devout student of Scottish philosophy, author of *Discourse on the Qualifications and Duties of an Historian,* contributor to the *Southern Quarterly Review,* and cofounder of both the South Carolina Historical Society and the Literary and Philosophical Society. He was inclined to Unionism, though, and along with Robert Barnwell, represented St. Philip's and St. Michael's (Charleston) at the Southern Rights Convention in 1852.

King encouraged his friend John Bachman (q.v.) to publish in 1842 his reflections on the agricultural survey (see Introduction). King's interest is indicated by his visit with Bachman to greet Ruffin on his arrival. His concern for agriculture was that of a public figure rather than a planter. It was probably rhetorical and legal "bigwiggery" that secured his invitation from the State Agricultural Society to deliver the Anniversary Address in November 1846. King insisted that he was undeserving of the honor. "The engrossing studies and practice of a jealous profession, have left me little leisure or opportunity for Agricultural inquiries." He congratulated the society on its success in getting the survey under way, "which, in the able hands by which it has been conducted, has already afforded much valuable information, and it is earnestly hoped, will be continued until all the objects contemplated by it be fully attained." He argued that the old seaboard cotton states urgently needed alternative staples.

It was in King's house, eighteen years later, that Ruffin abandoned his abstinence to toast the news of Virginia's secession. Two days later he was there again: "Judge King is one of the first men of S.C. for his intellect & his virtues, though now very old, & withdrawn from all public matters." Extensive remains of his personal library survive at the College of Charleston. [Bachman, *Inquiry into the Agricultural Survey;* King, *Address before St. Andrew's Society;* King, *Discourse on the Qualifications and Duties of an Historian;* King, *Anniversary Address of the State Agricultural Society;* Moltke-Hansen, private communication to editor; Ruffin, *Diary,* vol. 1; Wauchope, *Writers of South Carolina; Journal of the State Convention of South Carolina;* "Memoir of F. A. Porcher"]

LARTIGUE, Colonel Isadore, was a representative for the Beaufort District parish of St. Peter's. Active in the South Carolina Agricultural Society, he served on the Committee on Horses in 1844–45 and on another set up about the same time "to enquire into the means" of the society. He was a friend of the survey, voting supportively in the House, and also was involved with the parish agricultural society. [*Journal of the House of Representatives, 1843; Proceedings of the Agricultural Convention,* 1844]

LAWTON, Mr., must have been one of the prominent Lawtons of that time, the most notable of which was Winborn Lawton, former St. Andrew's parish representative and senator and James Island planter, then about age sixty. The man mentioned in the diary, however, almost certainly was William M. Lawton. Ruffin wrote in 1857 of a dinner meeting with "Wm. M. Lawton esq., a prominent & wealthy & very intelligent merchant, to whom I have before been indebted for much attention."

Lawton was a planter and trader, conducting most of his business in and around Charleston. He is cited in Robert Allston's (q.v.) *Essay on Sea Coast Crops* as an authority on cotton cultivation and as the man who drew the attention of Carolina planters to the existence of the superior McCarthy (or Florida) cotton gin. In 1870, he was elected to a vice-presidency of the State Agricultural Society, along with Sanford Barker (q.v.) and Isaac Jenkins Mikell (q.v.). He was still active in the society in 1878.

Immediately after South Carolina's secession in December 1860, Ruffin traveled south in anticipation of similar developments in Florida. Lawton was then president of a steamship line that ran to Fernandina and helped Ruffin on his way with a free passage ticket. [Allston, *Essay on Sea Coast Crops;* Reynolds and Faunt, *Biographical Directory of the Senate;* Ruffin, *Diary,* vol. 1; Walker, *History of the Agricultural Society*]

LEGARÉ, Dr. Thomas, one of the largest James Island farmers, was of Huguenot descent. He worked Light House Point plantation and owned around 150 slaves. He was about the same age as Ruffin and an M.D. graduate of the University of Pennsylvania. He was married to the former Mrs. Sarah Jenkins Mikell. Legaré gave what was later referred to as a "capital practical speech" at the anniversary meeting of the United Agricultural Society in Columbia in 1838, in which he complained of federal prejudice in favor of manufacturers and the resultant tariff burdens on planters. He died in 1855. [Davidson, *Last Foray;* Walker, *History of the Agricultural Society*]

LITTLEJOHN, Major John (Jack), born in 1794, was a large Union District planter, merchant, and owner of a gristmill on the Pacolet River. As a state representative he was contemporary with Roper, Middleton, Jamison, and many others who assisted Ruffin. He served as senator, 1836–40, and had been a member of the Nullification Convention. The diary gives his agricultural biography. In the report, Ruffin mentions him and William Davis (q.v.) of Fairfield as the only planters in South Carolina known to be engaged in the leguminous cropping described. Littlejohn did not marry and died in 1872. [Reynolds and Faunt, *Biographical Directory of the Senate;* Ruffin, *Report of Survey; Journal of the House of Representatives, 1843*]

MCCORD, Colonel David James, was a planter in South Carolina and Alabama. At the time of Ruffin's visit, he was forty-six years old. McCord was a prominent nullifier, free trader, and legal author of local distinction. He began both matrimonial and public life early, marrying in his teens, fathering fourteen children, working in Columbia after the early 1820s as a lawyer, editor of the *Telescope,* intendant of the city (receiving Lafayette in 1825), representative, and president of the local branch of the Bank of South Carolina. A year or so after his first wife's death in 1839 he married writer Louisa Cheves, daughter of Langdon Cheves, and moved in 1841 to her recently acquired plantation at Lang Syne on the Congaree thirty miles from Columbia, and living for part of the year in the capital.

McCord was the grandson of an Irishman who first appeared in the state in the mid-eighteenth century as a Catawba and Cherokee trader and later operated McCord's Ferry on the Congaree. David McCord had a reputation for impulsive, intemperate behavior and, in the words of Susan Smyth Bennett (a collateral descendant), "never hesitated to use a cane, or his fist." Ruffin and he got on well, being of the same generation. Except for banking, they would have agreed on most current issues. In late November 1843 Ruffin wrote him a personal letter from Charleston expressing his frustrated desire to visit McCord and his wife at Lang Syne and asking that his belongings be forwarded as well as any newly found fossils of interest. In his report, Ruffin acknowledged their help and named the *Cytheria Maccordia* shell in Louisa's honor. McCord died in 1855. [Davidson, *Last Foray;* Fraser, *Louisa C. McCord;* Longton, "Some Aspects of Intellectual Activity in South Carolina"; O'Brien, ed., *All Clever Men;* Ruffin, *Report of Survey; South Carolina Genealogies,* vol. 3; Edmund Ruffin to Col. D. M. [sic] McCord, 25 November 1843 (ms. in the Library of the University of South Carolina, Columbia]

MCKEE, Mr., would be Henry McKee (1811–75), owner of Ashdale plantation, three miles east of Beaufort, and a brick mansion in Prince Street, Beaufort town (still known as the McKee House). Currently he was in a legal dispute with Gabriel Capers (q.v.) over the road that crossed Capers's land (and which, as a result of accompanying drainage ditching, had severed one of Capers's lines of access to his own plantation). [Davidson, *Last Foray;* Hilton, *Old Homes and Churches of Beaufort;* Mills, *Atlas of South Carolina;* Rosengarten, *Tombee;* Ruffin, *Report of Survey*]

MANNING, Elizabeth Peyre, née Richardson (1801–69), was the widow of Governor Richard Manning, sister of Governor John P. Richardson (q.v.), and mother of

Governor John L. Manning (q.v.). [Reynolds and Faunt, *Biographical Directory of the Senate*]

MANNING, John Lawrence, son of former governor Richard Irvine Manning (died 1836) and Elizabeth Peyre (Richardson), and grandson on his father's side of General Richard Richardson, owned the house, probably Milford (still standing), built around 1840 and sometimes known as Manning's Folly. Davidson describes him as "princely," and Scarborough calls him "one of the largest slaveholders in the South," with land in Louisiana as well as Carolina. His slaves numbered 425 in 1850, most of them in Louisiana sugar growing. Like his father, Manning served as a member of the South Carolina legislature and later as governor (1852–54). Both men served in Washington, as congressman and senator respectively, and both were disunionists, one being a member of the Nullification Convention of 1832–33, the other head of a Committee of Safety in 1850 and delegate to the Secession Convention in 1860.

Manning supported the survey, though not to the extent of assisting the surveyor. The two became friendly just before the outbreak of the war, and the acquaintanceship, as with David Jamison (q.v.), helped Ruffin gain a front-row view of events. "When the Convention adjourned," he recorded on 20 December 1860, "it was to meet again at 7 P.M. at Secession Hall, a very large room, to sign & ratify the act of secession & independence. I was taken in by Ex-Gov. Manning, among the members of the Convention, in the centre of the hall." [Channing, *Crisis of Fear;* Davidson, *Last Foray;* Reynolds and Faunt, *Biographical Directory of the Senate;* Nicholes, *Historical Sketches of Sumter County,* vol. 2; Ruffin, *Diary,* vol. 1; private communication from Scarborough to editor; Ruffin, *Report of Survey*]

MAZYCK, Robert, was probably over sixty at the time and one of a prominent Huguenot family (originally Mazicq) that arrived from Liège by way of Holland, France, and Britain at the end of the seventeenth century. Owned Fair Spring plantation. He was another of the rare marlers in the state. F. A. Porcher's (q.v.) committee on manures visited his farm in August 1844 to look at his results from greensand marling, reporting to the Black Oak Agricultural Society: "The marl . . . is found in a ravine on the eastern side of Begin swamp. . . . It is to be regretted that Mr. Mazyck did not accurately observe the quantity applied to his land; and it is rather too early in the season for us to be furnished with the results of his experiment. At that late period of the summer, however, a practical eye can judge with tolerable accuracy, what the result will be. It required but a glance to convince us, that the cotton manured with green sand, was worth three fold the best portion of his crop not so manured." As with virtually every other case of apparently successful marling in Carolina, the exercise was not extended in subsequent years. [MacDowell, *Gaillard Genealogy;*

Holmes, "Notes on the Geology of Charleston"; Porcher, *Report on Manures*]

MEANS, General John Hugh, the thirteenth child of Sarah Milling and Thomas Means (the former of an old Carolina family, the latter originating in Boston), owned the large Oaklands plantation in Fairfield District. He was married to Susan Rebecca Stark. Then aged thirty, he held the title of "general" in the local militia. Means was yet another of the past and future governors Ruffin encountered. Robert Allston (q.v.) described him before Means's election as "an untried man, fond of popular favor, and very successful in commanding it." He was a determined secessionist, and his term (1850–52) was, by Yates Snowden's account, noted for states'-rights propaganda and many gubernatorial "speeches in favor of the withdrawal of the South from the Union." Means was president of the Southern Rights Convention in 1852 and prominent during the presecession period, seeing Ruffin again on several occasions.

Means took his planting seriously, but nothing around Oaklands was of particular interest to Ruffin. In 1839 he and his brother Edward represented Fairfield at the state Agricultural Convention in Columbia. He was killed in 1862 at the second Battle of Manassas. [Barnwell, *Love of Order;* Davidson, *Last Foray;* Ruffin, *Report of Survey; Journal of the State Convention of South Carolina; Proceedings of the Agricultural Convention; South Carolina Genealogies,* vol. 3]

MELLETT, Peter, had died two years earlier. The Melletts were a prominent Sumter family of Huguenot origin, dating back to the seventeenth century. "The death of Mr. Mellett," wrote Ruffin in his report, "and there being no person particularly acquainted with his experiments, prevented my obtaining more precise information," but there could be no doubting "the very great improvement" that had occurred.

Ruffin, however, did not reflect on the implications of Mellett's work. Such an instance of calcareous experimentation by an uneducated farmer was very rare. Not only was it extremely difficult to get the work started at the subgrandee level, but there were high risks, as here, of it being conducted inefficiently, of results being unrecorded, and of influence on the neighborhood being nonexistent. Ruffin never gave much attention to such problems. [Mathew, *Edmund Ruffin;* Nicholes, *Historical Sketches of Sumter County,* vol. 2; Ruffin, *Report of Survey*]

MIDDLETON, John Izard, should not be mistaken for his uncle John Izard Middleton (1785–1849), archaeologist and painter, formerly of Cedar Grove. This Middleton was a politically prominent planter of Crowfield, then aged forty-three. He was the fourth son of Governor Henry Izard Middleton of Middleton Place and a member of an English family that had been in the state since the 1670s. He married an Alston. Middleton graduated from the Colleges of South Carolina and New Jersey (later Princeton) and worked

(1822–24) as legation secretary in St. Petersburg, where his father was U.S. minister.

He was a strong disunionist, attending the Nullification Convention in 1832–33, becoming president of the Winyah and All Saints Southern Rights Association in 1850, talking secession throughout the 1850s, traveling north with the radical Carolina delegation to the Richmond Democratic Convention in 1860, and signing the ordinance at the Secession Convention in 1860. He was also elected representative, senator, and Speaker of the House, gaining abundant opportunity to express his forceful opinions.

Middleton was a planter on a grand scale. In 1860 he had 201 slaves at Crowfield and (counting shared property) 318 in total. He also had a reputation for innovative farming. As early as 1824 he was awarded a premium by the United Agricultural Society for achieving a rice yield of seventy bushels per acre. In the 1850s he more than doubled his output through the application of improved methods and advanced technology. Ruffin observed in his report, "The steam machine of Mr. J. I. Middleton thrashes from 600 to 1,000 bushels a day." The two men met again frequently in the months preceding the war. They also corresponded, Middleton briefing Ruffin in July 1860 on the vulnerability of the federal forts in Charleston Harbor and praising him for his "zeal in the good cause of Southern redemption." He died in 1877. [Channing, *Crisis of Fear;* Davidson, *Last Foray;* Heitzler, *Historic Goose Creek;* Joyner, *Down by the Riverside;* Rogers, *History of Georgetown County;* Reynolds and Faunt, *Biographical Directory of the Senate;* Ruffin, *Report of Survey;* Ruffin, *Diary,* vol. 1; private communication from Scarborough to editor; Walker, *History of the Agricultural Society of South Carolina; South Carolina Genealogies,* vol. 3; Ruffin Papers, Southern Historical Collection, University of North Carolina]

MIKELL, Isaac Jenkins, then thirty-four years old, of Peter's Point plantation and Charleston, was from old Edisto families. A graduate of the College of New Jersey, he was a bookish man, generally disinclined to involve himself with public affairs. He did, however, attend the Agricultural Convention of 1839 with his friend William Murray (q.v.) and in 1870 agreed to become a vice-president of the State Agricultural Society.

As corresponding secretary of St. John's, Colleton, Agricultural Society in 1840, he communicated the club's queries about liming of cotton ground to Edmund Ruffin in Virginia. Ruffin published his letter and his own reply in the *Farmers' Register.* The exchange had arisen from Professor C. U. Shepard's (q.v.) analysis of island soils, requested by planters who were becoming increasingly concerned over the deteriorating yields of sea-island cotton and their evident inability to halt the decline by manuring. Mikell's urgent interest in such matters was reflected in subsequent experiments with Peruvian guano. He was probably the first man in the state to publish an account of his trials with the fertilizer.

Mikell was a planter of great wealth, owning well over two hundred slaves and farming in a manner that clearly impressed his guest. His mansion was built in a commanding location by the sea around 1840 on an elevated foundation of brick and tabby (a mixture of shell lime and sand). It had, according to one of his sons, "twelve great rooms with white and colored marble for inside adornment, a spiral stairway, broad brown stone steps and double piazzas." There was a large fishpond outside for winter provision, within and around which lay an ornamental profusion of palmettos, live oaks, saltwater cedars, Venetian bridges, and Chinese tea gardens. A dozen years or so later Mikell built a spectacular classical revival house in Charleston, which still stands, at 94 Rutledge Avenue. He died in 1881, aged seventy-two. [Davidson, *Last Foray;* Graydon, *Tales of Beaufort;* Taylor, "Commercial Fertilizers"; Walker, *History of the Agricultural Society; Farmers' Register,* 1840; *Proceedings of the Agricultural Convention*]

MILNE, Mr., almost certainly was Andrew Milne (1783–1857), a Scotsman who had shown the canniness Ruffin mentions by profitable trading in Charleston and marriage into the prominent Seabrook family (q.v.). [Davidson, *Last Foray;* Mills, *Statistics of South Carolina*]

MOSER, Dr. Philip, a Pennsylvania physician, had practiced medicine in Charleston up to 1831, owning property in the city and representing it in the state legislature, 1808–16 and 1820–24. He died in Philadelphia in 1835. His account of boring was published in 1826 in Robert Mills's *Statistics of South Carolina.* City authorities had been eager to secure a good, regular supply of pure water. "An idea," wrote Mills, "seemed long to be entertained of the existence of subterraneous streams of water running to the ocean, from distant highlands; which if struck upon, by boring, would rise up and overflow the surface, in proportion to its original source." The city council acted on this suggestion, giving Moser authority to bore in the poorhouse yard in the summer of 1823. Although the work extended into 1824 and penetration was a very considerable 335.5 feet, no water was found except small quantities seeping in from surrounding strata. It was, though, a valuable exercise scientifically. Mills declared that there was "no instance . . . in the alluvial country of any part of the world, of such a depth being penetrated before." Moser's samples told Ruffin that probably the best marl stratum on the eastern seaboard lay beneath the city, dipping too deeply to be available to farmers, though close to the surface farther inland (and accessible in later years when mining companies applied advanced technology in search of phosphates). There was, Ruffin wrote in his report, "in all two hundred and forty-two feet of marl, rich, and not anywhere of stony hardness, except of a few very thin layers." [Reynolds and Faunt, *Biographical Directory of the Senate;* Mills, *Statistics of South Carolina;* Ruffin, *Report of Survey; De Bow's Review,* 1847]

MURRAY, Major William Meggett, born in Edisto in 1806, was in 1843 state senator for St. John's, Colleton. His political interests led to his attending the Nullification Convention of 1832–33, and his military education at Partridge Academy helped bring him commands in the Mexican and Civil wars. His life, however, was mainly that of an agriculturist, and his Jack Daw Hill plantation at Cedar Grove was one of the largest on Edisto, using about 150 slaves in its operation.

In 1840 he had been on a committee set up by the Agricultural Society of St. John's, Colleton, to examine Professor C. U. Shepard's (q.v.) analysis of island soils. This and the society's correspondence with the *Farmers' Register* familiarized him with the issue of calcareous manuring and the name of Ruffin. He had also represented St. John's parish at the 1839 Agricultural Convention in Columbia and five years later delivered its petition to the senate "praying for the continuation of the Agricultural Survey." In 1844–45 he served as chairman of the state Agricultural Society's Committee on Marl. He died on his sixtieth birthday in 1866. [Davidson, *Last Foray;* Reynolds and Faunt, *Biographical Directory of the Senate; Farmers' Register,* 1840; *Journal of the Senate, 1844; Proceedings of the Agricultural Convention*]

NELSON, Mr., probably was a relation (father?) of General Patrick Henry Nelson of Indigo Hill and Marston in Sumter District, the latter then aged eighteen and a student in Columbia. [Davidson, *Last Foray; Eminent and Representative Men of the Carolinas*]

NESBITT, Colonel, was a member of a noted up-country industrial family, probably the landowning Colonel Wilson Nesbitt mentioned in Mills's *Statistics* of 1826. The same man was, at the time of Ruffin's visit, a representative in Columbia and a member of the State Agricultural Society. [Lander, *Textile Industry in South Carolina;* Mills, *Statistics of South Carolina*]

NOTT, Dr., a local physician, planter, and mine owner with an experimental interest in liming, should not be confused with the better-known Dr. Josiah Clark Nott, an old college friend of James Hammond's, then resident in Georgia, or with the late Henry Junius Nott, David McCord's (q.v.) collaborator in legal scholarship. [Faust, *James Henry Hammond;* Ruffin, *Report of Survey; Garnet and Black,* vol. 1]

PALMER, Dr. John S., and PALMER, Colonel Samuel J., were brothers. John Saunders (1804–81) and Samuel Jerman (1807–53) were the sole issue of Thomas Palmer's (Gravel Hill) third marriage, to Harriet Jerman, and the grandsons of wealthy naval stores entrepreneur of English parentage "Turpentine John" (who had changed the family name from Pamor). Dr. John, a graduate of

New York College of Physicians and Surgeons and former medical practitioner, worked Balls Dam plantation above Lenud's Ferry on the Santee and subsequently Betow and Laurel Hill as well. Colonel Samuel, A.B. of the College of South Carolina, lived lower downriver at Mt. Moriah, owning the site of the deserted Huguenot village of Jamestown and also buying more land. Together they owned hundreds of slaves. Both were justices of the peace, justices of the quorum, commissioners of free schools, and, for spells, representatives and senators. In 1852 Samuel was St. James's, Santee, delegate to the Southern Rights Convention. In 1860 John was a delegate to the Secession Convention and met Ruffin again.

In 1843 the Palmer brothers were, with James Hammond, Ruffin's principal disciples in the state. Their enthusiasm and success must have been enormously gratifying to him. As elsewhere, though, Ruffin combined pleasure at innovative spirit with dissatisfaction over methods that made it impossible for the virtues of amelioration to be thoroughly realized. No doubt he took time to spell this out to his hosts.

In June 1843, a talk on marling by John Palmer, originally delivered to the Black Oak Agricultural Society, was published by the *Southern Agriculturist.* In it Palmer gave details of his own, his brother's, and Robert Gourdin's (q.v.) experiments. Ruffin's account in the diary appears to draw on the paper. Palmer concluded, with misplaced optimism: "In the course of a few years, we may look to see marl made an article of internal commerce, and those who are now afraid to carry it 100 yards upon their own soil, may, when they shall have covered every foot of arable land in their possession, be found supplying it at a profitable rate to their neighbors five miles distant." Ruffin secured Palmer's permission to reproduce the piece in the Appendix to the report. [Davidson, *Last Foray;* MacDowell, *Gaillard Genealogy;* Reynolds and Faunt, *Biographical Directory of the Senate;* Ruffin, *Report of Survey;* Ruffin, *Diary,* vol. 1; *Journal of the State Convention of South Carolina; Southern Agriculturist,* 1843]

PATTERSON, Major Angus, a large planter of Pine Forest in the Barnwell District, had been born in North Carolina and was mostly self-educated. In 1843 he was aged fifty-two. A man of extended political experience, he was then in the fifth of twelve years as president of the state senate and the fifteenth of thirty-two years as a member of the state legislature. As a Barnwell representative, he had been a strong supporter of nullification, insisting that "one of the avowed objects of the Tariff, is to favor free labor, as it is called, at the expense of slave labor," rendering the latter valueless and forcing Carolinians "to assent to a system of emancipation, through the agency of the General Government."

Patterson collected a noted library of law books. He was active in the Barnwell Agricultural Society, for a time serving as president. He died in 1854. [Davidson, *Last Foray;* Freehling, *Prelude to Civil War;* Reynolds and Faunt, *Biographical Directory of the Senate; Southern Agriculturist,* 1842]

PICKENS, Francis Wilkinson, owned the extensive Edgewood plantation, Edgefield District, on the Savannah. He had 165 slaves there and around 90 on other properties. Clearly Pickens had offended Ruffin, causing an angry response quite unlike anything else recorded in the diary. Pickens's social and political eminence gave him a powerful appeal in an area where grandees were few, so Ruffin was probably disappointed as well. Pickens had a reputation for curtness and reserve, and his behavior may have been affected by recent bereavements. In 1843 he was thirty-six years old; he was the son of Governor Andrew Pickens, grandson of General Andrew Pickens, and cousin of John C. Calhoun, "a public figure almost by inheritance," as Davidson writes. He had just left Washington after eight years in Congress, becoming involved in Calhoun's campaign for the Democratic nomination for president. Pickens was elected as Edgefield's senator in Columbia. He has been viewed as a place-seeker, with a provincial, Carolinian perspective on national politics. He held office in the Edgefield Agricultural Society and was active in the state society as well.

Later Pickens served as U.S. minister to Russia and, despite his attendance at the Nashville and Southern Rights conventions in the early 1850s, preserved a reputation for moderate Unionism. Ironically, he was elected governor the year of secession. "I would have preferred any of the other persons named," wrote Ruffin in Charleston. "I think he is an empty boaster." In 1861 Ruffin was playing tit for tat after fresh rebuffs, possibly with 1843 still in mind. The governor had three times confused the relationship by asking him to call. "On each occasion I politely returned my thanks, but did not promise to accept the invitation, nor did I intend to do so. The Governor's polite attentions come too late." Ruffin had no interest in "deferred civilities." Pickens died in 1869. [Channing, *Crisis of Fear;* Davidson, *Last Foray;* Edmunds, *Francis W. Pickens;* Faunt and Reynolds, *Biographical Directory of the Senate;* Faust, *James Henry Hammond;* Freehling, *Prelude to Civil War;* Ruffin, *Diary,* vol. 1; Wakelyn, *Biographical Dictionary of the Confederacy; Journal of the State Convention of South Carolina; Proceedings of the Agricultural Convention*]

POINSETT, Joel Roberts, born in 1779, was one of the best known Carolinians of his day. He was widely traveled and educated in Europe and the Americas in medicine, law, and military science. He was elected to both state and national legislatures, served as minister plenipotentiary in Mexico, 1825–30, and as Martin Van Buren's secretary of war, 1837–41. Poinsett was a founder of the National Institute for the Promotion of Science and the Useful Arts. He was a consistent and eloquent Unionist. In 1843 a full-time planter of some scale and repute at the Homestead, he took an active interest in local and state agricultural societies, working Edward Heriot's (q.v.) system of "rest, manuring and rotation" on rice land. He declared himself in support of the survey in an address to the South Carolina Agricultural Society in 1845. He died in 1851 on a visit to Dr. W. W. Anderson (q.v.) of Sumter. [Allston, *Essay on Sea Coast Crops;* Channing, *Crisis of Fear;* Freehling, *Prelude to Civil War;* Moltke-Hansen, "Expansion of Intellectual Life"; Nicholes, *Historical Sketches of Sumter County;* Poinsett, "Agricultural Address"; Rogers, *History of Georgetown County;* Reynolds and Faunt, *Biographical Directory of the Senate; Proceedings of the Agricultural Convention; Eminent and Representative Men of the Carolinas; Southern Agriculturist,* 1844]

POPE, Joseph James, should not be confused with his son Joseph Daniel Pope (1820–1908), later head of the Confederacy's Revenue Department. The Popes were an English family that had been in Carolina since the early eighteenth century. In St. Helena, they had joined the ranks of the great sea-island cotton producers, "that fortunate class," according to William Elliott (q.v.), "whose names gave title to the highest prices in the agricultural lottery." Like Elliott and Isaac Jenkins Mikell (q.v.) just up the coast, Pope had literary inclinations, purchasing books for his large, wide-ranging library. This may partly explain why Ruffin considered him "good company." He served as president of the St. Helena Agricultural Society (see entry for Gabriel Capers).

In 1846 Pope delivered a high-flown Fourth of July address, heavy with learned allusion, in which he warned his St. Helena audience against the danger of federal authority growing at the expense of state legislatures and said that war could be a means of bringing this about. "May we never surrender our jewels for some pitiful gewgaw, whose brilliancy is delusive, and vanishes as we approach!" In 1849, he gave the anniversary oration at Moultrieville. In 1850, he helped form the St. Helena branch of the Southern Rights Association, joining its council of safety. [Channing, *Crisis of Fear;* Davidson, *Last Foray;* Johnson, *Social History of the Sea Islands;* Pope, *Moral Influence of the American Government;* Pope, *Oration in the Fort at Moultrieville;* Reynolds and Faunt, *Biographical Directory of the Senate;* Rosengarten, *Tombee;* Wakelyn, *Biographical Dictionary of the Confederacy; Eminent and Representative Men of the Carolinas; Southern Agriculturist,* 1842]

PORCHER, Frederick Adolphus, of Somerton, was in his mid-thirties in 1843. He was perhaps the most urbane and scholarly member of his notable South Carolina Huguenot family. He was the great-great-grandson of the Isaac Porcher who emigrated from Berri in France, first cousin of the contemporary Isaac (q.v.) and Philip (q.v.), and second cousin of Major Samuel (q.v.). After attending Partridge Military Academy and Yale College and undergoing a serious bout of tubercular ill health, Porcher had traveled to Europe for eighteen months, where, according to an obituary, he devoted himself "to the cultivation of his mind and taste."

He was elected to the state legislature three times in the 1830s, despite his youth and frequent absence. He retired from politics in 1840, maintaining strong states'-rights convictions. In 1848 he won the chair of history and belles lettres at the College of Charleston and thereafter was primarily a literary figure, sending off sententious articles to *Russell's Magazine* and the *Southern Quarterly Review,* taking an interest in the local Library Society and South-Carolina Institute, contributing to the Literary and Philosophical Society of South Carolina, and helping found the state historical society in 1855 (later serving as its president).

Despite his scholarly city work and dislike of planting, Porcher was much involved with agriculture. The principal reason for Ruffin's visit to him probably lay in Porcher's reputation as an improving cotton farmer. He was then serving on the State Agricultural Society's three-man committee looking into "Cotton, Its Different Species and All Matters and Facts Relating to It." Whitemarsh Seabrook made use of Porcher's expertise in his 1844 *Memoir on Cotton* and thanked him in the Preface. Most notably, however, Ruffin's own *Farmers' Register* in March 1840 carried an account of successful experiments Porcher had undertaken in 1839 with a combination of salt, manure, and lime (using local lime rock smashed up with hammers). He failed, however, to persist with amelioration.

In 1844 Porcher presented the report of his Committee on Manures to the Black Oak Agricultural Society, declaring therein that the prosperity of the state depended on "judicious use of manures" and regretting the skepticism and tardy responses of farmers on matters concerning such improvement. But there had recently been notable change: "Our planters long since knew that lime was used in their very neighborhood with favorable results; but it required the energy and fire of a Ruffin to make the adoption of its use general." Porcher, like so many others, was being overoptimistic. He died in old age in 1889. [Longton, "Some Aspects of Intellectual Activity in South Carolina"; MacDowell, *Gaillard Genealogy;* O'Brien, ed., *All Clever Men;* Porcher, *Report on Manures;* Seabrook, *Memoir on Cotton;* Walker, *History of the Agricultural Society;* Wauchope, *Writers of South Carolina; Collections of the South-Carolina Historical Society,* vol. 1; *Farmers' Register,* 1840; "Memoir of Professor F. A. Porcher"]

PORCHER, Isaac, older brother of Philip Porcher (q.v.), also died young, in 1849. He married Elizabeth Gaillard and lived at Chapel Hill plantation. He is not to be confused with his father, Isaac Porcher of Oldfield (1778–1849), who had been politically prominent (q.v. Porcher, Philip Mazyck). [MacDowell, *Gaillard Genealogy;* Reynolds and Faunt, *Biographical Directory of the Senate*]

PORCHER, Philip Mazyck, aged about twenty-five, took an earnest interest in Ruffin's ideas. He was the brother of Isaac (q.v.). Their father, Isaac Porcher of Oldfield (1778–1849), was a former senator for St. Stephen's and first cousin of Major Samuel Porcher (q.v.). Philip died in 1850. [MacDowell, *Gaillard Genealogy;* Reynolds and Faunt, *Biographical Directory of the Senate*]

PORCHER, Major Samuel, of Mexico plantation, St. Stephen's, Charleston District, was born in 1768 and died in 1851. He married his first cousin Harriet Porcher (1772–1843). The Porcher genealogy shows much intermarriage within the French community. The first four generations established alliances with the Gendron, Mazyck, Cordes, Bonneau, Gignilliat, Peyre, Ravenel, DeVeaux, Cahusac, DuBose, and Couturier families. Two of Samuel's four children married Gaillards.

Ruffin's diary supplies an agricultural biography of Porcher. The *Farmers' Register* had cited him as a leading Carolinian improver. He was another of the planters who helped Whitemarsh Seabrook with his *Memoir on Cotton.* Robert Allston (q.v.) gave his achievement special posthumous acclaim (q.v. Porcher, William Mazyck). [Allston, *Essay on Sea Coast Crops;* MacDowell, *Gaillard Genealogy;* Seabrook, *Memoir on Cotton; Farmers' Register,* 1840]

PORCHER, Dr. Thomas William, should not be confused with Colonel Tom Porcher of Ophir, by 1843 deceased. Thomas Porcher, son of Major Samuel (q.v.) and brother of William Mazyk (q.v.), was then thirty-five years old and married to Elinor Cordes Gaillard. He held a degree from South Carolina Medical College. Clearly, Porcher was an improving planter in the style of Dr. Henry Ravenel (q.v.). He owned many slaves at Walworth, permitting extensive resort to labor-intensive compost manuring (described briefly in the *Farmers' Register*). [Davidson, *Last Foray;* MacDowell, *Gaillard Genealogy; Farmers' Register,* 1840]

PORCHER, William Mazyk, the youngest of Samuel Porcher's (q.v.) four children and brother of Thomas William Porcher (q.v.) of Walworth, unmarried and in his early thirties, managed the great embanked Mexico plantation for his aging father. Three years after Samuel's death, it was referred to in Robert Allston's (q.v.) pamphlet on coastal crops as belonging to Mazyk and being one of the few examples of successful swamp farming in the southeastern states. "None of the swamps on the great rivers," wrote Allston, "are under profitable cultivation . . . except those which, having been reclaimed, are protected by dams (levees) from the destructive influx of heavy freshets to which those rivers are annually subject." He was the St. Stephen's delegate at the Southern Rights Convention in 1852. His antebellum ways are recalled in a poem, "A Carolina Bourbon," by Yates Snowden: "Perhaps too easy life he led—Four hours afield, and ten abed." Porcher later visited Ruffin at Marlbourne. The two met again at William Cain's (q.v.) house in Pinopolis in 1863 and conversed about the war and Yankee destruction. Porcher survived until 1902. [Allston, *Essay on Sea Coast Crops;* MacDowell, *Gaillard Genealogy;* O'Brien, ed., *All Clever Men;*

Ruffin, *Diary,* vol. 3; *Journal of the State Convention of South Carolina*]

RAMSAY, Dr., probably was Dr. William G. Ramsay, 52 East Bay, Charleston, younger brother of Dr. James Ramsay, professor of surgery at the Medical College, and youngest son of Dr. David Ramsay, noted historian and physician. The family was of Pennsylvanian and Scotch-Irish origin. William was then about forty and died not many years later. He achieved considerable public notice for his work at Marine Hospital during the yellow fever outbreaks of 1834, 1835, and 1838, later communicating his observations on these and on remittent fever to learned journals. [Brunhose, *David Ramsay;* Reynolds and Faunt, *Biographical Directory of the Senate;* Waring, *History of Medicine in South Carolina; Charleston Medical Journal and Review,* 1849; *Directory of Charleston, 1837–38; Eminent and Representative Men of the Carolinas; Southern Journal of Medicine and Pharmacy*]

RAMSAY, John A., belonged to the same family as David Ramsay, the historian (q.v. Dr. Ramsay) but should not be confused with David Ramsay's nephew Dr. John Ramsay (1768–1828), state representative and senator. John A. Ramsay was an officeholder, with other notable planters, in the recently formed St. Andrew's, Ashley, and Stono Agricultural Association.

Shortly after Ruffin's visit, he began experimenting with marl. A patch of new cotton land was treated at the modest rate of 100 bushels per acre, and the results were decidedly favorable. Ramsay considered (through loose observation rather than precise measurement) that marling had at least doubled his yield, and Ruffin, looking at the remains of the crop at the end of the year, was inclined to agree. The plants generally were about ten days ahead of those left unaided. In February 1844 Ramsay tackled 25 acres, dosing them at a rate of 150 bushels. The members of his agricultural association were told in June that he had made the largest experiments in the parishes, and they were urged to examine the results. The *Southern Agriculturist,* in a brief notice on his work in September 1844, said "that any one riding along, will decide in a moment which rows are marled and which not, so great is the difference between them in the growth and appearance of the cotton-plant, the size, number, and maturity of its pods, etc." Ramsay reported drought damage in September but observed that his marled cotton had suffered "much less so than any other."

He also tried burning marl for nonagricultural lime, producing at the rate of 1 barrel of lime to 3.5 bushels of marl. His figures show that if he had aimed for annual marling of 50 acres and annual lime production of 200 barrels, he would have required 8,200 bushels of marl in all. To haul this along a rough, mile-long track at the current rate of 84 bushels a day, and making allowances for bad weather, he would either have had to begin preparatory operations at the end of the preceding October or divert much more manpower and equipment to the task. Ruffin never offered any publicized answer to such a problem, and it was one reason why so much otherwise successful experimentation—including Ramsay's—lapsed after Ruffin left the state. [Brunhose, *David Ramsay;* Reynolds and Faunt, *Biographical Directory of the Senate;* Ruffin, *Report of Survey; Southern Agriculturist,* 1844]

RAVENEL, Edmund, forty-five years old, came from a Breton Huguenot family resident in the low country since the seventeenth century. He was born in Charleston and later worked there as professor of chemistry and pharmacy at the Medical College of South Carolina. He held a medical degree from the University of Pennsylvania. Dissatisfaction with the Medical Society (original sponsor of the college) and indifferent health brought about his resignation in 1835. He purchased the Grove, on the Cooper River, that year, thereby gaining his first experience of planting.

Like John Bachman, Ravenel was one of those Carolinian men of science who enjoyed major local reputations but were not well known on a national level. Today his huge shell collection lies hidden away in dozens of storeroom drawers in the Charleston Museum. His 1834 conchological catalog, acclaimed as the first of its kind in the United States, was described by Longton as little more than "a convenient check list," the product of accumulatory instinct rather than analytical zeal. He was the author of a study of *Echinidae,* a handful of journal articles, and a pamphlet outlining the benefits of seaside residence. As both a chemist and conchologist, he was able to acquire some geological expertise. Louis Agassiz, John James Audubon, and Sir Charles Lyell visited him at his plantation (see Introduction).

Recent and fossil shells could be found at the Grove, and the latter were essential to the location of marl strata. Marine specimens could be gathered on Sullivan's Island, where Ravenel had maintained a summer house since 1823. His catalog shows that he also relied heavily on gifts and on the cooperation of traveling acquaintances.

He was interested in agricultural experimentation, if only temporarily in the case of marl. He owned 3,364 acres and 104 slaves at his main property and by the 1850s, after buying three neighboring plantations, had amassed over 7,500 acres. The brick manufacturing mentioned by Ruffin had been established on the estate when he moved in, but he became very attached to the business, seeking to monopolize trade by buying up South Carolina patents on both Sawyer and Brown brickmaking machines in the 1830s.

Ravenel remained generally aloof from politics, though he did chair a meeting convened in his home parish in November 1860 for the purpose of endorsing secession. He returned to Charleston when the war was over, by then virtually blind and much diminished commercially. He died of typhoid fever in 1871. [Davidson, *Last Foray;* Irving, *Day on Cooper River;* Longton, "Some Aspects of Intellectual Activity in South Carolina"; MacDowell, *Gaillard Genealogy; Catalogue of Recent and Fossil Shells;* Ravenel, *Advantages of a Sea-Shore Residence in the Treatment of Certain Diseases;* Ruffin, *Report of Survey;* Shaffer, *Carolina Gardens;* Waring, *History of Medicine in South Carolina; South Carolina Genealogies,* vol. 2]

RAVENEL, Dr. Henry, not to be confused with his son Henry William (q.v.), owned Pooshee plantation, St. John's, Berkeley, Charleston District, and over one hundred slaves. He was a graduate of New York College of Physicians and Surgeons and a distant relative of Edmund Ravenel (q.v.). Ravenel was a few years older than Ruffin. He married, successively, a Stevens, a Dwight, and a Porcher.

Ruffin would have known of Ravenel's reputation as an improving planter. He is mentioned in a *Farmers' Register* article of 1840. In July 1831 the *Southern Agriculturist* carried an extended "Account of the Management of Pushee [sic]." When Henry took over the farm "it was much exhausted, by long and severe culture. Not having land enough cleared to enable him to alternate his fields, he judiciously determined to endeavour to restore their fertility by the free use of manures," his efforts being "crowned with considerable and deserved success." A specialist force was kept permanently at work attending to "daily augmentation of the manure heap" (particularly with pine litter) and this, combined with clearings of stables and pens, gave Ravenel enough to treat all his subsistence land as well as half the cotton crop. Among other results, corn yields had doubled over eleven years. (How much greater the reward would be, Ruffin must have told him, if manuring could be accompanied by marling or liming.) Self-sufficiency apparently was an important objective. Ravenel also made considerable efforts to have slave clothing manufactured within the plantation, using home-produced raw materials.

Ruffin saw Ravenel again in 1863, at Pinopolis. "He is now very old, & so infirm that he rarely goes out." He died in 1867. [Davidson, *Last Foray;* MacDowell, *Gaillard Genealogy;* Ruffin, *Diary,* vol. 3; *Farmers' Register,* 1840; *Southern Agriculturist,* 1831]

RAVENEL, Henry William, son of Dr. Henry Ravenel (q.v.) and a graduate of Pineville Academy and South Carolina College, was in his twenties. He had recently begun botanical investigations that were to win him an international reputation. He would become the leading authority on American fungi and a vigorous collector and classifier of botanical specimens. "It is doubtful," wrote the mycologist W. G. Farlow in 1887, "whether any other American botanist has ever covered so wide a range of plants." One genus and fifty species were named after him. To Ruffin in 1843, however, Ravenel was probably just a young scientist-planter with an interest in agricultural improvement. "It had been my wish," Ravenel later recorded, "to study Medicine as a profession, but my father advised strongly against it (as too laborious

& liable to exposure for what he thought my weak constitution) that I was reluctantly obliged to give it up. He gave me a planting interest, & I commenced at Northampton." This plantation stood on the headwaters of the Cooper River just below the Santee Canal. The mansion was destroyed by fire at the end of 1843. Later, Ravenel, in poor health, sold his land and retreated to Aiken and a life of scientific inquiry.

Ravenel was a keen supporter of Ruffin's survey and had read his *Calcareous Manures.* In April 1843 (just after Ruffin had passed through), as secretary of the Black Oak Agricultural Society, he sent a copy of John Palmer's (q.v.) recent address on marling to the *Southern Agriculturist* as an "inducement" to other planters in the state. A year later, he presented four resolutions to the society, lauding Ruffin for his efforts and suggesting that citizens of South Carolina should tell him of their "entire satisfaction" and extend a "cordial well-done." As a Virginia planter, scientific author, and journal editor, Ruffin had manifested "untiring zeal . . . blending in a happy union, the lights of science with a sound practical judgement." He had labored "amid many difficulties and discouragements," winning through as "benefactor of his country." The resolutions were unanimously adopted. Ravenel also contributed to Ruffin's report by submitting a list of post-Pliocene fossils to be found in the neighborhood of Charleston. Ruffin asked him in 1843 to arrange for a group of Carolinians to visit the marled fields of Virginia, offering to pay his keep if he paid his own transport expenses; nothing resulted.

Ravenel developed stern racist and disunionist perspectives on South Carolina's political affairs, stating that blacks were people "whose physical instincts lead them to subjection and control." He died in 1887. [Channing, *Crisis of Fear;* Childs, ed., *Private Journal of Henry William Ravenel;* Haygood, *Henry William Ravenel;* Longton, "Some Aspects of Intellectual Activity in South Carolina"; MacDowell, *Gaillard Genealogy;* Ruffin, *Report of Survey; Southern Agriculturist,* 1843, 1844]

RICHARDSON, Colonel James, was a member of an English family that came to Sumter District by way of Virginia in the mid-eighteenth century. He was the brother of Governor John P. Richardson (q.v.) and brother-in-law, through his sister Elizabeth (q.v.), of the late governor Richard I. Manning (q.v.). He is referred to in the diary and the report as an eager hauler, burner, and spreader of lime. He is not to be confused with his better-known namesake (1770–1836), former president of the state senate. [Nicholes, *Historical Sketches of Sumter County,* vol. 2; Reynolds and Faunt, *Biographical Directory of the Senate;* Ruffin, *Report of Survey*]

RICHARDSON, John Peter, born in 1801, was a Sinkler (q.v.) on his mother's side and brother-in-law of the late governor Richard I. Manning (q.v.). He contested the governorship with James Hammond (q.v.) in 1840.

Old nullification men, wishing for a return of political deference and sharing John C. Calhoun's dislike for growth of party sentiment, selected Richardson as the unity candidate. Hammond presented himself as the true leader of the nullifiers but had to make do with whatever scraps of support could be mustered against the old guard. Animosities were intense, and several duels were fought between the two camps. Richardson won decisively, by 104 to 47. Hammond had his second chance in 1842.

Richardson, a noted horse breeder, had a large plantation at Manchester and was one of the Sumter representatives at the 1839 Agricultural Convention. He was governor when proposals for agricultural investigation first were widely mooted and the House Committee of Agriculture went to work under Robert Roper (q.v.) to formulate survey proposals (see Introduction). Richardson supported the idea when in office, commending it to the legislature in a Message on 29 November 1842. He observed, in clear reference to slavery, "Our domestic institutions are of a character so immutably agricultural, as to vibrate with all its reverses and vicissitudes." Five years earlier, he wrote to Richard Singleton (q.v.): "The wealth—the talents—the enterprise—and religion of the whole North are concentrating to effect the subversion of our Domestic Institutions." Richardson developed into an unequivocal secessionist. He was the Clarendon delegate at the Southern Rights Convention of 1852. [Channing, *Crisis of Fear;* Davidson, *Last Foray;* Faust, *James Henry Hammond;* Moltke-Hansen, private communication to the editor; Nicholes, *Historical Sketches of Sumter County,* vol. 2; Ruffin, *Report of Survey;* Richardson, *Governor's Message, No. 1* (1842); *Journal of the State Convention of South Carolina; Proceedings of the Agricultural Convention*]

RIVERS, John, was a member of a large James Island family descended from William Rivers, a settler in the 1690s. In 1843 he was aged fifty-six and a man of some political weight: he had been a representative in Columbia (1820–24), a member of the Nullification Convention (1832–33), and senator for St. Andrew's parish since 1840. He was the St. Andrew's delegate at the 1852 Southern Rights Convention. Rivers was active, along with Brisbane, Ramsay, and others, in the St. Andrew's, Ashley, and Stono Agricultural Association, representing his parish at annual meetings of the State Agricultural Society. In 1844, he served as chairman of the Committee on Long Cotton for the society and presented a petition to the senate from the St. Andrew's Police Society requesting the continuance of the survey. He died in 1857. [Reynolds and Faunt, *Biographical Directory of the Senate; Journal of the Senate, 1844; Journal of the State Convention of South Carolina; Proceedings of the Agricultural Convention,* 1844; *Southern Agriculturist,* 1844]

ROPER, Robert William, a wealthy Cooper River rice planter of Point

Comfort on the West Branch, was also a longtime resident of Charleston where he built a magnificent (surviving) house at 9 East Battery in 1838. He had been active in the State Agricultural Society and its precursor, the United Agricultural Society, for many years before his promotion of the survey, serving as orator in 1834, working on committees, and presenting papers. In 1844 he served on a committee to consider alterations to the society's constitution and on the Committee on Mules, as well as chairing the Committee on Rice. He was invited to deliver the 1842 anniversary oration at Fort Moultrie. And, of course, he was a representative in Columbia and chairman of the House Committee on Agriculture.

In 1834 Roper submitted a letter to the United Agricultural Society from Joseph Johnson (q.v.), declaring marl to be "fraught with valuable consequences to the whole State." In December 1843 he moved that the House authorize his committee to publish three thousand copies of Ruffin's report (this figure was raised to four thousand at the suggestion of J. I. Middleton [q.v.]). Nature, he observed in 1844, had "based the country below the falls of the rivers upon a calcareous formation, which, when spread upon the surface of the soil, imparts exhaustless fertility. . . . The Report of our late Agricultural Surveyor, Mr. Ruffin, has imparted a mass of information on this manure, which has kindled a spirit of inquiry, and given an impetus to the agriculture of the country, resulting in incalculable value." The latter claim was mistaken, but Roper died well before this became apparent. He did, however, merit his reputation with Seabrook, Poinsett, and many others as a man who had articulated major anxieties over the economic weakness of the state and, despite confusions, had gone some way toward acting on his perceptions.

On his death, Point Comfort was sold to the Laurens family by his wife, Martha Rutledge Laurens. [Johnson, *Social History of the Sea Islands;* Poinsett, "Agricultural Address"; Roper, "Address before the State Agricultural Society"; Roper, *Oration on Anniversary of Battle of Fort Moultrie;* Ruffin, *Report of Survey;* Seabrook, *Memoir on Cotton;* Shaffer, *Carolina Gardens;* Walker, *History of the Agricultural Society; Journal of the House of Representatives, 1843; Proceedings of the Agricultural Convention; Reports and Resolutions of the General Assembly, 1843*]

SEABROOK, Whitemarsh Benjamin, was born in 1793, only a few months before Ruffin. A major sea-island cotton planter, he was the great-grandson of an English merchant, Captain Richard Seabrook, who settled in the state in the 1670s. The family was one of great wealth and prominence on Edisto.

Seabrook was a law graduate of the College of New Jersey (later Princeton) in 1812. He served in the state House of Representatives (1814–20) and later the senate (1826–33). Despite his early doubts over slavery, Seabrook was a leading radical on the nullification issue in the legislature by 1828, remaining a determined secessionist for the remainder of his life. He was elected lieutenant governor (1834–36) and governor (1848–49).

Ruffin was already familiar with Seabrook as co-author of a piece in the 1840 *Farmers' Register.* As an opinionated young senator, he was invited to deliver the anniversary address to the United Agricultural Society of South Carolina in 1827. In a speech full of states'-rights and antitariff sentiment, he criticized his fellow planters for absenteeism, monoculture, and ignorance of elementary principles of agricultural science. Later he declared that planters' "repugnance to test a new project is a truth of daily observation." But he continued to publish farming papers and addresses and served as president of the State Agricultural Society. In late 1843 (at Ruffin's request) he presented his celebrated *Memoir on Cotton* to the state society, claiming incidentally that Ruffin's work had ushered in a new era for Carolina planters: "Before another year have [sic] passed, lime and marl will be the most common, and the most extensively employed, of all the natural means for resuscitating exhausted lands." In the same year he and John Belton O'Neall prepared a report on cotton for the society, advising a combination of augmented staple production and soil-relieving diversification. Like Ruffin, Seabrook was much concerned over the institutional consequences of agricultural complacency, believing that the survival of slavery depended upon plantation reform.

For most of his adult life, he viewed slavery as a positive good. When governor, he applauded the politicizing effects of the Wilmot Proviso: "The South has at last been aroused from its criminal lethargy to a knowledge of the dangers of its position." He was the St. John's, Colleton, delegate to the Southern Rights Convention in 1852. He died at age sixty-one, some years before his state seceded. [Allston, *Essay on Sea Coast Crops;* Barnwell, *Love of Order;* Eaton, *Growth of Southern Civilization;* Freehling, *Prelude to Civil War;* Graydon, *Tales of Edisto;* Johnson, *Social History of the Sea Islands;* O'Neall and Seabrook, "Report on the Cotton Crop"; Seabrook, *Address at the United Agricultural Society;* Seabrook, *Memoir on Cotton;* Seabrook, *Message to the Legislature of South-Carolina;* Walker, *History of the Agricultural Society; Journal of the State Convention of South Carolina; South Carolina Genealogies,* vol. 1]

SHEPARD, Charles Upham, a scientist of national repute, was aged thirty-eight in 1843. He had already published his *Treatise on Mineralogy,* co-authored the recent *Geological Survey of Connecticut,* and discovered two species of mineral microlite. A native of Rhode Island, he was at this time both professor of chemistry at the Medical College in Charleston and lecturer in natural history at Yale. In 1845 he also became professor of chemistry and natural history at Amherst College, where he taught until 1852; he dropped the Yale post in 1847. Shepard retained his Charleston chair until the outbreak of war in 1861, when he chose to move north and resume employment at Yale and Amherst for the duration of hostilities. He was prevailed upon to return to South Carolina later in the 1860s, and he died in Charleston in 1886.

Shortly after his first Carolina appointment, Shepard was described by Samuel Henry Dickson as "quite a prize." He not only brought proven ability but also what was to become the finest mineralogical collection in the United States. Using his own money, he converted the campus grounds into what Middleton Michel called "a fairyland" of hothouses and flower displays. The Moser (q.v.) specimens gave him useful preliminary insights into the geological substrata of the Charleston area, which he developed later into an accurate appraisal of phosphate resources underlying the city.

Interest in the Moser samples gave Shepard and Ruffin common ground for mutually beneficial exchanges. Additional interests were the James River analyses referred to in the diary and Shepard's experience on the Connecticut survey. At the request of the secretary of the treasury, he had undertaken a study of sugar cultivation and manufacture in the early 1830s. He also was currently involved with soil analysis. Of particular interest to Ruffin was his examination of Edisto Island soil specimens in 1840 for the Agricultural Society of St. John's, Colleton, and his discovery that the best land for sea-island cotton had a natural lime component of about 2 percent. Shepard had concluded, in language that could have been Ruffin's own, that the "peculiar fertility of the new sea-island cotton land may be owing to the proportion of comminuted shells, natural to such soils, and the deterioration of these lands under long cultivation, ascribable to the exhaustion of carbonate of lime." Local planters, represented by Isaac Jenkins Mikell (q.v.), wrote to Ruffin at the *Farmers' Register* for advice about liming of cotton. Ruffin admitted to limited experience with the crop but suggested that it was as responsive as any to such treatment. He sent the society twenty complimentary copies of his *Calcareous Manures*. This point of intellectual contact apparently was not developed during the survey, but Shepard did oblige Ruffin by sending him "Analyses of Soils from the Tide-Swamp Plantation of Col. R. F. W. Allston" (q.v.) in December 1843 for inclusion in the Appendix to the report. [Bachman, *Inquiry into the Agricultural Survey;* Johnson, *Social History of the Sea Islands;* Longton, "Some Aspects of Intellectual Activity in South Carolina"; Porcher, *Report on Manures;* Ruffin, *Report of Survey;* Seabrook, *Memoir on Cotton;* Smith, *Economic Readjustment of an Old Cotton State;* Waring, *History of Medicine in South Carolina; Appleton's Cyclopedia of American Biography,* vol. 5; *Farmers' Register,* 1840; *Southern Agriculturist,* 1844]

SIMONS, Keating, son of Keating Simons of revolutionary war prominence and Sarah Lewis, owned Lewisfield plantation on the West Branch of the Cooper. This Huguenot family had been of some consequence in low-country military, political, and farming affairs since the end of the seventeenth century. Several of them bore the same name. [Channing, *Crisis of Fear;* Davidson, *Last Foray;* Irving, *Day on Cooper River;* Rogers, *History of Georgetown County;* Shaffer, *Carolina Gardens*]

SINGLETON, Richard, of True Blue, came from a long-established Charleston, Sumter, and Richland family of great wealth and was a horse breeder. In 1850 he owned 549 slaves. He is not to be mistaken for the long-deceased Senator Richard Singleton. His reputation as a cotton planter is indicated in his being thanked by Whitemarsh Seabrook in his study of the staple. [Davidson, *Last Foray;* Reynolds and Faunt, *Biographical Directory of the Senate;* Rogers, *History of Georgetown County;* Scarborough, communication with the editor; Seabrook, *Memoir on Cotton;* Walker, *History of the Agricultural Society*]

SINKLER, William, owned the large Eutaw plantation. Like Ruffin, he was descended from Scottish nobility and bore a phonetically adjusted surname (Sinkler from Sinclair, Ruffin from Ruthven). He was the grandson of the original settler and was married to Elizabeth Broun, daughter of Captain Archibald Broun, the noted revolutionary war soldier. One of the Sinkler children married a Manning and another a Huger.

Sinkler clearly had experimental inclinations; he was the first planter in the state whom Ruffin saw marling. He also gave Whitemarsh Seabrook material for his *Memoir of Cotton*. In his mid-fifties at the time of the survey, he died in 1853. [MacDowell, *Gaillard Genealogy;* Mathew, Edmund Ruffin; Seabrook, *Memoir on Cotton*]

SMITH, Dr. John Lawrence, was a strong publicist for calcareous manuring after Ruffin returned to Virginia. "The importance of the presence of lime," he wrote in the *Southern Agriculturist* in 1844, "has been fully settled; and it may be considered indispensable to the healthy growth of the plant." South Carolina had "the richest marls that have ever been discovered," and Charleston, according to a letter from Professor Jacob Whitman Bailey of West Point, was "built upon a bed of animalcules several hundred feet in thickness." Smith's article mentioned certain features of marling which help account for its almost total neglect in South Carolina, its virtues notwithstanding. First, planters could "not expect the maximum benefit . . . for several years." Second, because of soil variations, no standardized advice could be given on dosage: "Planters should experiment upon this." Third, there was "diversity of opinion" concerning frequency of marling. Fourth, marl was best when used with compost or manure: applied alone it could damage soil. Smith gained an early reputation abroad, delivering a paper to the Academy of Science in Paris and undertaking a four-year geological and mineralogical survey at the request of the sultan of Turkey. [*Charleston Medical Journal and Review; Southern Agriculturist*, 1844]

STARK, Dr. Thomas, of Belville, Orangeburg District, on the Congaree about four miles above the point where it joins the Wateree to form the Santee, is not to be confused with James Hammond's old friend Theodore Stark or the "Mr. Starke" Ruffin also met in May. [Faust, *James Henry Hammond;* Mills, *Atlas of South Carolina*]

STROMAN, Jacob, was a substantial planter, then in his mid-fifties, working Rocky Swamp plantation on the creek of the same name twenty-odd miles west of Orangeburg. He was the Orange and St. Matthew's delegate to the state Agricultural Convention of 1839. He died in 1877. [Davidson, *Last Foray;* Mills, *Atlas of South Carolina;* Ruffin, *Report of Survey; Proceedings of the Agricultural Convention*]

TUCKER, John Hyrne, of Litchfield, Willbrook, and Charleston, another of the great Waccamaw Neck planters, was then aged sixty-two. He also owned rice land on the lower Cooper and was a graduate of Rhode Island College (later Brown). His father had been in trade with the Heriots (q.v.), and his second wife, Elizabeth Ann, was a sister of R. F. W. Allston (q.v.). On his death in 1859, he owned around 350 slaves, all in Georgetown District.

As Ruffin indicates, Tucker was an improver. He was a founding member and current president of the Pee Dee Planters' Club, president of the State Agricultural Society for periods in the 1840s and 1850s, and a member of the society's Committee on Corn in 1843. In 1850 his Georgetown plantations produced 1,140,000 pounds of rice, and in the following year his yield of 888.5 bushels on 10.5 acres won him the society's annual premium. Reputedly he was an inelegant, pox-pitted man with a dislike for books. It was said that "planting was his sole delight. He lived for and in rice. It was the first and the last thought of his mind." [Davidson, *Last Foray;* Irving, *Day on Cooper River;* Joyner, *Down by the Riverside;* Rogers, *History of Georgetown County;* Scarborough, private communication to the editor; Walker, *History of the Agricultural Society*]

VERDIER, Dr., cannot be identified with certainty. There were numerous Verdiers in the south of the state. Two of the possible "doctors" were A. G. Verdier, a St. Luke's Beaufort representative in the State Agricultural Society, and James Robert Verdier, father of the future Beaufort senator William Johnson Verdier (1841–1902). He is probably not the better-known Simon Verdier, wealthy planter and former state senator of Colleton District (1781–1853), for he had no title, and it is unlikely that Ruffin had any plans to backtrack into Colleton. [Davidson, *Last Foray;* Reynolds and Faunt, *Biographical Directory of the Senate;* Rosengarten, *Tombee; Proceedings of the Agricultural Convention*]

Bibliography

Books and Articles

Akehurst, B. C. *Tobacco.* London: Longmans, 1968.

Allmendinger, David F. "The Early Career of Edmund Ruffin, 1810–1840." *Virginia Magazine of History and Biography* 93 (1985).

Allston, R. F. W. *Essay on Sea Coast Crops; Read before the Agricultural Association of the Planting States . . . December 3d, 1853.* Charleston: A. E. Miller, 1854.

———. *Memoir of the Introduction and Planting of Rice in South-Carolina.* Charleston: Miller & Browne, 1843.

Anderson, Ralph V., and Robert E. Gallman. "Slaves as Fixed Capital: Slave Labor and Southern Economic Development." *Journal of American History* 64 (1977).

Appleton's Cyclopædia of American Biography. New York: D. Appleton, 1888.

Arden, Daniel D. "Charles Lyell's Observations on Southeastern Geology." In James X. Corgan, ed., *The Geological Sciences in the Antebellum South.* University, Ala.: University of Alabama Press, 1982.

Bachman, C. L. *John Bachman.* Charleston: Walker, Evans, & Cogswell, 1888.

Bachman, Rev. J. *An Address Delivered before the Horticultural Society of Charleston at the Anniversary Meeting, July 10th, 1833.* Charleston: A. E. Miller, 1833.

———. *A Discourse, Delivered on the Forty-Third Anniversary of His Ministry in Charleston.* Charleston: A. Burke, 1858.

———. *The Doctrine of the Human Race Examined on the Principles of Science.* Charleston: C. Canning, 1850.

———. *An Inquiry into the Nature and Benefits of an Agricultural Survey of the State of South-Carolina.* Charleston: Miller & Browne, 1843.

Barnwell, John. *Love of Order: South Carolina's First Secession Crisis.* Chapel Hill: University of North Carolina Press, 1982.

Barnwell, Stephen B. *The Story of an American Family.* Marquette: N.p., 1969.

Bartram, John. *Diary of a Journey through the Carolinas, Georgia, and*

Florida from July 1, 1765, to April 10, 1766. Annotated by Francis Harper. *Transactions of the American Philosophical Society* n.s., 33 (1942).

Bleser, Carol, ed. *The Hammonds of Redcliffe*. New York: Oxford University Press, 1981.

Brady, Nyle C. *The Nature and Properties of Soils*. New York: Macmillan, 1974.

Brunhose, Robert L. *David Ramsay, 1749–1815: Selections from His Writings. Transactions of the American Philosophical Society* n.s., 55 (1965).

Buol, S. W., F. J. McCracken, and F. D. Hole. *Soil Genesis and Classification*. Ames: Iowa State University Press, 1973.

Catalogue of Recent and Fossil Shells in the Cabinet of the Late Edmund Ravenel, M.D. Charleston: Walker, Evans & Cogswell, 1875.

Catalogue of the Trustees, Faculty and Students of the South Carolina College. Columbia: Morgan, 1843, 1844.

Channing, Steven A. *Crisis of Fear: Secession in South Carolina*. New York: Norton, 1970.

Chernov, V. A. *The Nature of Soil Acidity*. Madison, Wisc.: Soil Science Society of America, 1964.

Childs, Arney Robinson, ed. *The Private Journal of Henry William Ravenel, 1859–1887*. Columbia: University of South Carolina Press, 1947.

Clay, James W., Douglas M. Orr, Jr., and Alfred W. Stuart. *North Carolina Atlas: Portrait of a Changing Southern State*. Chapel Hill: University of North Carolina Press, 1975.

Coclanis, Peter A. *The Shadow of a Dream: Economic Life and Death in the South Carolina Low Country, 1670–1920*. New York: Oxford University Press, 1989.

Collections of the South-Carolina Historical Society. Vol. 1. Charleston, 1857.

Collings, Gilbert H. *Commercial Fertilizers*. London: H. K. Lewis, 1947.

Collins, Bruce. *White Society in the Antebellum South*. London: Longmans, 1985.

Colquhoun, Donald J. *Geomorphology of the Lower Coastal Plain of South Carolina*. Columbia, S.C.: State Development Board, 1969.

Craven, Avery O. *Edmund Ruffin, Southerner*. 1932. Reprint. Baton Rouge: Louisiana State University Press, 1966.

Cruickshank, Helen Gere, ed. *John and William Bartram's America: Selections from the Writings of the Philadelphia Naturalists*. New York: Devin-Adair, 1957.

Cuthbert, James. "An Address Delivered in Charleston, before the Agricultural Society of South-Carolina, at Its Anniversary Meeting, 18th August, 1830." *Southern Agriculturist* 3 (November 1830).

Dahl, Robert Alan. *Democracy in the United States: Promise and Performance*. 2d ed. Chicago: Rand McNally, 1972.

Darwin, Charles. *Journal of Researches into the Natural History and Geology of the Countries Visited during the Voyage of H.M.S. "Beagle" Round the World*. 7th ed. London: Ward, Lock and Co., 1890.

Davidson, Charles Gaston. *The Last Foray: The South Carolina Planters of 1860, A Sociological Study*. Columbia: University of South Carolina Press, 1976.

Dickson, S. H. "Essay on Malaria. Read before the State Agricultural Society of South Carolina." *Proceedings of the Agricultural Convention and of the State Agricultural Society, 1839 to 1845*. Columbia: Summer & Carroll, 1846.

Directory of Charleston, 1837–38. Charleston, 1838.

Donahue, Roy L., John C. Shickluna, and Lynn S. Robertson. *Soils*. Englewood Cliffs, N.J.: Prentice-Hall, 1971.

Drayton, John. *A View of South-Carolina, as Respects Her Natural and Civil Concerns*. 1802. Reprint. Spartanburg: Reprint Company, 1972.

Easterby, J. H., ed. *The South Carolina Rice Plantation as Revealed in the Papers of Robert F. W. Allston*. Chicago: University of Chicago Press, 1945.

Eaton, Clement. *The Growth of Southern Civilization, 1790–1860*. New York: Harper and Brothers; 1961.

Edmunds, John B., Jr. *Francis W. Pickens and the Politics of Destruction*. Chapel Hill: University of North Carolina Press, 1986.

Elliott, William. *Address Delivered by Special Request before the St. Paul's Agricultural Society, May, 1850*. Charleston: Walker & James, 1850.

———. *Carolina Sports, by Land and Water; Including Incidents of Devil-Fishing*. Charleston: Burges and James, 1846.

Faust, Drew Gilpin. *James Henry Hammond and the Old South: A Design for Mastery*. Baton Rouge: Louisiana State University Press, 1982.

———. "The Rhetoric and Ritual of Agriculture in Antebellum South Carolina." *Journal of Southern History* 45 (1979).

———. *A Sacred Circle: The Dilemma of the Intellectual in the Old South, 1840–1860*. Baltimore: Johns Hopkins University Press, 1977.

Fergus, E. N., Carsie Hammonds, and Hayden Rogers. *Southern Field Crops Management*. Chicago: J. B. Lippincott, 1944.

Fippin, Elmer O. *Address Delivered at the Summer Meeting of the Agricultural Society of South Carolina, July 14, Nineteen-fifteen*. Charleston: Walker, Evans and Cogswell, 1915.

Flanders, Ralph Betts. *Plantation Slavery in Georgia*. 1933. Reprint. Cos Cob, Conn.: John E. Edwards, 1967.

Ford, Lacy K. *Origins of Southern Radicalism: The South Carolina Upcountry, 1800–1860*. New York: Oxford University Press, 1988.

Foth, H. D., and J. W. Schafer. *Soil Geography and Land Use*. New York: N.p., 1980.

Fraser, Jessie Melville. *Louisa C. McCord*. Columbia: University of South Carolina Press, 1919.

Freehling, William W. *Prelude to Civil War: The Nullification Controversy in South Carolina, 1816–1836*. New York: Harper & Row, 1965.

Genovese, Eugene D. *The Political Economy of Slavery*. 1965. Reprint. New York: Vintage Books, 1967.

Gray, Lewis Cecil. *History of Agriculture in the Southern States to 1860*. 2 vols. 1933. Reprint. Gloucester, Mass.: Peter Smith, 1958.

Graydon, Nell S. *Tales of Beaufort*. Beaufort, S.C.: Beaufort Bookshop.

———. *Tales of Edisto*. Columbia: R. L. Bryan.

Greene, Mott T. *Geology in the Nineteenth Century: Changing Views of a Changing World*. Ithaca: Cornell University Press, 1982.

Hamilton, James, Jr. "An Address on the Agriculture and Husbandry of the South." *Southern Agriculturist* n.s., 4 (August 1844).

Hammond, James H. *An Address Delivered before the South-Carolina Institute, at Its First Annual Fair, on the 20th November, 1849*. Charleston: Walker and James, 1849.

———. "Anniversary Oration, of the State Agricultural Society of South Carolina; . . . 25th November, 1841." *Proceedings of the Agricultural Convention and of the State Agricultural Society of South Carolina, 1839 to 1845*. Columbia: Summer & Carroll, 1846.

———. *Governor's Message, No. 1, to the Senate and House of Representatives of the State of South-Carolina, November 28th, 1843*. Columbia: Executive Department, 1842.

Hart, John Fraser. *The Southeastern United States*. Princeton: D. Van Nostrand, 1967.

Haygood, Tamara Miner. *Henry William Ravenel, 1814–1887: South Carolina Scientist in the Civil War Era*. Tuscaloosa: University of Alabama Press, 1987.

Heitzler, Michael J. *Historic Goose Creek, South Carolina, 1670–1980*. Edited by Richard N. Cote. Easley: South Carolina Historical Press, 1983.

Hennig, Helen Kohn, ed. *Columbia: Capital City of South Carolina, 1786–1936*. Columbia: R. L. Bryan, 1936.

Henry, Robert. *Eulogy on the Late Honorable John Caldwell Calhoun, Delivered at Columbia, South Carolina, on Thursday May 16, 1850*. Columbia: I. C. Morgan, 1850.

Hilton, Mary Kendall. *Old Homes and Churches of Beaufort County, South Carolina*. Columbia: State Printing Company, n.d.

Holmes, Francis S. "Notes on the Geology of Charleston." *Charleston Medical Journal and Review* 4 (1849).

Hunt, Charles B. *Natural Regions of the United States and Canada*. San Francisco: 1967.

Irving, John B. *A Day on Cooper River*. 1842. Reprint. Edited by L. C. Stone. Columbia: R. L. Bryan, 1969.

Jackson, Andrew. *Correspondence of Andrew Jackson*. Edited by John

Spencer Bassett. Washington, D.C.: Carnegie Institution of Washington, 1926–35.

James, Marquis. *The Life of Andrew Jackson*. Indianapolis: Bobbs-Merrill, 1937.

Johnson, Guion Griffis. *A Social History of the Sea Islands*. Chapel Hill: University of North Carolina Press, 1930.

Johnson, Henry S. *Geology in South Carolina*. Miscellaneous Report 3. Columbia: Division of Geology State Development Board, 1964, as revised by E. James Clay, 1971.

Journal of the House of Representatives of the State of South-Carolina. Being the Annual Session of 1843. Columbia: A. H. Pemberton, 1843.

Journal of the Senate of the State of South-Carolina. Being the Annual Session of 1844. Columbia: A. H. Pemberton, 1844.

Journal of the State Convention of South Carolina. Columbia, 1852.

Joyner, Charles. *Down by the Riverside: A South Carolina Slave Community*. Urbana: University of Illinois Press, 1984.

King, Mitchell. *The Anniversary Address of the State Agricultural Society of South Carolina*. Columbia: I. C. Morgan, 1846.

———. *A Discourse on the Qualifications and Duties of an Historian; Delivered before the Georgia Historical Society*. Savannah: Georgia Historical Society, 1843.

Lachicotte, Alberta Morel. *Georgetown Rice Plantations*. Columbia: State Printing Company, 1955.

Lander, Ernest McPherson, Jr. "Manufacturing in South Carolina, 1815–60." *Business History Review* 28 (March 1954).

———. *The Textile Industry in Antebellum South Carolina*. Baton Rouge: Louisiana State University Press, 1969.

Longton, William Henry. "Some Aspects of Intellectual Activity in Ante-Bellum South Carolina, 1830–1860: An Introductory Study." Ph.D. dissertation, University of North Carolina, 1969.

Low, Sir Alfred Maurice. *Protection in the United States*. London: King, 1904.

Lyell, Charles. *Principles of Geology*. 3 vols. London: Murray, 1830–33.

———. *Travels in North America in the Years 1841–2; with Geological Observations on the United States, Canada, and Nova Scotia*. New York: Wiley & Putnam, 1845.

MacDowell, Dorothy Kelly. *Gaillard Genealogy: Descendants of Joachim Gaillard and Esther Paparel*. Aiken.

McDuffie, Gen. George. "Anniversary Oration, of the State Agricultural Society of South Carolina; . . . 26th November 1840." *Proceedings of the Agricultural Convention and of the State Agricultural Society of South Carolina, 1839 to 1845*. Columbia: Summer & Carroll, 1846.

Mathew, W. M. "Agricultural Adaptation and Race Control in the American South: The Failure of the Ruffin Reforms." *Slavery and Abolition* 7 (September 1986).

———. *Edmund Ruffin and the Crisis of Slavery in the Old South: The Failure of Agricultural Reform.* Athens: University of Georgia Press, 1988.

———. "Edmund Ruffin and the Demise of the *Farmers' Register.*" *Virginia Magazine of History and Biography* 94 (January 1986).

———. "Planter Entrepreneurship and the Ruffin Reforms in the Old South, 1820–60." *Business History* 27 (July 1985).

———. "Slave Skills, Plantation Schedules, and Net Returns in Southern Marling, 1830–60." *Plantation Society in the Americas* 2 (December 1986).

Memoir of Professor F. A. Porcher. Charleston: Historical Society of South Carolina, 1889.

Merrens, Harry Roy, ed. *The Colonial South Carolina Scene, 1697–1774.* Columbia, 1977.

Midgely, A. R. "Lime—Its Importance and Efficient Use in Soils." *Proceedings of the Soil Science Society of America* 8 (1943).

Millar, C. E., L. M. Turk, and H. D. Foth. *Fundamentals of Soil Science.* New York: Wiley, 1965.

Millbrooke, Anne. "South Carolina State Geological Surveys of the Nineteenth Century." In James X. Corgan, ed., *The Geological Sciences in the Antebellum South.* University, Ala.: University of Alabama Press, 1982.

Mills, Robert. *Mills' Atlas of the State of South Carolina, 1825.* Introduction by Gene Waddell. Reprint. Easley: Southern Historical Press, 1980.

———. *Statistics of South Carolina, Including a View of Its Natural, Civil, and Military History, General and Particular.* 1826. Reprint. Spartanburg, 1972.

Mitchell, Betty. *Edmund Ruffin: A Biography.* Bloomington: Indiana University Press, 1981.

Moltke-Hansen, David. "The Expansion of Intellectual Life: A Prospectus." In Michael O'Brien and David Moltke-Hansen, eds., *Intellectual Life in Antebellum Charleston.* Knoxville: University of Tennessee Press, 1987.

———. "Protecting Interests, Maintaining Rights, Emulating Ancestors: U.S. Constitution Bicentennial Reflections on 'The Problem of South Carolina,' 1787–1860." *South Carolina Historical Magazine* 90 (1989).

Nicholes, Cassie. *Historical Sketches of Sumter County: Its Birth and Growth.*

O'Brien, Michael, ed. *All Clever Men Who Make Their Way: Critical Discourse in the Old South.* Fayetteville: University of Arkansas Press, 1982.

O'Brien, Michael, and David Moltke-Hansen, eds. *Intellectual Life in Antebellum Charleston.* Knoxville: University of Tennessee Press, 1987.

Ochenkowski, J. P. "The Origins of Nullification in South Carolina." *South Carolina Historical Magazine* 83 (1982).

O'Neall, John Belton. "An Agricultural Address Delivered before the State Agricultural Society, 29th Dec., 1842." *Proceedings of the Agricultural Convention and of the State Agricultural Society of South Carolina, 1839 to 1845.* Columbia: Summer & Carroll, 1846.

O'Neall, John Belton, and W. B. Seabrook. "Report, on the Reduction of the Cotton Crop." *Proceedings of the Agricultural Convention and of the State Agricultural Society of South Carolina, 1839 to 1845*. Columbia: Summer & Carroll, 1846.

Paden, W. R., and W. H. Garman. "Yield and Composition of Cotton and Kobe Lespedeza Grown at Different pH Levels." *Proceedings of the Soil Science Society of America* 11 (1947).

Pease, Jane H., and William H. Pease. "The Economics and Politics of Charleston's Nullification Crisis." *Journal of Southern History* 47 (1981).

Poinsett, Joel R. "An Agricultural Address Delivered before the State Agricultural Society, 27th Nov. 1845." *Proceedings of the Agricultural Convention and of the State Agricultural Society of South Carolina, 1839 to 1845*. Columbia: Summer & Carroll, 1846.

Pope, J. J., Jr., *The Moral Influence of the American Government: An Oration, Delivered on St. Helena Island, on the Fourth of July, 1846*. Charleston: Walker & Burke, 1846.

———. *An Oration Delivered in the Fort at Moultrieville, on the Twenty-Eighth of June, 1849*. Charleston: James S. Burgess, 1849.

Porcher, F. A. *The History of the Santee Canal*, 1903. Reprint. Moncks Corner, 1950.

———. *Report on Manures Read before the Black Oak Agricultural Society*. Charleston: Miller & Browne, 1844.

Ravenel, Edmund. *The Advantages of a Sea-Shore Residence in the Treatment of Certain Diseases, and the Therapeutic Employment of Sea-Water*. Charleston: Walker & James, 1850.

———. "Mr. [H. W.] Ravenel's Letter on Marling" [to R. W. Roper]. *Proceedings of the Agricultural Convention and of the State Agricultural Society of South Carolina, 1839 to 1845*. Columbia: Summer & Carroll, 1846.

Proceedings of the Agricultural Convention and of the State Agricultural Society of South Carolina, 1839 to 1845. Columbia: Summer & Carroll, 1846.

Reports and Resolutions of the General Assembly of South-Carolina, Passed at Its Regular Session of 1843. Columbia: A. H. Pemberton, 1844.

Reynolds, Emily Bellinger, and John Reynolds Faunt. *Biographical Directory of the Senate of the State of South Carolina, 1776–1964*. Columbia: South Carolina Archives Department, 1964.

Richardson, John P. *Governor's Message, No. 1, to the Senate and House of Representatives of the State of South-Carolina*. Columbia: Executive Department, 1842.

Rogers, G. S. "The Phosphate Deposits of South Carolina." *Geological Survey Bulletin*, no. 580 (1915).

Rogers, George C., Jr. *The History of Georgetown County, South Carolina*. Columbia: University of South Carolina Press, 1970.

Roper, R. W. "Address Delivered before the State Agricultural Society,

November, 1844." *Proceedings of the Agricultural Convention and of the State Agricultural Society of South Carolina, 1839 to 1845*. Columbia: Summer & Carroll, 1846.

———. *An Oration Delivered in the Fort at Moultrieville, on the Twenty-eighth of June, 1842*. Charleston: James S. Burgess, 1842.

Rosengarten, Theodore. "*The Southern Agriculturist* in an Age of Reform." In Michael O'Brien and David Moltke-Hansen, eds., *Intellectual Life in Antebellum Charleston*. Knoxville: University of Tennessee Press, 1987.

———. *Tombee: Portrait of a Cotton Planter, with the Journal of Thomas B. Chaplin (1822–1890)*. New York: William Morrow, 1986.

Ruffin, Edmund. *An Address on the Opposite Results of Exhausting and Fertilizing Systems of Agriculture, Read before the South-Carolina Institute . . . November 18th, 1852*. Charleston: Walker and James, 1853.

———. *The Diary of Edmund Ruffin*. Edited by William Kauffman Scarborough. Vol. 1, *Toward Independence, October, 1856–April, 1861*. Vol. 2, *The Years of Hope, April, 1861–June, 1863*. Vol. 3, *A Dream Shattered, June, 1863–June, 1865*. Baton Rouge: Louisiana State University Press, 1972–89.

———. *An Essay on Calcareous Manures* (1832). Edited by J. Carlyle Sitterson. Cambridge, Mass.: Belknap Press of Harvard University Press, 1961.

———. "Marling Facts and Estimates." *Proceedings of the Agricultural Convention and of the State Agricultural Society of South Carolina, 1839 to 1845*. Columbia: Summer & Carroll, 1846.

———. *Report of the Commencement and Progress of the Agricultural Survey of South-Carolina, for 1843*. Columbia: A. H. Pemberton, 1843.

Seabrook, Whitemarsh B. *An Address Delivered at the First Anniversary Meeting of the United Agricultural Society of South-Carolina*. Charleston: A. E. Miller, 1828.

———. *A Memoir on the Origin, Cultivation and Uses of Cotton, from the Earliest Ages to the Present Time, with Especial Reference to the Sea-Island Cotton Plant*. Charleston: Miller & Browne, 1844.

———. *Message of His Excellency W. B. Seabrook, Read to Both Houses of the Legislature of South Carolina, on Tuesday, November 27, 1849*. Columbia: I. C. Morgan, 1849.

———. "To the Planters and Farmers of South Carolina." *Proceedings of the Agricultural Convention and of the State Agricultural Society of South Carolina, 1839 to 1845*. Columbia: Summer & Carroll, 1846.

Shaffer, E. T. H. *Carolina Gardens*. New York: Devin-Adair, 1963.

Smith, Alfred Glaze. *Economic Readjustment of an Old Cotton State: South Carolina, 1820–1860*. Columbia: University of South Carolina Press, 1958.

South Carolina Genealogies. Charleston: South Carolina Historical Society, 1983.

Stuck, W. M. *Soil Survey of Beaufort and Jasper Counties, South Carolina*. Washington, D.C.: U.S. Department of Agriculture, 1973.

Taylor, Rosser H. "Commercial Fertilizers in South Carolina." *South Atlantic Quarterly* 29 (1930).

Taylor, William Robert. *Cavalier and Yankee: The Old South and the American National Character.* New York: Braziller, 1961.

Tuomey, Michael. *Report on the Geological and Agricultural Survey of the State of South Carolina, 1844.* Columbia: A. S. Johnston, 1844.

———. *Report on the Geology of South Carolina.* Columbia: A. S. Johnston, 1848.

Van Deusen, Glyndon Garlock. *The Life of Henry Clay.* Boston: Little, Brown, 1937.

Vanuxem, Lardner, and S. G. Morton. "Geological Observations on the Secondary, Tertiary, and Alluvial Formations of the Atlantic Coast of the United States of America." *Journal of the Philadelphia Academy of Sciences* 6, pt. 1 (1829).

Wakelyn, Jon L. *Biographical Dictionary of the Confederacy.* Westport, Conn.: Greenwood Press, 1977.

Walker, C. Irvine. *History of the Agricultural Society of South Carolina.* Charleston, 1919.

Waring, Joseph Ioor. *A History of Medicine in South Carolina, 1825–1900.* Charleston [?]: South Carolina Medical Association, 1964.

Wauchope, George Armstrong. *The Writers of South Carolina.* Columbia: State Publishing Company, 1910.

Wiltse, Charles M. *John C. Calhoun, Nullifier, 1829–1839.* Indianapolis: Bobbs-Merrill, 1949.

Wright, Gavin. *The Political Economy of the Cotton South.* New York: Norton, 1978.

Newspapers and Journals

American Farmer. Baltimore and Washington.

American Journal.

Charleston Medical Journal and Review. Charleston.

Courier. Charleston.

De Bow's Review. New Orleans.

Farmer and Planter. Pendleton and Columbia.

Farmers' Register. Shellbanks and Petersburg.

Garnet and Black. Vol. 1. South Carolina College student paper. 1899.

Mercury. Charleston.

Southern Agriculturist. Charleston.

Southern Cultivator. Augusta, Athens, and Atlanta.

Southern Journal of Medicine and Pharmacy. Charleston (precursor of *Charleston Medical Journal and Review,* above).

Manuscripts

Hammond, James Henry. Papers. Microfilm. Library of Congress.

Ruffin, Edmund. Papers. Microfilm. Southern Historical Collection, University of North Carolina, Chapel Hill; originals in Virginia Historical Society, Richmond.

Index